Astrophysics and Space Science

Volume 466

Series Editor
Steven N. Shore, Dipartimento di Fisica "Enrico Fermi", Università di Pisa, Pisa, Italy

The Astrophysics and Space Science Library is a series of high-level monographs and edited volumes covering a broad range of subjects in Astrophysics, Astronomy, Cosmology, and Space Science. The authors are distinguished specialists with international reputations in their fields of expertise. Each title is carefully supervised and aims to provide an in-depth understanding by offering detailed background and the results of state-of-the-art research. The subjects are placed in the broader context of related disciplines such as Engineering, Computer Science, Environmental Science, and Nuclear and Particle Physics. The ASSL series offers a reliable resource for scientific professional researchers and advanced graduate students.

Series Editor:
STEVEN N. SHORE, Dipartimento di Fisica "Enrico Fermi", Università di Pisa, Pisa, Italy

Advisory Board:
F. BERTOLA, University of Padua, Italy
C. J. CESARSKY, Commission for Atomic Energy, Saclay, France
P. EHRENFREUND, Leiden University, The Netherlands
O. ENGVOLD, University of Oslo, Norway
E. P. J. VAN DEN HEUVEL, University of Amsterdam, The Netherlands
V. M. KASPI, McGill University, Montreal, Canada
J. M. E. KUIJPERS, University of Nijmegen, The Netherlands
H. VAN DER LAAN, University of Utrecht, The Netherlands
P. G. MURDIN, Institute of Astronomy, Cambridge, UK
B. V. SOMOV, Astronomical Institute, Moscow State University, Russia
R. A. SUNYAEV, Max Planck Institute for Astrophysics, Garching, Germany

More information about this series at https://link.springer.com/bookseries/5664

Katia Biazzo · Valerio Bozza · Luigi Mancini ·
Alessandro Sozzetti
Editors

Demographics of Exoplanetary Systems

Lecture Notes of the 3rd Advanced School on Exoplanetary Science

Editors
Katia Biazzo
INAF—Astronomical Observatory of Rome
Rome, Italy

Luigi Mancini
Department of Physics
University of Rome "Tor Vergata"
Rome, Italy

Valerio Bozza
Department of Physics
University of Salerno
Fisciano, Italy

Alessandro Sozzetti
INAF—Turin Astrophysical Observatory
Turin, Italy

ISSN 0067-0057   ISSN 2214-7985 (electronic)
Astrophysics and Space Science Library
ISBN 978-3-030-88126-9   ISBN 978-3-030-88124-5 (eBook)
https://doi.org/10.1007/978-3-030-88124-5

© Springer Nature Switzerland AG 2022
This work is subject to copyright. All rights are reserved by the Publisher, whether the whole or part of the material is concerned, specifically the rights of translation, reprinting, reuse of illustrations, recitation, broadcasting, reproduction on microfilms or in any other physical way, and transmission or information storage and retrieval, electronic adaptation, computer software, or by similar or dissimilar methodology now known or hereafter developed.
The use of general descriptive names, registered names, trademarks, service marks, etc. in this publication does not imply, even in the absence of a specific statement, that such names are exempt from the relevant protective laws and regulations and therefore free for general use.
The publisher, the authors and the editors are safe to assume that the advice and information in this book are believed to be true and accurate at the date of publication. Neither the publisher nor the authors or the editors give a warranty, expressed or implied, with respect to the material contained herein or for any errors or omissions that may have been made. The publisher remains neutral with regard to jurisdictional claims in published maps and institutional affiliations.

This Springer imprint is published by the registered company Springer Nature Switzerland AG
The registered company address is: Gewerbestrasse 11, 6330 Cham, Switzerland

To our mentors

Acknowledgements

The organizing committee of the 3rd Advanced Scool on Exoplanetary Science (ASES-3) would like to thank the Department of Physics of the University of Salerno, the University of Rome "Tor Vergata" and the Italian National Institute for Astrophysics (INAF) for their financial support. ASES-3 was also supported by the "Progetto Premiale INAF 2015 *Frontiera*" of the Italian Ministry of the University and Research (MIUR), and by Europlanet 2020 RI Work Package NA1—Innovation through Science Networking, Task 5 (the EU-project Europlanet 2020 RI has received funding from the European Union's Horizon 2020 research and innovation programme under grant agreement No. 654208). The organizing committee kindly thanks the Max Planck Institute for Astronomy for hosting the website of the school and the director of the International Institute for Advanced Scientific Studies (IIASS) in Vietri sul Mare, *Prof. Emeritus Ferdinando Mancini*, for hosting the event. Finally, the organizing committee would also like to acknowledge the great help offered by *Mrs. Tina Nappi* for the secretariat of the school.

Contents

Part I Planet Formation

1 **Planet Formation: Key Mechanisms and Global Models** 3
Sean N. Raymond and Alessandro Morbidelli

Part II Star-Planet Interactions

2 **The Role of Interactions Between Stars and Their Planets** 85
A. F. Lanza

Part III Close-In Exoplanets

3 **The Demographics of Close-In Planets** 143
K. Biazzo, V. Bozza, L. Mancini, and A. Sozzetti

Part IV Wide-Orbit Exoplanets

4 **The Demographics of Wide-Separation Planets** 237
B. Scott Gaudi

Contributors

K. Biazzo INAF—Rome Astronomical Observatory, Monte Porzio Catone, RM, Italy

V. Bozza Department of Physics "E.R. Caianiello", University of Salerno, Fisciano, SA, Italy

A. F. Lanza INAF-Osservatorio Astrofisico di Catania, Catania, Italy

L. Mancini Department of Physics, University of Rome "Tor Vergata", Rome, Italy

Alessandro Morbidelli Laboratoire Lagrange, Observatoire de la Cote d'Azur, Nice, France

Sean N. Raymond Laboratoire d'Astrophysique de Bordeaux, CNRS and Université de Bordeaux, Pessac, France

B. Scott Gaudi Department of Astronomy, The Ohio State University, Columbus, OH, USA;
Jet Propulsion Laboratory, California Institute of Technology, Pasadena, CA, USA

A. Sozzetti INAF—Turin Astrophysical Observatory, Pino Torinese, TO, Italy

Part I
Planet Formation

Chapter 1
Planet Formation: Key Mechanisms and Global Models

Sean N. Raymond and Alessandro Morbidelli

Abstract Models of planet formation are built on underlying physical processes. In order to make sense of the origin of the planets we must first understand the origin of their building blocks. This review comes in two parts. The first part presents a detailed description of six key mechanisms of planet formation:

- The structure and evolution of protoplanetary disks
- The formation of planetesimals
- Accretion of protoplanets
- Orbital migration of growing planets
- Gas accretion and giant planet migration
- Resonance trapping during planet migration.

While this is not a comprehensive list, it includes processes for which our understanding has changed in recent years or for which key uncertainties remain. The second part of this review shows how global models are built out of planet formation processes. We present global models to explain different populations of known planetary systems, including close-in small/low-mass planets (i.e., *super-Earths*), giant exoplanets, and the Solar System's planets. We discuss the different sources of water on rocky exoplanets, and use cosmochemical measurements to constrain the origin of Earth's water. We point out the successes and failings of different models and how they may be falsified. Finally, we lay out a path for the future trajectory of planet formation studies.

S. N. Raymond (✉)
Laboratoire d'Astrophysique de Bordeaux, CNRS and Université de Bordeaux, Pessac, France

A. Morbidelli
Laboratoire Lagrange, Observatoire de la Cote d'Azur, Nice, France
e-mail: morby@oca.eu

© Springer Nature Switzerland AG 2022
K. Biazzo et al. (eds.), *Demographics of Exoplanetary Systems*,
Astrophysics and Space Science Library 466,
https://doi.org/10.1007/978-3-030-88124-5_1

1.1 Observational Constraints on Planet Formation Models

If planet building is akin to cooking, then a review of planet formation is a cookbook. Planetary systems—like dishes—come in many shapes and sizes. Just as one cooking method cannot produce all foods, a single growth history cannot explain all planets. While the diversity of dishes reflects a range of cooking techniques and tools, they are all drawn from a common set of cooking methods. Likewise, the diversity of planetary systems can be explained by different combinations of processes drawn from a common set of physical mechanisms. Our goal in this review is first to describe the key processes of planet formation and then to show how they may be combined to generate global models, or recipes, for different types of planetary systems.

To illustrate the processes involved, Fig. 1.1 shows a cartoon picture of our current vision for the growth of Earth and Jupiter. Both planets are thought to have formed from planetesimals in different parts of the Solar System. In our current understanding, the growth tracks of these planets diverge during the pebble accretion process, which is likely to be much more efficient past the snow line (Lambrechts and Johansen 2014; Morbidelli et al. 2015). There exists a much larger diversity of planets than just Jupiter and Earth, and many vital processes are not included in the figure, yet it serves to illustrate how divergent formation pathways can contribute to planetary diversity.

We start this review by summarizing the key constraints on planet formation models. Constraints come from Solar System measurements (e.g., meteorites), observations of other planetary systems (e.g., exoplanets and protoplanetary disks), as well as laboratory measurements (e.g., to measure the sticking properties of small grains).

Solar System Constraints

Centuries of human observation have generated a census of the Solar System, albeit one that is still not 100% complete. The most important constraints for planet formation include our system's orbital architecture as well as compositional and timing information gleaned from in-situ measurements. An important but challenging exercise is to distill the multitude of existing constraints into just a few large-scale factors to which resolution-limited models can be compared.

The central Solar System constraints are:

- **The masses and orbits of the terrestrial planets.**[1]. The key quantities include their number, their absolute masses and mass ratios, and their low-eccentricity, low-inclination orbits. These have been quantified in studies that attempted to match their orbital distribution. For example the normalized angular momentum deficit AMD is defined as (Laskar 1997; Chambers 2001):

$$AMD = \frac{\sum_j m_j \sqrt{a_j} \left(1 - \cos(i_j)\sqrt{1 - e_j^2}\right)}{\sum_j m_j \sqrt{a_j}}, \quad (1.1)$$

---

[1] The terms "terrestrial" and "rocky" planet are interchangeable: the Solar System community generally uses the term *terrestrial* and the exoplanet community uses *rocky* We use both terminologies in this review to represent planets with solid surfaces that are dominated (by mass) by rock and iron

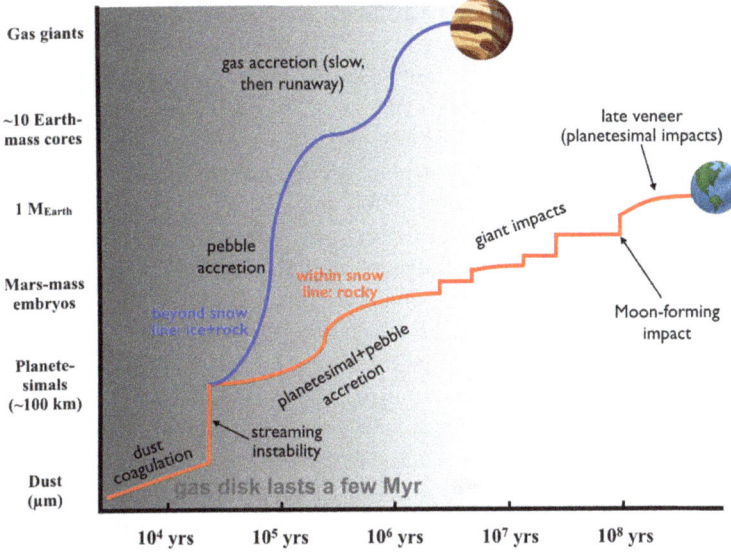

Fig. 1.1 Schematic view of some of the processes involved in forming Jupiter and Earth. This diagram is designed to present a broad view of the relevant mechanisms but still does not show a number of important effects. For instance, we know from the age distribution of primitive meteorites that planetesimals in the Solar System formed in many generations, not all at the same time. In addition, this diagram does not depict the large-scale migration thought to be ubiquitous among any planets more massive than roughly an Earth-mass (see discussion in text). Adapted from Meech and Raymond (2019)

where a_j, e_j, i_j, and m_j correspond to planet j's semimajor axis, eccentricity, orbital inclination, and mass. The Solar System's terrestrial planets have an AMD of 0.0018.

The radial mass concentration statistic RMC (called S_c by Chambers 2001) is a measure of the radial mass profile of the planets. It is defined as:

$$RMC = \max\left(\frac{\sum m_j}{\sum m_j [\log_{10}(a/a_j)]^2}\right). \quad (1.2)$$

The function in brackets is calculated sweeping a across all radii, and the RMC represents the maximum. For a one-planet system RMC is infinite. The RMC is higher when the planets' masses are concentrated in narrow radial zones (as is the case in the terrestrial planets, with two large central planets and two small exterior ones). The RMC becomes smaller for systems that are more spread out and systems in which all planets have similar masses. The Solar System's terrestrial planets' RMC is 89.9.

Confronting distributions of simulated planets with these empirical statistics (as well as other ones) has become a powerful and commonly-used discriminant of terrestrial planet formation models (Chambers 2001; O'Brien et al. 2006; Raymond et al. 2006a, 2009b, 2014; Clement et al. 2018; Lykawka and Ito 2019).

- **The masses and orbits of the giant planets.** As for the terrestrial planets, the number (two gas giant, two ice giant), masses and orbits of the giant planets are the central constraints. The orbital spacing of the planets is also important, for instance the fact that no pair of giant planets is located in mean motion resonance. An important, overarching factor is simply that the Solar System's giant planets are located far from the Sun, well exterior to the orbits of the terrestrial planets.

- **The orbital and compositional structure of the asteroid belt.** While spread over a huge area, the asteroid belt contains only $\sim 4.5 \times 10^{-4} \, M_\oplus$ in total mass (Krasinsky et al. 2002; Kruijer and Folkner 2013; DeMeo and Carry 2013), orders of magnitude less than would be inferred from models of planet-forming disks such as the very simplistic minimum-mass solar nebula model (Weidenschilling 1977a; Hayashi 1981). The orbits of the asteroids are excited, with eccentricities that are roughly evenly distribution from zero to 0.3 and inclinations evenly spread from zero to more than 20° (a rough stability limit given the orbits of the planets). While there are a number of compositional groups within the belt, the general trend is that the inner main belt is dominated by S-types and the outer main belt by C-types (Gradie and Tedesco 1982; DeMeo and Carry 2013, 2014). S-type asteroids are associated with ordinary chondrites, which are quite dry (with water contents less than 0.1% by mass), and C-types are linked with carbonaceous chondrites, some of which (CI, CM meteorites) contain $\sim 10\%$ water by mass (Robert et al. 1977; Kerridge 1985; Alexander et al. 2018).

- **The cosmochemically-constrained growth histories of rocky bodies in the inner Solar System.** Isotopic chronometers have been used to constrain the accretion timescales of different solid bodies in the Solar System. Ages are generally measured with respect to CAIs (Calcium and Aluminum-rich Inclusions), mm-sized inclusions in chondritic meteorites that are dated to be 4.568 Gyr old (Bouvier and Wadhwa 2010). Cosmochemical measurements indicate that chondrules, which are similar in size to CAIs, started to form at roughly the same time (Connelly et al. 2008; Nyquist et al. 2009). Age dating of iron meteorites suggests that differentiated bodies—large planetesimals or planetary embryos—were formed in the inner Solar System within 1 Myr of CAIs (Halliday and Kleine 2006; Kruijer et al. 2014; Schiller et al. 2015). Isotopic analyses of Martian meteorites show that Mars was fully formed within 5–10 Myr after CAIs (Nimmo and Kleine 2007; Dauphas and Pourmand 2011), whereas similar analyses of Earth rocks suggest that Earth's accretion did not finish until much later, roughly 100 Myr after CAIs (Touboul et al. 2007; Kleine et al. 2009).

  There is evidence that two populations of isotopically-distinct chondritic meteorites—the so-called carbonaceous and non-carbonaceous meteorites—have similar age distributions (Kruijer et al. 2017). Given that chondrules are expected to undergo very fast radial drift within the disk (Weidenschilling 1977b; Lambrechts and Johansen 2012), this suggests that the two populations were kept apart and radially segregated, perhaps by the early growth of Jupiter's core (Kruijer et al. 2017).

Constraints from Observations of Planet-Forming Disks Around Other Stars

Gas-dominated protoplanetary disks are the birthplaces of planets. Disks' structure and evolution plays a central role in numerous processes such as how dust drifts (Birnstiel et al. 2016), where planetesimals form (Drążkowska et al. 2016; Drążkowska and Alibert 2017), and what direction and how fast planets migrate (Bitsch et al. 2015a).

We briefly summarize the main observational constraints from protoplanetary disks for planet formation models (see also dedicated reviews: Williams and Cieza 2011; Armitage 2011; Alexander et al. 2014):

- **Disk lifetime**. In young clusters, virtually all stars have detectable hot dust, which is used as a tracer for the presence of gaseous disks (Haisch et al. 2001; Briceño et al. 2001). However, in old clusters very few stars have detectable disks. Analyses of a large number of clusters of different ages indicate that the typical timescale for disks to dissipate is a few Myr (Haisch et al. 2001; Briceño et al. 2001; Hillenbrand et al. 2008; Mamajek 2009). Figure 1.2 shows this trend, with the fraction of stars with disks decreasing as a function of cluster age. It is worth noting that observational biases are at play, as the selection of stars that are members of clusters can affect the interpreted disk dissipation timescale (Pfalzner et al. 2014).

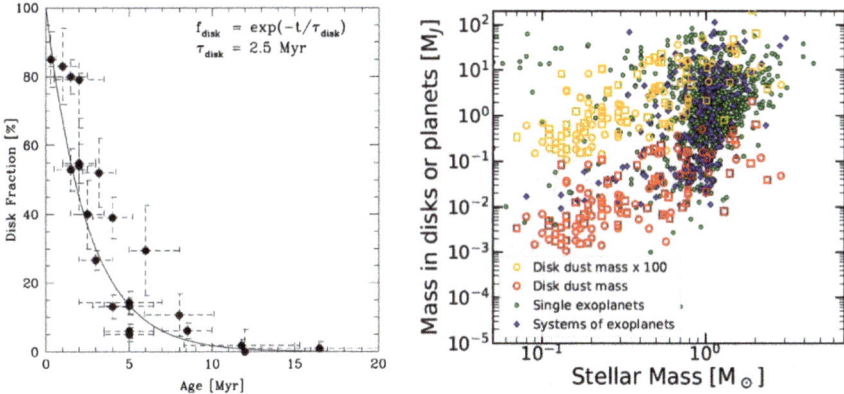

Fig. 1.2 Two observational constraints on planet-forming disks. *Left-hand panel:* The fraction of stars that have detectable disks in clusters of different ages. This suggests that the typical gaseous planet forming disk only lasts a few Myr (Haisch et al. 2001; Briceño et al. 2001; Hillenbrand et al 2008). Figure courtesy of Eric Mamajek, from Mamajek (2009). *Right-hand panel:* A comparison between inferred disk masses and the mass in planets in different systems, as a function of host star mass. The dust mass (red) is measured using sub-mm observations (and making the assumption that the emission is optically thin), and the gas mass is inferred by imposing a 100:1 gas to dust ratio. There is considerable tension, as the population of disks does not appear massive enough to act as the precursors of the population of known planets. The solution to this problem is not immediately obvious. Perhaps disk masses are systematically underestimated (Greaves and Rice 2011), or perhaps disks are continuously re-supplied with material from within their birth clusters via Bondi-Hoyle accretion (Throop and Bally 2008; Moeckel and Throop 2009). Figure courtesy of Carlo Felice Manara, from Manara et al. (2018)

- **Disk masses**. Most masses of protoplanetary disks are measured using sub-mm observations of the outer parts of the disk in which the emission is thought to be optically thin (Williams and Cieza 2011). Disk masses are commonly found to be roughly equivalent to 1% of the stellar mass, albeit with a 1–2 order of magnitude spread (Eisner and Carpenter 2003; Andrews and Williams 2005, 2007a; Andrews et al. 2010; Williams and Cieza 2011). It has recently been pointed out that there is tension between the inferred disk masses and the masses of exoplanet systems, as a large fraction of disks do not appear to contain enough mass to produce exoplanet systems (Manara et al. 2018; Mulders et al. 2018), even assuming a very high efficiency of planet formation (see Fig. 1.2).
- **Disk structure and evolution**. ALMA observations suggest that disks are typically 10–100 au in scale (Barenfeld et al. 2017), similar to the expected dimensions of the Sun's protoplanetary disk (Hayashi 1981; Gomes et al. 2004; Kretke and Lin 2012). Sub-mm observations at different radii indicate that the surface density of dust Σ in the outer parts of disks follows a roughly $\Sigma \sim r^{-1}$ radial surface density slope (Mulders et al. 2000; Looney et al. 2003; Andrews and Williams 2007b; Andrews et al. 2009), consistent with simple models for accretion disks. Many disks observed with ALMA show ringed substructure (ALMA Partnership 2015; Andrews et al. 2016, 2018). Disks are thought to evolve by accreting onto their host stars, and the accretion rate itself has been measured to vary as a function of time; indeed, the accretion rate is often used as a proxy for disk age (Hartmann et al. 1998; Armitage 2011). As disks age, they evaporate from the inside-out by radiation from the central star (Alexander and Armitage 2006; Owen et al. 2011; Alexander et al. 2014) and, depending on the stellar environment, may also evaporate from the outside-in due to external irradiation (Hollenbach et al. 1994; Adams et al. 2004).
- **Dust around older stars**. Older stars with no more gas disks often have observable dust, called *debris disks* (recent reviews: Matthews et al., 2014; Hughes et al., 2018). Roughly 20% of Sun-like stars are found to have dust at mid-infrared wavelengths (Bryden et al. 2006; Trilling et al. 2008; Montesinos et al. 2016). This dust is thought to be associated with the slow collisional evolution of outer planetesimal belts akin to our Kuiper belt, but generally containing much more mass (Wyatt 2008; Krivov 2010). The occurrence rate of dust is observed to decrease with the stellar age (Meyer et al. 2008; Carpenter et al. 2009). More [less] massive stars have significantly higher [lower] occurrence of debris disks (Su et al. 2006; Lega et al. 2012). There is no clear observed correlation between debris disks and planets (Moro-Martín et al. 2007, 2015; Marshall et al. 2014). A significant fraction of old stars have been found to have warm or hot *exo-zodiacal* dust (Absil et al. 2010; Ertel et al. 2014; Kral et al. 2017). The origin of this dust remains mysterious as there is no clear correlation between the presence of cold and hot dust (Ertel et al. 2014).

Constraints from Extra-Solar Planets

With a catalog of thousands of known exoplanets, the constraints from planets around other stars are extremely rich and constantly being improved. Figure 1.3 shows the orbital architecture of a (non-representative) selection of known exoplanet systems. While there exist biases in the detection methods used to find exoplanets (Winn and Fabrycky 2015), their sheer number form the basis of a statistical framework with which to confront planet formation theories.

We can grossly summarize the exoplanet constraints as follows:

- **Occurrence and demographics**. Over the past few decades it has been shown using multiple techniques that exoplanets are essentially ubiquitous (Mayor et al. 2011; Cassan et al. 2012; Batalha et al. 2013). Despite the observational biases, a huge diversity of planetary systems has been discovered. Yet when drawing analogies with the Solar System, it is worth noting that, if our Sun were to have been observed with present-day technology, Jupiter is the only planet that could have been detected (Morbidelli and Raymond 2016; Raymond et al. 2018d). This makes the Solar System unusual at roughly the 1% level. In addition, the Solar System is borderline unusual in *not* containing any close-in low-mass planets (Martin and Livio 2015; Mulders et al. 2018). For the purposes of this review we focus on two categories of planets: gas giants and close-in low-mass planets, made up of high-density 'super-Earths' and puffy 'mini-Neptunes'.
- **Gas giant planets: occurrence and orbital distribution**. Radial velocity surveys have found giant planets to exist around roughly 10% of Sun-like stars (Cumming et al. 2008; Mayor et al. 2011). Roughly one percent of Sun-like stars have *hot Jupiters* on very short-period orbits (Howard et al. 2010; Wright et al. 2011), very few have *warm Jupiters* with orbital radii of up to 0.5–1 au (Butler et al. 2006; Udry et al. 2007), and the occurrence of giant planets increases strongly and plateaus between 1 to several au, and there are hints that it decreases again farther out (Mayor et al. 2011; Fernandes et al. 2019); see Fig. 1.16. Direct imaging surveys have found a dearth of giant planets on wide-period orbits, although only massive young planets tend to be detectable (Bowler and Nielsen 2018). Microlensing surveys find a similar overall abundance of gas giants as radial velocity surveys and have shown that ice giant-mass planets appear to be far more common than their gas giant counterparts (Gould et al. 2010; Suzuki et al. 2016b). Giant planet occurrence has also been shown to be a strong function of stellar metallicity, with higher metallicity stars hosting many more giant planets (Gonzalez et al. 1997; Santos et al. 2001; Laws et al. 2003; Fischer and Valenti 2005; Dawson and Murray-Clay 2013). More in-depth discussions of the above trends in giant planet demographics at short, intermediate, and wide separations are presented in Chap. 3 and 4.
- **Close-in low-mass planets: occurrence and orbital distribution**. Perhaps the most striking exoplanet discovery of the past decade was the amazing abundance of close-in small planets. Planets between roughly Earth and Neptune in size or mass with orbital periods shorter than 100 days have been shown to exist around roughly 30 − 50% of all main sequence stars (Mayor et al. 2011; Howard et al. 2012; Fressin et al. 2013; Dong and Zhu 2013; Petigura et al. 2013; Winn and Fabrycky 2015). Both the masses and radii have been measured for a subset of planets (Marcy et al. 2014) and analyses have shown that the smaller planets tend to have high

densities and the larger ones have low densities, which has been interpreted as a transition between rocky 'super-Earths' and gas-rich 'mini-Neptunes' with a transition size or mass of roughly $1.5-2\,R_\oplus$ or $\sim 3-5\,M_\oplus$ (Weiss et al. 2013; Weiss and Marcy 2014; Rogers 2015; Wolfgang et al. 2016; Chen and Kipping 2017). For the purposes of this review we generally lump together all close-in planets smaller than Neptune and call them *super-Earths* for simplicity. The super-Earth population has a number of intriguing characteristics that constrain planet formation models. While they span a range of sizes, within a given system super-Earths tens to have very similar sizes (Millholland et al. 2017; Weiss et al. 2018). Their period ratios form a broad distribution and do not cluster at mean motion resonances (Lissauer et al. 2011; Fabrycky et al. 2014). Finally, in the Kepler survey the majority of super-Earth systems only contain a single super-Earth (Batalha et al. 2013; Rowe et al. 2014), which contrasts with the high-multiplicity rate found in radial velocity surveys (Mayor et al. 2011). For detailed reviews and discussions on the demographics of close-in small planets we refer the reader to Chap. 3.

Outline of This Review

The rest of this chapter is structured as follows.

In Sect. 1.2 we will describe six essential mechanisms of planet formation. These are:

- The structure and evolution of protoplanetary disks (Sect. 1.2.1)
- The formation of planetesimals (Sect. 1.2.2)
- Accretion of protoplanets (Sect. 1.2.3)
- Orbital migration of growing planets (Sect. 1.2.4)
- Gas accretion and giant planet migration (Sect. 1.2.5)
- Resonance trapping during planet migration (Sect. 1.2.6).

This list does not include every process related to planet formation. These processes have been selected because they are both important and are areas of active study.

Next we will build global models of planet formation from these processes (Sect. 1.3). We will first focus on models to match the intriguing population of close-in small/low-mass planets: the *super-Earths* (Sect. 1.3.1). Next we will turn our attention to the population of giant planets, using the population of known giant exoplanets to guide our thinking about the formation of our Solar System's giant planets (Sect. 1.3.2). Next we will turn our attention to matching the Solar System itself (Sect. 1.3.3), starting from the classical model of terrestrial planet formation (Sect. 1.3.3.1) and then discussing newer ideas: the Grand Tack (Sect. 1.3.3.2), Low-mass Asteroid belt (Sect. 1.3.3.3) and Early Instability (Sect. 1.3.3.4) models. Then we will discuss the different sources of water on rocky exoplanets, and use cosmochemical measurements to constrain the origin of Earth's water (Sect. 1.3.4). Finally, in Sect. 1.4 we will lay out a path for the future trajectory of planet formation studies.

1 Planet Formation: Key Mechanisms and Global Models 11

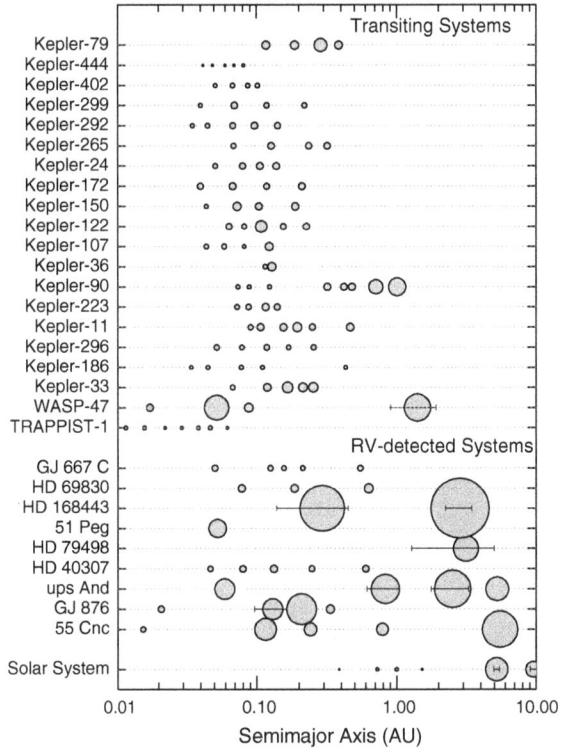

Fig. 1.3 A sample of exoplanet systems selected by hand to illustrate their diversity (adapted from Raymond et al., 2018d). The systems at the top were discovered by the transit method and the bottom systems by radial velocity (RV). Of course, some planets are detected in both transit and RV (e.g. 55 Cnc e; Demory et al., 2011). The planet's size is proportional to its actual size (but is not to scale on the x-axis). For RV planets without transit detections we used the $M \sin i \propto R^{2.06}$ scaling law derived by Lissauer et al. (2011). For giant planets ($M > 50\,M_\oplus$) on eccentric orbits ($e > 0.1$; also for Jupiter and Saturn), the horizontal error bar represents the planet's pericenter to apocenter orbital excursion. The central stars vary in mass and luminosity; e.g., TRAPPIST-1 is an ultracool dwarf star with mass of only $0.09\,M_\odot$ (Gillon et al. 2017). A handful of systems have ~Earth-sized planets in their star's habitable zones, such as Kepler-186 (Quintana et al. 2014), TRAPPIST-1 (Gillon et al. 2017), and GJ 667 C (Anglada-Escudé et al. 2013). Some planetary systems—for example, 55 Cancri (Fischer et al. 2008)—are found in multiple star systems

1.2 Key Processes in Planet Formation

In this section we review basic properties that affect the disk's structure, planetesimals and planet formation as well as dynamical evolution. These processes build the skeleton of our understanding of how planetary systems are formed. As pieces of a puzzle, they will then be put together to develop models on the origin of the different observed structures of planetary systems in Sect. 1.3.

1.2.1 Protoplanetary Disks: Structure and Evolution

Planet formation takes place in gas-dominated disks around young stars. These disks were inferred by Laplace (1822) from the near-perfect coplanarity of the orbits of the planets of the Solar System and of angular momentum conservation during the process of contraction of gas towards the central star. Disks are now routinely observed (imaged directly or deduced through the infrared excess in the spectral energy distribution) around young stars. The largest among protoplanetary disks are now resolved by the ALMA mm-interferometer (Andrews et al. 2018). Here we briefly review the viscous-disk model and the wind-dominated model. For more in depth reading we recommend Armitage (2011) and Turner et al. (2014).

1.2.1.1 Viscous-Disk Model(s)

The simplest model of a protoplanetary disk is a donut of gas and dust in rotation around the central star evolving under the effect of its internal viscosity. This is hereafter dubbed the *viscous-disk model*. Because of Keplerian shear, different rings of the disk rotate with different angular velocities, depending on their distance from the star. Consequently, friction is exerted between adjacent rings. The inner ring, rotating faster, tends to accelerate the outer ring (i.e. it exerts a positive torque) and the outer ring tends to decelerate the inner ring (i.e. exerting a negative torque of equal absolute strength). It can be demonstrated (see for instance Hahn 2009) that such a torque is

$$T = 3\pi \Sigma \nu r^2 \Omega, \quad (1.3)$$

where Σ is the surface density of the disk at the boundary between the two rings, ν is the viscosity, and Ω is the rotational frequency at the distance r from the star.

A fundamental assumption of a viscous-disk model is that it is in steady state, which means the the mass flow of gas $\dot{M}$ is the same at any distance r. Under this assumption it can be demonstrated that the gas flows inwards with a radial speed

$$v_r = -\frac{3}{2}\frac{\nu}{r} \quad (1.4)$$

and that the product $\nu \Sigma$ is independent of r. That is, the radial dependence of Σ is the inverse of the radial dependence of ν. Of course the steady-state assumption is valid only in an infinite disk. In a more realistic disk with a finite size, this assumption is good only in the inner part of the disk, whereas the outer part expands into the vacuum under the effect of the viscous torque (Lynden-Bell and Pringle 1974).

If viscosity rules the radial structure of the disk, pressure rules the vertical structure. At steady state, the disk has to be in hydrostatic equilibrium, which means that the vertical component of the gravitational force exerted by the star has to be equal and opposite to the pressure force, i.e.:

1 Planet Formation: Key Mechanisms and Global Models

$$\frac{GM_\star}{r^3}z = -\frac{1}{\rho}\frac{dP}{dz}, \tag{1.5}$$

where $M_\star$ is the mass of the star, z is the height over the disk's midplane, ρ is the volume density of the disk and P is its pressure. Using the perfect gas law $P = \mathcal{R}\rho T/\mu$ (where $\mathcal{R}$ is the gas constant, μ is the molecular weight of the gas and T is the temperature) and assuming that the gas is vertically isothermal (i.e. T is a function of r only), Eq. (1.5) gives the solution:

$$\rho(z) = \rho(0)\exp\left(-\frac{z^2}{2H^2}\right), \tag{1.6}$$

where $H = \sqrt{\mathcal{R}r^3T/\mu}$ is called the *pressure scale-height* of the disk (the gas extends to several scale-heights, with exponentially vanishing density).

We now need to compute $T(r)$. The simplest way is to assume that the disk is solely heated by the radiation from the central star (passive disk assumption, see Chiang and Goldreich 1997). Most of the disk is opaque to radiation, so the star can illuminate and deposit heat only on the surface layer of the disk, here defined as the layer where the integrated optical depth along a stellar ray reaches unity. For simplicity, we assume that the stellar radiation hits a hard surface, whose height over the midplane is proportional to the pressure scale-height H. Then, the energy deposited on this surface between r and $r + \delta r$ from the star is:

$$E_+^{Irr} = \left(\frac{L_\star}{4\pi r^2}\right)(2\pi r)(r\delta h), \tag{1.7}$$

where δh is the change in aspect ratio H/r over the range δr, namely $d(H/r)/dr\,\delta r$; the parentheses have been put in (1.7) to regroup the terms corresponding respectively to (1) stellar brightness $L_\star$ at distance r, (2) circumference of the ring and (3) projection of $H(r + \delta r) - H(r)$ on the direction orthogonal to the stellar ray hitting the surface. On the other had, the same surface will cool by black-body radiation in space at a rate

$$E_-^{Irr} = 2\pi r \delta r \sigma T^4, \tag{1.8}$$

where σ is Boltzmann's constant. Equating (1.7) and (1.8) and remembering the definition of H as a function of r and T leads to

$$T(r) \propto r^{-3/7} \quad \text{and} \quad H/r \propto r^{2/7}. \tag{1.9}$$

The positive exponent in the dependence of the aspect ratio H/r on r implies that the disk is *flared*. Notice that neither quantities in Eq. (1.9) depend on disk's surface density, opacity or viscosity.

However, because we are dealing with a viscous disk, we cannot neglect the heat released by viscous dissipation, i.e. in the friction between adjacent rings rotating at different speeds. Over a radial width δr, this friction dissipates energy at a rate

(Armitage 2011)

$$E_+^{Visc} = \frac{9}{8}\Sigma\nu\Omega^2 2\pi r\,\delta r. \quad (1.10)$$

This heat is dissipated mostly close to the midplane, where the disk's volume density is highest. This changes the cooling with respect to Eq. (1.8). The energy cannot be freely irradiated in space; it has first to be transported from the midplane through the disk, which is opaque to radiation, to the "surface" boundary with the optically thin layer. Thus the cooling term in Eq. (1.8) has to be divided by $\kappa\Sigma$, where κ is the disk's opacity. Again, by balancing heating and cooling and the definition of H we find:

$$H/r \propto (\dot{M}^2\kappa/\nu r)^{1/8}, \quad (1.11)$$

where we have used that $\dot{M} = 2\pi r v_r \Sigma = -3\pi\nu\Sigma$.

To know the actual radial dependence of this expression, we need to know the radial dependences of κ and ν (remember that $\dot{M}$ is assumed to be independent of r). The opacity depends on temperature, hence on r in a complicated manner, with abrupt transitions when the main chemical species (notably water) condense (Bell and Lin 1994). Let's ignore this for the moment. Concerning $\nu(r)$, Shakura and Sunyaev (1973) proposed from dimensional analysis that the viscosity is proportional to the square of the characteristic length of the system and is inversely proportional to the characteristic timescale. At a distance r from the star, the characteristic length of a disk is $H(r)$ and the characteristic timescale is the inverse of the orbital frequency Ω. Thus they postulated $\nu = \alpha H^2 \Omega$, where α is an unknown coefficient of proportionality. If one adopts this prescription for the viscosity, the viscous-disk model is qualified as an α-*disk model*. Injecting this definition of ν into Eq. (1.11), one obtains

$$H/r \propto \left(\frac{\kappa\dot{M}^2}{\alpha}\right)^{1/10} r^{1/20}. \quad (1.12)$$

This result implies that the aspect ratio of a viscously heated disk is basically independent of r (and $T \propto 1/r$), in sharp contrast with the aspect ratio of a passive disk. Because the disk is both heated by viscosity and illuminated by the star, its aspect ratio at each r will be the maximum between Eqs. (1.12) and (1.9): it will be flat in the inner part and flared in the outer part. Because Eq. (1.12) depends on opacity, accretion rate and α, the transition from the flat disk and the flared disk will depend on these quantities. In particular, given that $\dot{M}$ decreases with time as the disk is consumed by accretion of gas onto the star (Hartmann et al. 1998), this transition moves towards the star as time progresses (Bitsch et al. 2015a). The effects of non-constant opacity introduce wiggles of H/r over this general trend (Bitsch et al. 2015a).

The viscous disk model is simple and neat, but its limitation is in the understanding of the origin of the disk's viscosity. The molecular viscosity of the gas is by orders of magnitude insufficient to deliver the observed accretion rate $\dot{M}$ onto the central star (Hartmann et al. 1998) given a reasonable disk's density, comparable to that of the Minimum Mass Solar Nebula model (MMSN: Weidenschilling 1977a; Hayashi

1981). It was thought that the main source of viscosity is turbulence and that turbulence was generated by the magneto-rotational instability (see Balbus and Hawley 1998). But this instability requires a relatively high ionization of the gas, which is prevented when grains condense in abundance, at a temperature below ∼1000 K (Desch and Turner 2015). Thus, only the very inner part of the disk is expected to be turbulent and have a high viscosity. Beyond the condensation line of silicates, the viscosity should be much lower. Remembering that $\nu\Sigma$ has to be constant with radius, the drop of ν at the silicate line implies an abrupt increase of Σ. As we will see, this property has an important role in the drift of dust and the migration of planets. It was expected that near the surface of the disk, where the gas is optically thin and radiation from the star can efficiently penetrate, enough ionization may be produced to sustain the Magneto Rotational Instability (MRI; Stone et al. 1998). However, these low-density regions are also prone to other effects of non-ideal magneto-hydrodynamics (MHD), like ambipolar diffusion and the Hall effect (Bai and Stone 2013), which are expected to quench turbulence. Thus, turbulent viscosity does not seem to be large enough beyond the silicate condensation line to explain the stellar accretion rates that are observed. This has promoted an alternative model of disk structure and evolution, dominated by the existence of disk winds, as we review next.

1.2.1.2 Wind-Dominated Disk Models

Unlike viscous disk models, which can be treated with simple analytic formulae, the emergence of disk winds and their effects are consequence of non-ideal MHD. Thus their mathematical treatment is complicated and the results can be unveiled only through numerical simulations. Therefore, this section will remain at a phenomenological level. For an in-depth study, we refer the reader to Bai and Stone (2013), Turner et al. (2014) and Bai (2017).

As we have seen above, the ionized regions of the disk have necessarily a low density. If the disk is crossed by a magnetic field, the ions atoms in these low-density regions can travel along the magnetic field lines without suffering collision with neutral molecules. This is the essence of *ambipolar diffusion*.

Consider a frame co-rotating with the disk at radius r_0 from the central star. In this frame, fluid parcels feel an effective potential combining the gravitational and centrifugal potentials. If poloidal (r, z) magnetic field lines act as rigid wires for fluid parcels (which happens as long as the poloidal flow is slower than the local Alfvén speed[2].), then a parcel initially at rest at $r = r_0$ can undergo a runaway if the field line to which it is attached is more inclined than a critical angle. Along such a field line, the effective potential decreases with distance, leading to an acceleration of magnetocentrifugal origin. This yields the inclination angle criterion $\theta > 30°$ for the

---

[2] The Alfvén speed is defined as the ratio between the magnetic field intensity B and $\sqrt{\mu_0 \rho}$, where μ_0 is the permeability of vacuum and ρ is the total mass density of the charged plasma particles. Apart from relativity effects, the Alfvén speed is the phase speed of the Alfvén wave, which is a magnetohydrodynamic wave in which ions oscillate in response to a restoring force provided by an effective tension on the magnetic field lines.

disk-surface poloidal field with respect to the vertical direction. Here fluid parcels rotate at constant angular velocity and so increase their specific angular momentum. Angular momentum is thus extracted from the disk and transferred to the ejected material. As the disk loses angular momentum, some material has to be transferred towards the central star, driving the stellar accretion. The efficiency of this process is directly connected to the disk's magnetic field strength, with a stronger field leading to faster accretion.

In wind-dominated disk models the viscosity can be very low and the disk's density can be of the order of that of the MMSN. The observed accretion rate onto the central star is not due to viscosity (the small value of $\nu\Sigma$ provides only a minor contribution) but is provided by the radial, fast advection of a small amount of gas, typically at 3-4 pressure scale heights H above the disk's midplane (Gressel et al. 2015). The global structure of the disk can be symmetric (top panel of Fig. 1.4) or asymmetric (bottom panel of Fig. 1.4) depending on simulation parameters and the inclusion of different physical effects (e.g. the Hall effect). The origin of the asymmetry is not fully understood. In some cases, the magnetic field lines can be concentrated in narrow radial bands that can fragment the disk into concentric rings (Béthune 2017; see bottom panel of Fig. 1.4). This effect is intriguing in light of the recent ALMA observations of the ringed structure of protoplanetary disks (ALMA Partnership 2015; Andrews et al. 2016, 2018). It should be stressed, however, that there is currently no consensus on the origin of the observed rings. An alternative possibility is that they are the consequence of planet formation (Zhang 2018) or of other disk instabilities (Tominaga et al. 2019). Understanding whether ring formation in protoplanetary disks is a prerequisite for, or a consequence of, planet formation is an essential goal of current research.

Depending on assumptions on the radial gradient of the magnetic field, and hence of the strength of the wind, the gas density of the disk may be partially depleted in its inner part (Suzuki et al. 2016a) or preserve a global power-law structure similar to that of viscous-disk models (Bai 2016) (see Fig. 1.5). A positive surface density gradient, as in Suzuki et al. (2016a), has implication for the radial drift of dust and planetesimals (Ogihara et al. 2018a) and for the migration of protoplanets (Ogihara et al. 2018b). In addition, in Suzuki et al. (2016a) the maximum of the surface density (where dust and migrating planets tend to accumulate) moves outwards as time progresses and the disk evolves.

Disk winds do not generate an appreciable amount of heat. In wind-dominated models, the disk temperature is close to that of a passive disk (Mori et al. 2019; Chambers 2006b), which has a snowline inward of 1 au (Bitsch et al. 2015a). The deficit of water in inner solar system bodies (the terrestrial planets and the parent bodies of enstatite and ordinary chondrites) demonstrates that the protoplanetary solar disk inwards of 2–3 au was warm, at least initially (Morbidelli et al. 2016). This implies that either the viscosity of the disk was quite high or another form of heating—for instance the adiabatic compression of gas as it fell onto the disk from the interstellar medium—was operating early on.

1 Planet Formation: Key Mechanisms and Global Models

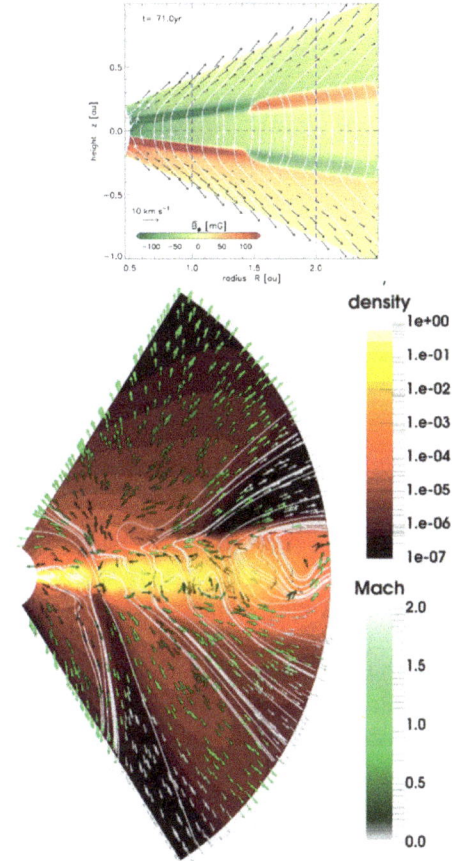

Fig. 1.4 Disk structure in two different wind-dominated disk global models. The top panel (figure courtesy of Oliver Gressel) from Gressel et al. (2015), shows the intensity and polarity of the magnetic field, the field lines and the velocity vectors of the wind. This disk model is symmetric relative to the mid-plane (anti-symmetric in polarity). The bottom panel (figure courtesy of William Béthune) from Béthune et al. (2017), shows the gas density, the magnetic field lines and the wind velocity vectors. This model has no symmetry relative to the midplane. Moreover, the disk is fragmented in rings by the accumulation of magnetic field lines in specific radial intervals

1.2.1.3 Dust Dynamics

The dynamics of dust particles is largely driven by gas drag. Any time that there is a difference in velocity between the gas and the particle, a drag force is exerted on the particle which tends to erase the velocity difference. The *friction time* t_f is defined as the coefficient that relates the accelerations felt by the particle to the gas-particle velocity difference, namely,

$$\dot{\mathbf{v}} = -\frac{1}{t_f}(\mathbf{v} - \mathbf{u}), \tag{1.13}$$

where $\mathbf{v}$ is the particle velocity vector and $\mathbf{u}$ is the gas velocity vector, while $\dot{\mathbf{v}}$ is the particle's acceleration. The smaller a dust particle the shorter its t_f. In the Epstein regime, where the particle size is smaller than the mean free path of a gas molecule, t_f is linearly proportional to the particle's size R. In the Stokes regime the particle size is larger than the mean free path of a gas molecule and $t_f \propto \sqrt{R}$. It is convenient

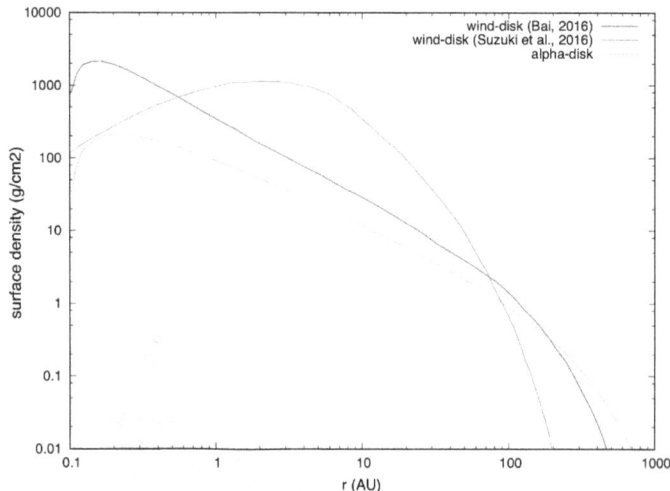

Fig. 1.5 Comparison between the radial surface density distribution of the wind-dominated disk models of Suzuki et al. (2016a), Bai (2016) and an α-disk model

to introduce a dimensionless number, called the *Stokes number*, defined as $\tau_s = \Omega t_f$, which represents the ratio between the friction time and the orbital timescale.

The effects of gas drag are mainly the sedimentation of dust towards the midplane and its radial drift towards pressure maxima.

To describe an orbiting particle as it settles in a disk, cylindrical coordinates are the natural choice. The stellar gravitational force can be decomposed into a radial and a vertical component. The radial component is canceled by the centrifugal force due to the orbital motion. The vertical component, $F_{g,z} = -m\Omega^2 z$, where m is the mass of the particle and z its vertical coordinate, instead accelerates the particle towards the midplane, until its velocity v_{settle} is such that the gas drag force $F_D = m v_{\text{settle}}/t_f$ cancels $F_{g,z}$. This sets $v_{\text{settle}} = \Omega^2 z t_f$ and gives a settling time $T_{\text{settle}} = z/v_{\text{settle}} = 1/(\Omega^2 t_f) = 1/(\Omega \tau_s)$. Thus, for a particle with Stokes number $\tau_s = 1$, the settling time is the orbital timescale. However, turbulence in the disk stirs up the particle layer, which therefore has a finite thickness. Assuming an α-disk model, the scale height of the particle layer is (Chiang and Youdin 2010):

$$H_p = \frac{H}{\sqrt{1 + \tau_s/\alpha}}. \tag{1.14}$$

Dust particles undergo radial drift due to a small difference of their orbital velocity relative to the gas. The gas feels the gravity of the central star and its own pressure. The pressure radial gradient exerts a force $F_r = -(1/\rho) dP/dr$ which can oppose or enhance the gravitational force. As we saw above, $P \propto \rho T$, and $\rho \sim \Sigma/H$. Because Σ and T in general decrease with r, $dP/dr < 0$. The pressure force opposes to the gravity force, diminishing it. Consequently, the gas parcels orbit the star at a speed that is slightly slower than the Keplerian speed at the same location. The difference between the Keplerian speed v_k and the gas orbital speed is ηv_K, where

$$\eta = -\frac{1}{2}\left(\frac{H}{r}\right)^2 \frac{d \log P}{d \log r}. \tag{1.15}$$

The radial velocity of a particle is then:

$$v_r = -2\eta v_K \tau_s/(\tau_s^2 - 1) + u_r/(\tau_s^2 - 1), \tag{1.16}$$

where u_r is the radial component of the gas velocity. Except for extreme cases, u_r is very small and hence the second term in the right-hand side of Eq. (1.16) is negligible relative to the first term. Consequently, the direction of the radial drift of dust depends on the sign of η, i.e., from Eq. (1.15), on the sign of the pressure gradient. If the gradient is negative as in most parts of the disk, the drift is inwards. But in the special regions where the pressure gradient is positive, the particle's drift is outwards. Consequently, dust tends to accumulate at pressure maxima in the disk. We have seen above that the MHD dynamics in the disk can create a sequence of rings and gaps, where the density is alternatively maximal and minimal (bottom panel of Fig. 1.4). Each of these rings therefore features a pressure maximum along a circle. In absence of diffusive motion, the dust would form an infinitely thin ring at the pressure maximum. Turbulence produces diffusion of the dust particles in the radial direction, as it does in the vertical direction. Thus, as dust sedimentation produces a layer with thickness given by Eq. (1.14), dust migration produces a ring with radial thickness $w_p = w/\sqrt{1 + \tau_s/\alpha}$ around the pressure maximum, where w is the width of the gas ring assuming that it has a Gaussian profile (Dullemond et al. 2018).

Consequently, observations of the dust distribution in protoplanetary disks can provide information on the turbulence in the disk. The fact that the width of the gaps in the disk of HL Tau appears to be independent of the azimuth despite the fact that the disk is viewed with an angle smaller than 90°, suggests that the vertical diffusion of dust is very limited such that α in Eq. (1.14) must be be 10^{-4} or less (Pinte et al. 2016). In contrast, the observation that dust is quite broadly distributed in each ring of the disks, suggest that α could be as large as 10^{-3}, depending on the particles Stokes number τ_s, which is not precisely known (Dullemond et al. 2018). These observations therefore suggest that turbulence in the disks is such that the vertical diffusion it produces is weaker than the radial diffusion. It is yet unclear which mechanism could generate turbulence with this property.

1.2.2 Planetesimal Formation

Dust particles orbiting within a disk often collide. If collisions are sufficiently gentle, they stick through electrostatic forces, forming larger particles (Blum and Wurm 2008). One could imagine that this process continues indefinitely, eventually forming macroscopic bodies called planetesimals. However, as we have seen above, particles drift through the disk at different speeds depending on their size (or Stokes number). Thus, there is a minimum speed at which particles of different sizes can collide.

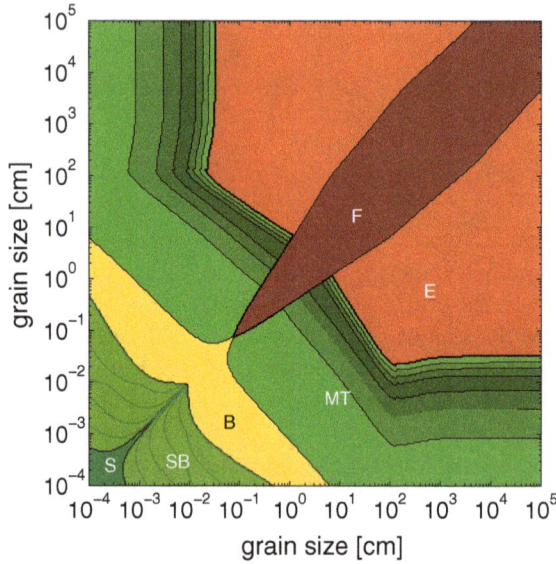

Fig. 1.6 Map of collisional outcome in the disk (figure courtesy of Fredrik Windmark, from Windmark et al., 2012). The sizes of colliding particles are reported on the axes. The colours denote the result of each pair-wise collision. Green denotes growth, red denotes erosion and yellow denotes neither of the above (i.e. a bounce). The label S stands for sticking, SB for stick and bounce, B for bounce, MT for mass transfer, E for erosion and F for fragmentation. This map is computed for compact (silicate) particles, at 3 au

Particles of equal size also have a distribution of impact velocities due to turbulent diffusion.

Figure 1.6 shows a map of the outcome of dust collisions within a simple disk model, from Windmark et al. (2012). Using laboratory experiments on the fate of collisions as a function of particle sizes and mutual velocities, and considering a disk with turbulent diffusion $\alpha = 10^{-3}$ and drift velocities as in a MMSN disk, Windmark et al. (2012) computed the growth/disruption maps for different heliocentric distances. The one from Fig. 1.6 is for a distance of 3 au. The figure shows that, in the inner part of the disk, particles cannot easily grow beyond a millimeter in size. A bouncing barrier prevents particles to grow beyond this limit. If a particle somehow managed to grow to ∼10 cm, its growth could potentially resume by accreting tiny particles. But as soon as particles of comparable sizes hit each other, erosion or catastrophic fragmentation occurs, thus preventing the formation of planetesimal-size objects.

The situation is no better in the outer parts of the disk. In the colder regions, due to the lower velocities and the sticking effect of water ice, particles can grow to larger sizes. But this size is nevertheless limited to a few centimeters due to the so-called *drift barrier* (i.e. large enough particles start drifting faster than they grow: Birnstiel et al. 2016). It has been proposed that if particles are very porous, they could absorb better the collisional energy, thus continuing to grow without bouncing or breaking (Okuzumi et al. 2012). Very porous planetesimals could in principle form this way and their low densities would make them drift very slowly through the disk. But eventually these planetesimals would become compact under the effect of their own gravity and of the ram pressure of the flowing gas (Kataoka et al. 2013). This formation mechanism for planetesimals is still not generally accepted in the community. At best, it could work only in the outer part of the disk, where icy monomers have the tendency to form very porous structures, but not in the inner part of the disk, dominated by silicate particles. Moreover, meteorites show that the interior structure of asteroids is made mostly of compact particles of 100 microns to a millimeter in size, called chondrules, which is not consistent with the porous formation mode.

A mechanism called the *streaming instability* (Youdin and Goodman 2005) can bypass these growth bottlenecks to form planetesimals. Although originally found to be a linear instability (see Jacquet et al. 2011), this instability raises even more powerful effects, which can be qualitatively explained as follows. This instability arises from the speed difference between gas and solid particles. As the differential makes particles feel drag, the friction exerted from the particles back onto the gas accelerates the gas toward the local Keplerian speed. If there is a small overdensity of particles, the local gas is in a less sub-Keplerian rotation than elsewhere; this in turn reduces the local headwind on the particles, which therefore drift more slowly towards the star. Consequently, an isolated particle located farther away in the disc, feeling a stronger headwind and drifting faster towards the star, eventually joins this overdense region. This enhances the local density of particles and reduces further its radial drift. This process drives a positive feedback, i.e. an instability, whereby the local density of particles increases exponentially with time.

Particle clumps generated by the streaming instability can become self-gravitating and contract to form planetesimals. Numerical simulations of the streaming instability process (Johansen et al. 2015; Simon et al. 2016, 2017; Schäfer et al. 2017; Abod et al. 2019) show that planetesimals of a variety of sizes are produced, but those that carry most of the final total mass are those of $\sim$100 km in size. This size is indeed prominent in the observed size-frequency distributions of both asteroids and Kuiper-belt objects. Thus, these models suggest that planetesimals form (at least preferentially) big, in stark contrast with the collisional coagulation model, in which planetesimals would grow progressively from pair wise collisions. If the amount of solid mass in small particles is large enough, even Ceres-size planetesimals can be directly produced from particle clumps (Fig. 1.7).

While Fig. 1.7 shows that the streaming instability can clearly form planetesimals, a concern arises from the initial conditions of such simulations. Simulations find that quite large particles are needed for optimal concentration, corresponding to at least decimeters in size when applied to the asteroid belt. Chondrules—a ubiquitous component of primitive meteorites—typically have sizes from 0.1 to 1 mm but such small particles are hard to concentrate in vortices or through the streaming instability. High-resolution numerical simulations (Carrera et al. 2015; Yang et al. 2017) show that chondrule-size particles can trigger the streaming instability only if the initial mass ratio between these particles and the gas is larger than about 4%. The initial solid/gas ratio of the Solar-System disk is thought to have been $\sim$1%. At face value,

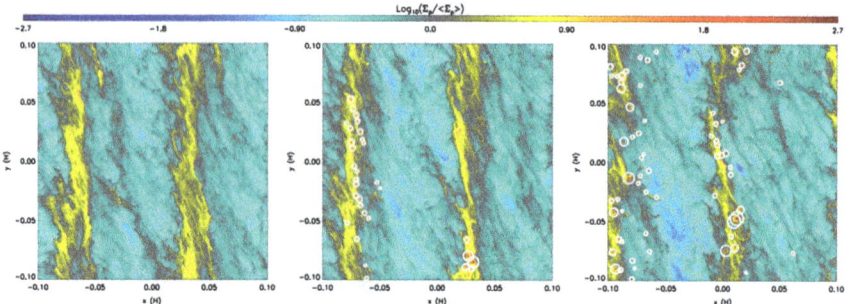

Fig. 1.7 Snapshots in time of a simulation of the streaming instability (figure courtesy of Jacob B. Simon, from Simon et al. 2016). The color scale shows the vertically integrated particle surface density normalized to the average particle surface density (log scale). Time increases from left to right. The left panel shows the clumping due to the streaming instability in the absence of self-gravity, but right before self-gravity is activated ($t = 110\,\Omega^{-1}$). The middle panel corresponds to a point shortly after self-gravity was activated ($t = 112.5\,\Omega^{-1}$), and the right panel corresponds to a time in which most of the planetesimals have formed ($t = 117.6\,\Omega^{-1}$). In the middle and right panel, each planetesimal is marked via a circle of the size of the Hill sphere

planetesimals should not have formed as agglomerates of chondrules. A possibility is that future simulations with even-higher resolution and run on longer timescales will show that the instability can occur for a smaller solid/gas ratio, approaching the value measured in the Sun.

Certain locations within the disk may act as preferred sites of planetesimal formation. Drifting particles may first accumulate at distinct radii in the disc where their radial speed is slowest and then, thanks to the locally enhanced particle/gas ratio, locally trigger the streaming instability. Two locations have been identified for this preliminary radial pile-up. One is in the vicinity of the snowline, where water transitions from vapor to solid form (Ida and Guillot 2016; Armitage et al. 2016; Schoonenberg and Ormel 2017). The other is in the vicinity of 1 au (Drążkowska et al. 2016). These could be the two locations where planetesimals could form very early in the protoplanetary disk (Drążkowska and Dullemond 2018). Elsewhere in the disk, the conditions for planetesimal formation via the streaming instability would only be met later on, when gas was substantially depleted by photo-evaporation from the central star, provided that the solids remained abundant (Throop and Bally 2008; Carrera et al. 2017).

At least at the qualitative level, this picture is consistent with available data for the Solar System. The meteorite record reveals that some planetesimals formed very early, in the first few 10^5 yr (Kleine et al. 2009; Kruijer et al. 2014; Schiller et al. 2015). Because of the large abundance of short-lived radioactive elements present at the early time (Grimm and McSween 1993; Monteux et al. 2018), these first planetesimals melted and differentiated, and are today the parent bodies of iron meteorites. But a second population of planetesimals formed 2 to 4 Myr later (Villeneuve et al. 2009). These planetesimals did not melt and are the parent bodies of the primitive meteorites called the chondrites. We can speculate that differentiated planetesimals formed at the two particle pile-up locations mentioned above, whereas the undifferentiated planetesimals formed elsewhere, for instance in the asteroid belt while the gas density was declining. Yet these preferred locations were certainly themselves evolving in time (Drążkowska and Dullemond 2018).

Strong support for the streaming instability model comes from Kuiper belt binaries. These binaries are typically made of objects of similar size and identical colors (see Noll et al. 2008). It has been shown (Nesvorný et al. 2010) that the formation of a binary is the natural outcome of the gravitational collapse of the clump of pebbles formed in the streaming instability, if the angular momentum of the clump is large. Simulations of this process can reproduce the typical semi-major axes, eccentricities and size ratios of the observed binaries. The color match between the two components is a natural consequence of the fact that both are made of the same material. This is a big strength of the model because such color identity cannot be explained in any capture or collisional scenario, given the observed intrinsic difference in colors between any random pair of Kuiper belt objects (KBOs—this statement holds even restricting the analysis to the cold population, which is the most homogeneous component of the Kuiper belt population). Additional evidence for the formation of equal-size KBO binaries by streaming instability is provided by the spatial orientation of binary orbits. Observations (Noll et al. 2008) show a broad distribution of binary inclinations with $\simeq$80% of prograde orbits ($i_b < 90°$) and $\simeq$20% of retrograde orbits ($i_b > 90°$). To explain these observations, Nesvorný et al. (2019) analyzed high-resolution simulations and determined the angular momentum vector of the gravitationally bound clumps produced by the streaming instability. Because the orientation of the angular momentum vector is approximately conserved during collapse, the distribution obtained from these simulations can be compared with known binary inclinations. The comparison shows that the model and observed distributions are indistinguishable. This clinches an argument in favor of the planetesimal formation by the streaming instability and binary formation by gravitational collapse. No other planetesimal formation mechanism has been able so far to reproduce the statistics of orbital plane orientations of the observed binaries.

1.2.3 Accretion of Protoplanets

Once planetesimals appear in the disk they continue to grow by mutual collisions. Gravity plays an important role by bending the trajectories of the colliding objects, which effectively increases their collisional cross-section by a factor

$$F_g = 1 + V_{\text{esc}}^2 / V_{\text{rel}}^2, \tag{1.17}$$

where V_{esc} is the mutual escape velocity defined as $V_{\text{esc}} = [2G(M_1 + M_2)/(R_1 + R_2)]^{1/2}$, M_1, M_2, R_1, R_2 are the masses and radii of the colliding bodies, V_{rel} is their relative velocity before the encounter and G is the gravitational constant. F_g is called the gravitational focusing factor (Safronov 1972).

The mass accretion rate of an object becomes

$$\frac{dM}{dt} \propto R^2 F_g \propto M^{2/3} F_g, \tag{1.18}$$

where the bulk density of planetesimals is assumed to be independent of their mass, so that the planetesimal physical radius $R \propto M^{1/3}$. These equations imply two distinct growth modes called runaway and oligarchic growth.

1.2.3.1 Runaway Growth

If one planetesimal (of mass M) grows quickly, then its escape velocity V_{esc} becomes much larger than its relative velocity V_{rel} with respect to the rest of the planetesimal population. Then, one can approximate F_g as V_{esc}^2/V_{rel}^2. Notice that the approximation $R \propto M^{1/3}$ makes $V_{esc}^2 \propto M^{2/3}$.

Substituting this expression into Eq. (1.18) leads to:

$$\frac{dM}{dt} \propto \frac{M^{4/3}}{V_{rel}^2}, \tag{1.19}$$

or, equivalently:

$$\frac{1}{M}\frac{dM}{dt} \propto \frac{M^{1/3}}{V_{rel}^2}. \tag{1.20}$$

This means that the relative mass-growth rate is a growing function of the body's mass. In other words, small initial differences in mass among planetesimals are rapidly magnified, in an exponential manner. This growth mode is called *runaway growth* (Greenberg et al. 1978; Wetherill and Stewart 1989, 1993; Kokubo and Ida 1996, 1998).

Runaway growth occurs as long as there are objects in the disk for which $V_{esc} \gg V_{rel}$. While V_{esc} is a simple function of the largest planetesimals' masses, V_{rel} is affected by other processes. There are two dynamical damping effects that act to decrease the relative velocities of planetesimals. The first is gas drag. Gas drag not only causes the drift of bodies towards the central star, as seen above, but it also tends to circularize the orbits, thus reducing their relative velocities V_{rel}. Whereas orbital drift vanishes for planetesimals larger than about 1 km in size, eccentricity damping continues to influence bodies up to several tens of kilometers across. However, in a turbulent disk gas drag cannot damp V_{rel} down to zero: in presence of turbulence the relative velocity evolves towards a size-dependent equilibrium value (Ida and Lin 2008). The second damping effect is that of collisions. Particles bouncing off each other tend to acquire parallel velocity vectors, reducing their relative velocity to zero. For a given total mass of the planetesimal population, this effect has a strong dependence on the planetesimal size, roughly $1/r^4$ (Wetherill and Stewart 1993).

Meanwhile, relative velocities are excited by the largest growing planetesimals by gravitational scattering, whose strength depends on those bodies' escape velocities. A planetesimal that experiences a near-miss with the largest body has its trajectory permanently perturbed and will have a relative velocity $V_{rel} \sim V_{esc}$ upon the next return. Thus, the planetesimals tend to acquire relative velocities of the order of the escape velocity from the most massive bodies, and when this happens runaway growth is shut off (see below).

To have an extended phase of runaway growth in a planetesimal disk, it is essential that the bulk of the solid mass is in small planetesimals, so that the damping effects are important. Because small planetesimals collide with each other frequently and either erode into small pieces or grow by coagulation, this condition may not hold for long. Moreover, if planetesimals really form with a preferential size of ~ 100 km, as in the streaming instability scenario, the population of small planetesimals would have been insignificant and therefore runaway growth would have only lasted a short time if it happened at all.

1.2.3.2 Oligarchic Growth

When the velocity dispersion of planetesimals becomes of the order of the escape velocity from the largest bodies, the gravitational focusing factor (Eq. 1.17) becomes of order unity. Consequently the mass growth equation (Eq. 1.18) becomes

$$\frac{1}{M}\frac{dM}{dt} \propto \frac{M^{-1/3}}{V_{rel}^2}. \tag{1.21}$$

In these conditions, the relative growth rate of the large bodies slows with the bodies' growth. Thus, the mass ratios among the large bodies tend to converge to unity.

In principle, one could expect that the small bodies also narrow down their mass difference with the large bodies. But in reality, the large value of V_{rel} prevents the small bodies from accreting each other. Small bodies only contribute to the growth of the large bodies (i.e. those whose escape velocity is of the order of V_{rel}). This phase is called *oligarchic growth* (Kokubo and Ida 1998, 2000).

In practice, oligarchic growth leads to the formation of a group of objects of roughly equal masses, embedded in the disk of planetesimals. The mass gap between oligarchs and planetesimals is typically of a few orders of magnitude. Because of dynamical friction—an equipartition of orbital excitation energy (Chandrasekhar 1943)—planetesimals have orbits that are much more eccentric than the oligarchs. The orbital separation among the oligarchs is of the order of 5 to 10 mutual Hill radii R_H, where:

$$R_H = \frac{a_1 + a_2}{2}\left(\frac{M_1 + M_2}{3M_\star}\right)^{1/3}, \tag{1.22}$$

and a_1, a_2 are the semi-major axes of the orbits of the objects with masses M_1, M_2, and $M_\star$ is the mass of the star.

1.2.3.3 The Need for an Additional Growth Process

In the classic view of planet formation (Wetherill 1992; Lissauer 1987, 1993), the processes of runaway growth and oligarchic growth convert most of the planetesimals mass into a few massive objects: the protoplanets (sometimes called *planetary embryos*). However, this picture does not survive close scrutiny.

In the Solar system, two categories of protoplanets formed within the few Myr lifetime of the gas component of the protoplanetary disk (see Fig. 1.2). In the outer system, a few planets of multiple Earth masses formed and were massive enough to be able to gravitationally capture a substantial mass of H and He from the disk and become the observed giant planets, from Jupiter to Neptune. In the inner disk, instead, the protoplanets only reached a mass of the order of the mass of Mars and eventually formed the terrestrial planets after the disappearance of the gas (see Sect. 1.3.3.1). Thus, the protoplanets in the outer part of the disk were 10-100 times more massive of those in the inner disk. This huge mass ratio is even more surprising if one considers that the orbital periods, which set the natural clock for all dynamical processes including accretion, are ten times longer in the outer disk.

The snowline represents a divide between the inner and the outer disk. The surface density of solid material is expected to increase beyond the snowline due to the availability of water ice (Hayashi 1981). However, this density-increase is only of a factor of ~ 2 (Lodders 2003), which is insufficient to explain the huge mass ratio between protoplanets in the outer and inner parts of the disc (Morbidelli et al. 2015).

In addition, whereas in the inner disk oligarchic growth can continue until most of the planetesimals have been accreted by protoplanets, the situation is much less favorable in the outer disc. There, when the protoplanets become sufficiently massive (about 1 Earth mass), they tend to scatter the planetesimals away, rather than accrete them. In doing this, they clear their neighboring regions, which in turn limits their own growth (Levison et al. 2010). In fact, scattering dominates over growth when the ratio $V_{esc}^2/2V_{orb}^2 > 1$, where V_{esc} is the escape velocity from the surface of the protoplanet and V_{orb} is its orbital speed (so that $\sqrt{2}V_{orb}$ is the escape velocity from the stellar potential well from the orbit of the protoplanet). This ratio is much larger in the outer disc than in the inner disc because $V_{orb}^2 \propto 1/a$, where a is the orbital semi-major axis.

Consequently, understanding the formation of the multi-Earth-mass cores of the giant planets and their huge mass ratio with the protoplanets in the inner Solar System is a major problem of the runaway/oligarchic growth models, and it has prompted the elaboration of a new planet growth paradigm, named *pebble accretion*.

1.2.3.4 Pebble Accretion

Let's take a step back to what seems to be most promising planetesimal formation model: that of self-gravitating clumps of small particles (hereafter called pebbles even though in the inner disc they are expected to be at most mm-size, so that

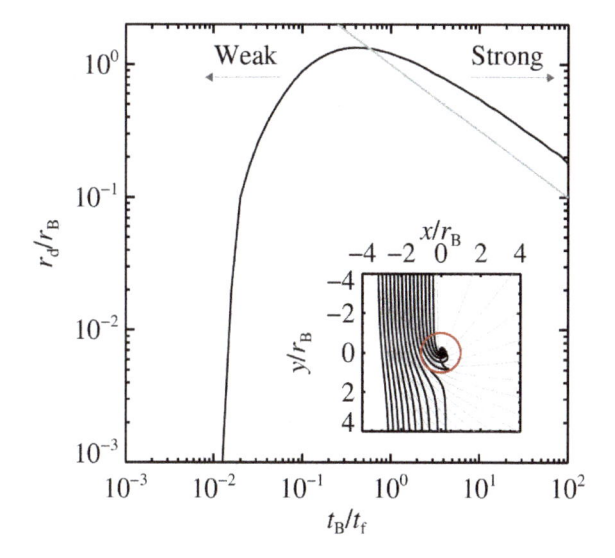

1 Planet Formation: Key Mechanisms and Global Models

Fig. 1.8 Efficiency of pebble accretion. The outer plot shows the accretion radius r_d, normalized to the Bondi radius r_B, as a function of t_B/t_f, where t_f is the friction time and t_B is the time required to cross the Bondi radius at the encounter velocity $v_{\rm rel}$. The smaller is the pebble the larger is t_B/t_f. The inset shows pebble trajectories (black curves) with $t_B/t_f = 1$, which can be compared with those of objects with $t_B/t_f \to 0$ (grey curves). Clearly the accretion radius for the former is much larger. A circle of Bondi radius is plotted in red. Figure courtesy of M. Lambrechts, from Lambrechts and Johansen (2012)

grains would be a more appropriate term). Once a planetesimal forms, it remains embedded in the disk of gas and pebbles and it can keep growing by accreting individual pebbles. This process was first envisioned by Ormel and Klahr (2010) and then studied in detailed by Lambrechts and Johansen (2012, 2014) (see also Johansen and Lambrechts, 2017). To avoid confusion, we call below the accreting body a *protoplanet* and we denote the accreted body as a pebble or a planetesimal, depending on whether it feels strong gas drag.

Pebble accretion is more efficient than planetesimal accretion for two reasons. First, the accretion cross-section for a protoplanet-pebble encounter is much larger than for a protoplanet-planetesimal encounter. As seen above, in a protoplanet-planetesimal encounter the accretion cross-section is $\pi R^2 F_g$, where R is the physical size of the protoplanet and F_g is the gravitational focusing factor. But in a protoplanet pebble encounter it can be as large as πr_d^2, where r_d is the distance at which the protoplanet can deflect the trajectories of the incoming objects. This is because, as soon as the pebble's trajectory starts to be deflected, its relative velocity with the gas increases and gas-drag becomes very strong. Thus, the pebble's trajectory spirals towards the protoplanet. This is shown in the inlet of Fig. 1.8, whereas the outer panel of the figure shows the value of r_d as a function of the pebble's friction time, normalized to the Bondi radius $r_B = GM/v_{\rm rel}^2$ ($v_{\rm rel}$ being the velocity of the pebble relative to the protoplanet, typically of order ηv_K).

The second reason that pebble accretion is more efficient than planetesimal accretion is that pebbles drift in the disk. Thus, the orbital neighborhood of the protoplanet cannot become empty. Even if the protoplanet accretes all the pebbles in its vicinity, the local population of pebbles is renewed by particles drifting inward from larger distances. This does not happen for planetesimals because their radial drift in the disk is negligible.

Provided that the mass-flux of pebbles through the disk is large enough, pebble accretion can grow protoplanets from about a Moon-mass up to multiple Earth-masses, i.e. to form the giant planets cores within the disc's lifetime (Lambrechts and Johansen 2012, 2014). The large mass ratio between protoplanets in the outer vs. inner parts of the disc can be explained by remembering that icy pebbles can be relatively large (a few centimeters in size), whereas in the inner disc the pebble's size is limited to sub-millimeter by the bouncing silicate barrier (chondrule-size particles) and by taking into account that pebble accretion is more efficient for large pebbles than for chondrule-size particles (Morbidelli et al. 2015).

For all these reasons, while some factors remain unknown (particularly the pebble flux and its evolution during the disk lifetime), pebble accretion is now considered to the dominant process of planet formation.

An important point is that pebble accretion cannot continue indefinitely. When a planet grows massive enough it starts opening a gap in the disk. This eventually creates a pressure bump at the outer edge of the gap which stops the flux of pebbles. The mass at which this happens is called *pebble isolation mass* (Morbidelli et al. 2012; Lambrechts et al. 2014) and depends on disk's viscosity and scale height (Bitsch et al. 2018). Once a planet reaches the pebble isolation mass, it stops accreting pebbles. Given that it blocks the inward pebble flux, this means that all protoplanets on interior orbits are starved of pebbles, regardless of their masses. Turbulent diffusion can allow some pebbles to pass through the pressure bump (Weber et al. 2018), particularly the smallest ones, because the effects of diffusion are proportional to $\sqrt{\alpha/\tau_s}$.

1.2.4 Orbital Migration of Planets

Once a massive body forms in the disk, it perturbs the distribution of the gas in which it is embedded. We generically denote the perturbing body as a *planet*. In this section we consider planets smaller than a few tens of Earth masses. The case of giant planets will be discussed in the next section.

Analytic and numerical studies have shown that a planet generates a spiral density wave in the disk, as shown in Fig. 1.9 (Goldreich and Tremaine 1979, 1980; Lin and Papaloizou 1979; Ward 1986; Tanaka et al. 2002). The exterior wave trails the planet. The gravitational attraction that the wave exerts on the planet produces a negative torque that slows the planet down. The interior wave leads the planet and exerts a positive torque. The net effect on the planet depends on the balance between these two torques of opposite signs. It was shown by Ward (1986) that for axis-symmetric disks with any power-law radial density profiles, the negative torque exerted by the wave

Fig. 1.9 The spiral density wave launched by a planet in the gas disk. The color brightness is proportional to the gas surface density. Courtesy of F. Masset

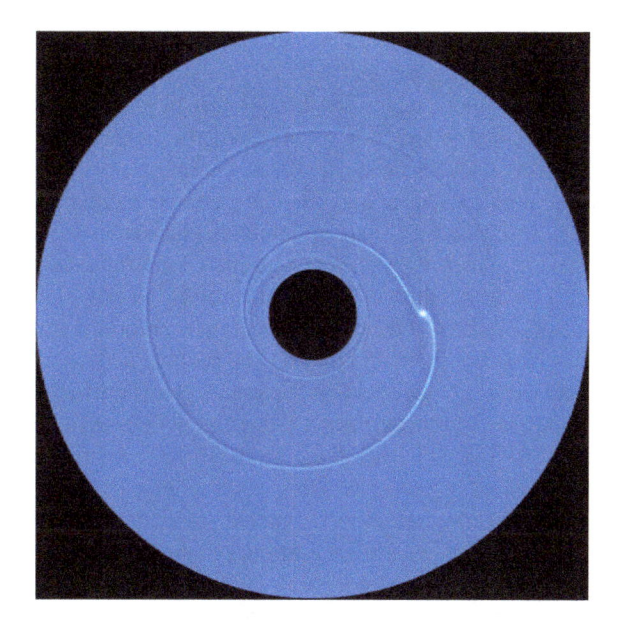

in the outer disk wins. This is because a power-law disk is in slight sub-keplerian rotation, so that the gravitational interaction of the planet with a disk's ring located at $a_p + \delta a$ (a_p being the orbital radius of the planet) is stronger than with the ring located at $a_p - \delta a$, given that the relative velocity with the former is smaller. As a consequence of this imbalance, the planet must lose angular momentum and its orbit shrinks: the planet migrates towards the central star. This process is called *Type-I migration*. The planet migration speed is:

$$\frac{da}{dt} \propto M_p \Sigma_g \left(\frac{a}{H}\right)^2, \qquad (1.23)$$

where a is the orbital radius of the planet (here assumed to be on a circular orbit), M_p is its mass, Σ_g is the surface density of the gas disk and H is its height at the distance a from the central star. A precise migration formula, function of the power-law index of the density and temperature radial profiles, can be found in Paardekooper et al. (2010, 2011). The planet-disk interaction also damps the planet's orbital eccentricity and inclination if these are initially non-zero. These damping timescales are a factor $(H/a)^2$ smaller than the migration timescale (Tanaka and Ward 2004; Cresswell et al. 2007).

Precise calculations show that an Earth-mass body at 1 au, in a Minimum Mass Solar Nebula ($\Sigma_g = 1700$ g/cm^2) with scale height $H/a = 5\%$, migrates into the star in 200,000 yr. For different planets or different disks, the migration time can be scaled using the relationship reported in Eq. (1.23). So, Lunar- to Mars-mass protoplanets are only mildly affected by Type-I migration because their migration timescales exceed the few Myr lifetime of the gas disk. Conversely, for more massive planets, migration should be substantial and should bring them close to the star before that the disk disappears.

Planet-disk interactions through the spiral density wave are only part of the story. An important interaction occurs along the planet's orbit due to fluid elements that are forced to do horseshoe-like librations in a frame corotating with the planet. Along these librations, as a fluid element passes from inside the planet's orbit to outside, it receives a positive angular momentum kick and exerts an equivalent but negative kick onto the planet. The opposite happens when a fluid element passes from outside of the planet's orbit to inside. It can be proven (Masset et al. 2006) that, if the radial surface density gradient at the planet's location is proportional to $1/r^{3/2}$ (i.e. the *vortensity* of the disk is constant with radius), the positive and negative kicks cancel out perfectly, and there is no net effect on the planet. But for different radial profiles, there is a net torque on the planet, named the *vortensity-driven corotation torque* (Paardekooper et al. 2010, 2011). If the disk's profile is shallower than $1/r^{3/2}$ this corotation torque is positive and it slows down migration relative to the rate from Eq. (1.23). Moreover, if the disk's radial surface density gradient is positive and sufficiently steep, the corotation torque (positive) can exceed the (negative) torque exerted by the wave and reverse migration (Masset et al. 2006). This implies the existence of a location in the disk—typically near the density maximum—where migration stops, dubbed *planet trap* (Lyra et al. 2010). Positive surface density gradients could exist at the inner edge of the protoplanetary disk, where the disk is truncated by the stellar magnetic torque (Chang et al. 2010), or at transition from the MRI-active to the MRI-inactive parts of the disk (Flock et al. 2017, 2019)—also very close to the central star—or at the inner edge of each ring observed in MHD simulations (see bottom panel of Fig. 1.4). Therefore, there can be several *planet traps* in the disk (Hasegawa and Pudritz 2012; Baillié et al. 2015).

The corotation region can also exert a positive torque on the planet in a region of the disk where the radial temperature gradient is steeper that $1/r$ (Paardekooper and Mellema 2006). This torque is called *entropy-driven corotation torque* (Paardekooper et al. 2010, 2011). Steep temperature gradients exist behind the "bumps" of the disk's aspect ratio that are generated by opacity transitions (Bitsch et al. 2015a). However, because the disk evolves over time towards a passive disk, with a temperature gradient shallower than $1/r$, the outward migration regions generated by the entropy-driven corotation torque exist only temporarily (Bitsch et al. 2015a).

Other torques can act on the planet and affect its migration in specific cases. If the viscosity of the disk is very small, *dynamical torques* are produced as a feedback of planet migration (Paardekooper 2014; Pierens 2015; Pierens and Raymond 2016). The feedback is negative, i.e. it acts to decelerate the migration, if the disk's surface density profile is shallower than $1/r^{3/2}$ and migration is inwards, or if the profile is steeper than $1/r^{3/2}$ and migration is outwards. In the opposite cases, the dynamical torque accelerates the migration.

Low-viscosity disks are also prone to a number of instabilities generating vortices when submitted to the perturbation of a planet. As a result, the migration of the planet can become stochastic, due to the interaction with these variable density structures (McNally et al. 2019).

As it approaches the planet gas is compressed then decompressed so that its temperature first increases then decreases. Because hot gas loses energy by irradiation, the situation is not symmetric and the gas is colder (i.e. denser) after the conjunction with the planet than it was before conjunction. This generates a negative torque (Lega et al. 2014). On the other hand, if a planet is accreting solids, gravitational energy is released as heat. This source of heat modifies the density of the gas in the vicinity of the planet. In some conditions, this *heating torque* can exceed the previous effect, so that the net effect is positive and can even overcome the negative torque exerted by the wave (Benítez-Llambay et al. 2015). This torque, however, also enhances the orbital eccentricity of the planet (Eklund and Masset 2017), which in turn reduces its accretion rate. Thus some self-regulated regime can be achieved (Masset 2017).

Finally, even the steady-state dust distribution can be perturbed by the presence of the planet, acquiring asymmetries that can exert torques on the planet (Benítez-Llambay and Pessah 2018).

To summarize, although the migration of a small-mass planet is typically inward and fast, there can be locations in the disk where migration is halted, as well as a number of temporary mechanisms that can reduce or enhance the migration rate. Therefore, the actual migration of a planet must be investigated in a case-by-case basis and requires a realistic modeling of the disk, given that its density and temperature gradients, opacity, viscosity and dust distribution play a key role. Unfortunately, so far our limited theoretical and observational knowledge of disks hampers our ability to model planet migration quantitatively.

1.2.5 Gas Accretion and Giant Planet Migration

A massive planet immersed in a gas disk can attract gas by gravity and build up an atmosphere. To distinguish between the solid part of the planet from its atmosphere, we will call the formed the *core*.

The closed set of equations that govern the distribution of gas in the atmosphere are:

$$\frac{dP}{dr} = \frac{GM(r)\rho(r)}{r}, \tag{1.24}$$

where $\rho(r)$ is the density of the gas at a distance r from the center of the planet, which describes hydrostatic equilibrium (gravity balanced by the internal pressure gradient);

$$\frac{dM}{dr} = 4\pi r^2 \rho(r), \tag{1.25}$$

which describes the planet's mass-radius relationship $M(r)$ from $M(r_c) = M_c$, where r_c is the radius of the core and M_c is its mass;

$$\frac{dT}{dr} = -\frac{3\kappa L \rho}{64\pi\sigma r^2 T^3}, \tag{1.26}$$

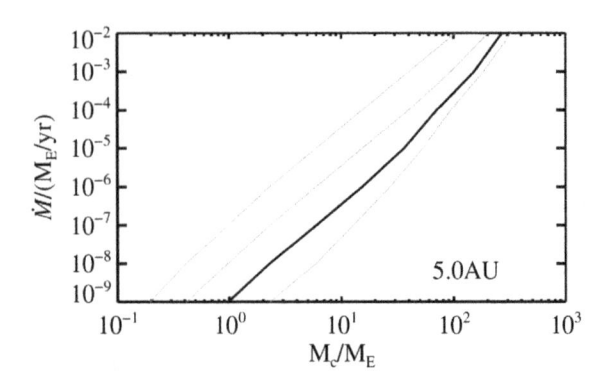

Fig. 1.10 The value of M_{crit} (on the horizontal axis) as a function of core's accretion rate (on the vertical axis), for different values of the opacity—increasing from left to right. The atmosphere can be in hydrostatic equilibrium only if $M_c < M_{crit}$. Figure courtesy of M. Lambrechts, from Lambrechts and Johansen (2014)

where σ is Boltzmann's constant, κ the gas opacity and $L \propto M_c \dot{M}_c / r_c$ is the luminosity of the core, due to the release of the gravitational energy delivered by the accretion of solids at a rate $\dot{M}_c$;

$$P = \frac{\mathcal{R}}{\mu \rho T}, \tag{1.27}$$

which is the equation of state, here for a perfect gas ($\mathcal{R}$ being the perfect gas constant and μ the molecular weight).

One can attempt to solve this set of equations using boundary condition $\rho(r_b) = \rho_0$ and $T(r_b) = T_0$, where ρ_0 and T_0 are the disk's values for gas density and temperature, respectively, and r_b is the disk-planet boundary, typically the Bondi radius $r_b = 2GM/c_s^2$ (c_s being the sound speed). A solution exists only for $M_c < M_{crit}(\kappa, L)$ where $M_{crit}(\kappa, L)$ is a threshold value depending on opacity and luminosity (and disk's properties), as shown in Fig. 1.10 (Piso and Youdin 2014; Lambrechts and Johansen 2014).

M_{crit} tends to zero as the core's accretion rate tends to zero. If $L = 0$, no hydrostatic solution can exist. Recall from the previous section that when a planet reaches the pebble isolation mass the accretion pebbles effectively stops (Morbidelli et al. 2012; Bitsch et al. 2018). This drastically changes the value of L and hence M_{crit}. If the atmosphere of the planet was in hydrostatic equilibrium up to that point, it may be out of equilibrium. As a rule of thumb, when M_c approaches M_{crit} the mass of the atmosphere in hydrostatic equilibrium approaches that of the core. This triggers runaway gas accretion (Lambrechts and Johansen 2014).

When the atmosphere is no longer in hydrostatic equilibrium, it contracts under the effect of gravity. The compression of gas releases energy, so the atmosphere can only contract on the Kelvin-Helmholtz timescale, which is effectively the atmosphere's cooling timescale through irradiation. As the atmosphere contracts, new gas can be captured within the Bondi radius r_b. This increases the mass of the atmosphere and hence the gravity of the full planet. This triggers a positive feedback on the accretion

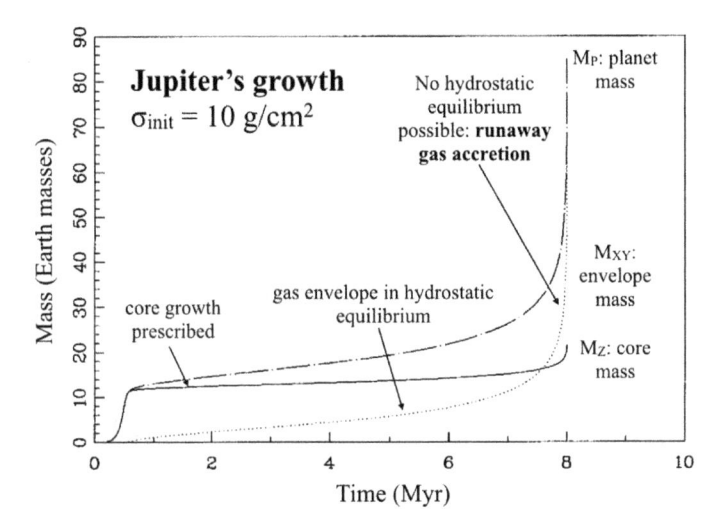

1 Planet Formation: Key Mechanisms and Global Models

Fig. 1.11 Runaway growth of a giant planet near Jupiter's orbital radius. The solid curve shows the mass of the solid core, the dotted curve the mass of the atmosphere and the dashed curve the total mass of the planet. Here the accretion of solids onto the core is prescribed according to a now obsolete planetesimal accretion model. The approximate boundary in time between the hydrostatic regime and the runaway regime is at ∼7.5 Myr. While the runaway gas accretion regime is not in hydrostatic equilibrium it can be modeled as a series of equilibrium states. Adapted from Pollack et al. (1996)

rate, so that the mass of the planet's atmosphere increases exponentially with time (see Fig. 1.11) (Pollack et al. 1996; Hubickyj et al. 2005; Lissauer et al. 2009).

A number of studies have modeled atmospheric accretion during the runaway phase (Ikoma et al. 2000; Coleman et al. 2017; Lambrechts et al. 2019b) using different approaches. All have confirmed the runaway gas accretion with a mass-doubling timescales for Jupiter-mass planets of order $10^4 - 10^5$ yr in a MMSN disk. This raises the question of why Saturn-mass exoplanets outnumber Jupiter-mass ones, which drastically outnumber super-Jupiters (Butler et al. 2006; Udry et al. 2007). One possibility is that giant planets enter in their runaway phase late, as the disk is disappearing (Pollack et al. 1996; Bitsch et al. 2015b). Given that the mass doubling timescale in the runaway phase is so much shorter than the disk's lifetime, this appears to be a surprising coincidence. The other possibility is that the growth of a planet is limited by the ability of the disk to transport gas radially. We know from observations, however, that the gas accretion rate onto the central star is typically of the order of 10^{-8} $M_\odot$/yr (Hartmann et al. 1998), which means that a Jupiter-mass of gas passes through the orbit of a giant planet in only 10^5 yr, again much shorter than the disk's lifetime. Perhaps the study of giant planet growth in low-viscosity disks dominated by winds can bring a solution to the problem, given the different geometry and mechanism of transport for the accreting gas relative to a classic, viscous disk.

1.2.5.1 Gap Opening and Type-II Migration

As we have seen in Sect. 1.2.4, a planet embedded in a disk exerts a positive torque on the outer part of the disk and a negative torque on the inner part. The torque is proportional to the planet's mass. If the planet is small, its torque is easily overcome by the viscous torque that the annuli of the disk exert on each other. The global surface density profile of the disk is not changed and only the spiral density wave appears. But if the planet is massive enough the torque it exerts on the disk overwhelms the disk's viscous torque. In this case, the planet effectively pushes gas away from its orbit: outer gas outwards and inner gas inwards. A gap opens in the gas distribution around the orbit of the planet.

As gas is removed, the gap becomes deeper and wider. But as the gradients of the disk's surface density distribution become steeper at the edges of the gap, the disk's viscous torque is enhanced. In fact, the viscous torque acting on elementary rings can be computed by differentiation of Eq. (1.3) to give:

$$\delta T_\nu = -\frac{3}{2}\nu\Omega \left[\frac{r}{\Sigma}\frac{d\Sigma}{dr} + \frac{1}{2} \right](2\pi r \Sigma). \tag{1.28}$$

Once the density gradient becomes steep enough, the viscous torque balances the torque that the planet exerts on the same annulus of the disk (Varnière et al. 2004).

The depth of a gap—defined as the ratio between the surface density of the disk perturbed by the planet Σ_p and the original value Σ_u—is (Kanagawa et al. 2018):

$$\frac{\Sigma_p}{\Sigma_u} = \frac{1}{1 + 0.04K}, \tag{1.29}$$

where

$$K = \left(\frac{M_p}{M_\star}\right)^2 \left(\frac{H}{r}\right)^{-5} \alpha^{-1} \tag{1.30}$$

and M_p, $M_\star$ are the masses of the planet and the star, respectively, the aspect ratio of the disk H/r is taken at the planet's location, and as usual $\alpha = \nu/(H^2\Omega)$. This formula holds up to $K \sim 10^4$ (see Fig. 11 in Kanagawa et al. 2018). However, it cannot hold indefinitely, in particular in the limit of low α. This is because, if the density gradients at the edges of the gap become too steep, the rotational properties of the gas change so much under the effect of the pressure gradient that the specific angular momentum of the disk $r^2\Omega(r)$ is no-longer a growing function of r (Kanagawa et al. 2015). When this happens the disk becomes Rayleigh unstable and develops local turbulence, in turn enhancing the local viscosity. This effectively limits the steepness of the gap "walls" and the depth of the gap. Because the pressure gradient is also proportional to $(H/r)^2$, this narrative implies that, in the limit of vanishing viscosity, the denser the disk the shallower the gap. Thus, a gap opening criterion—i.e. the minimal mass of a planet to cause a depletion of 90% of the gas in the gap—must depend not only on viscosity but also on the disk's aspect ratio. An often-quoted criterion is the following:

$$\frac{3}{4}\frac{H}{R_H} + \frac{50}{qR_e} < 1, \tag{1.31}$$

where $q = M_p/M_\star$, $R_H = (q/3)^{1/3}$ and $R_e = r_p^2\Omega_p/\nu$ (Crida et al. 2006).

The formation of a gap profoundly changes a planet's migration. This migration mode has been dubbed *Type-II migration*. The gap must migrate along with the planet. In particular, as the planet moves inwards, the disk has to refill the portion of the gap "left behind" by the planet's radial motion. Because the radial velocity of the gas in a viscous unperturbed disk is $v_r = -(3/2)\nu/r$, this was the expected migration speed of the planet, independent of the planet's mass and disk's scale aspect ratio (Ward 1997).

However, the planet's migration rate is not so simple (Duffell et al. 2014; Dürmann and Kley 2015). It depends on the ratio between the disk's density and the mass of the planet, exemplified by the dimensionless ratio $r_p^2\Sigma/M_p$ (see Fig. 1.12), and also on the disk's aspect ratio. Depending on these quantities, the radial velocity of the planet can be smaller or larger than the "idealized" Type-II migration speed of $-(3/2)\nu/r$.

It is easy to understand that a planet's migration speed is slower than the idealized speed for the case of a low disk mass. A light disk obviously cannot push a heavy planet. This is the inertial limit: the planet is an obstacle to the flow of the gas. But how can a planet migrate faster than the radial speed of the gas? In principle the gas should remain behind and the planet, losing contact from the outer disk, should feel a weaker negative torque, eventually slowing down its migration until it moves at the same speed of the gas. The explanation given by Duffell et al. (2014) and Dürmann and Kley (2015) was that, as the planet migrates inwards, the gas from the inner disk can pass through the gap, refilling the left-behind part of the gap. In this way, the planet does not have to wait for the gas to drift-in at viscous speed. But the passage of gas through the gap is insignificant if the gap is significantly wider than the planet's horseshoe region (Robert et al. 2018), which is the case for massive planets in low-viscosity disks. In that case, as the planet moves inwards, the gap must be refilled from the outer disk. However, the steep density gradient at the edge of the gap enhances the viscous torque, as discussed above, so that the gas radial speed can be several times faster than $-(3/2)\nu/r$. This explains why the planet can exceed the idealized Type-II migration rate. Nevertheless, the migration rate of the planet must linearly proportional to the viscosity of the disk (Robert et al. 2018).

The dependence of the migration rate of giant planets on disk's viscosity opens the possibility that, in low-viscosity disks, the migration timescale can exceed the disk's lifetime. This would be convenient to explain why most giant planets are warm/cold Jupiters and not hot Jupiters[3] (Butler et al. 2006; Udry et al. 2007). However, recall

---

[3] We define a hot Jupiter as a giant planet within 0.1 au from its star and warm/cold Jupiters as those that are beyond 0.5 au. Very few giant planets fall between 0.1 and 0.5 au so the exact values of these boundaries are not important. Debiased observations suggest that hot Jupiters are about 1/10th or less as abundant as warm/cold Jupiters (Fernandes et al. 2019; Wittenmyer et al. 2016, 2020).

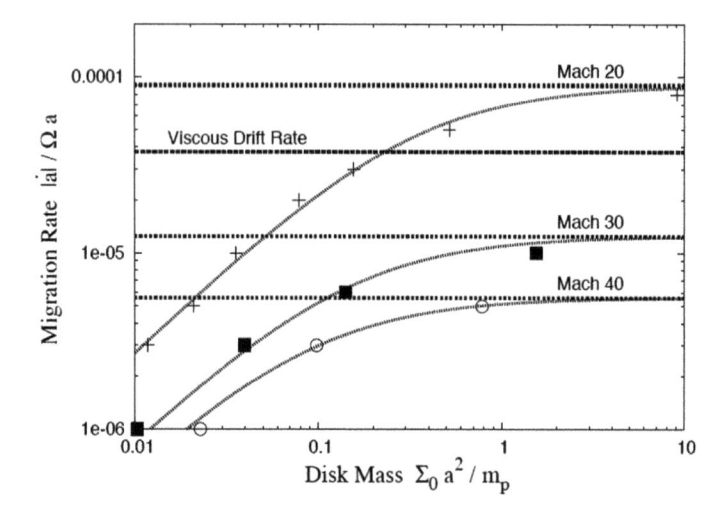

Fig. 1.12 The migration rate as a function of the ratio $r_p^2 \Sigma / M_p$ for a giant planet in a disk with $\alpha = 3 \times 10^{-3}$, for three different disk's aspect ratios. Notice that the migration rate can be smaller or larger than the viscous drift rate of the gas, i.e. the idealized Type-II migration speed. Figure courtesy of P. C. Duffell, from From Duffell et al. (2014)

from Sect. 1.2.1.1 that low-viscosity disks cannot explain the accretion rates observed for the central stars. If viscosity is low, there must be some additional mechanism for the radial transport of gas, possibly induced by angular momentum removal in disk winds. The migration rate of giant planets in these kind of disks has not yet been studied.

1.2.6 Resonance Trapping During Planet Migration

Numerical simulations show that multiple planets migrating in a disk have the tendency to lock into mutual mean motion resonance, where the period ratio is very close to the ratio of integer numbers (Terquem and Papaloizou 2007; Pierens and Nelson 2008; Izidoro et al. 2017). Typically, these numbers differ by just one such that the period ratios are 2:1, 3:2, 4:3 etc. These are called *first-order* resonances.

To understand this propensity to form resonant chains we need to dive into the complex world of dynamical planet-planet interactions. To stay simple, we will consider the planar three body problem, where two planets orbit a star on coplanar orbits. This is the simplest example of dynamics of a planetary system, yet it already captures most of the complexities of real systems. The best mathematicians made huge efforts to find an analytic solution of this problem, until Poincaré (1892) demonstrated that this was impossible. A general analytic solution does not exist. The system can exhibit chaotic behavior (Henon and Heiles 1964). Yet, some description

1 Planet Formation: Key Mechanisms and Global Models

can be provided, such as for instance for the dynamics of resonant planets with small libration amplitude, which is that of interest to understand the formation of resonant chains. This is what we attempt to do in this section.

To study the three body problem, the most effective approach is to use the Hamiltonian formalism. A system of first-order differential equations

$$\frac{d\mathbf{x}}{dt} = \mathbf{f}(\mathbf{x}, \mathbf{y}), \quad \frac{d\mathbf{y}}{dt} = \mathbf{g}(\mathbf{x}, \mathbf{y}) \tag{1.32}$$

is said to be in Hamiltonian forms, if there exists a scalar function $\mathcal{H}(\mathbf{x}, \mathbf{y})$ such as

$$f_i(\mathbf{x}, \mathbf{y}) = \frac{\partial \mathcal{H}}{\partial y_i}, \quad g_i(\mathbf{x}, \mathbf{y}) = -\frac{\partial \mathcal{H}}{\partial x_i}, \tag{1.33}$$

for each component $i = 1, \ldots, N$ of the vectors $\mathbf{x}, \mathbf{y}, \mathbf{f}, \mathbf{g}$. The function $\mathcal{H}$ is called the *Hamiltonian* of the problem; $\mathbf{x}$ is called the vector of *coordinates* and $\mathbf{y}$ the vector of *momenta*. If H is periodic on $\mathbf{x}$, the coordinates are also called *angles* and the momenta *actions*.

An important property of Hamiltonian dynamics is that $\mathcal{H}$ is constant over the flow $\mathbf{x}(t), \mathbf{y}(t)$ that is solution of the equations of motion. This means that, in the special case where $\mathcal{H}$ is a function of only one component of $\mathbf{x}$, namely $\mathcal{H} \equiv \mathcal{H}(x_1, \mathbf{y})$, the evolution of the system can be easily computed: $y_2, \ldots, y_N$ are constant of motion from (1.32) and (1.33); the motion of x_1, y_1 can be obtained from the level curve of $\mathcal{H}(x_1, y_1, y_2, \ldots, y_N) = \mathcal{H}(\mathbf{x}(0), \mathbf{y}(0))$, which is a 1D curve in a 2D x_1, y_1 space (called *phase space*). In this case, the problem is integrable (i.e. the solution is provided with analytic functions of time).

In studying a problem written in Hamiltonian form, a typical goal is to find a transformation of variables $\mathbf{x} \rightarrow \mathbf{x}', \mathbf{y} \rightarrow \mathbf{y}'$ that transforms the Hamiltonian function into one that is independent of $x'_2, \ldots, x'_N$. However, only transformations that preserve the form of Hamilton equations (Eqs. 1.32 and 1.33), called *canonical transformations*, are allowed. There are many forms of canonical transformations. In this section we will use only linear transformations $\mathbf{x}' = A\mathbf{x}, \mathbf{y}' = B\mathbf{y}$ where A, B are $N \times N$ matrices. It can be proven that the transformation is canonical if and only if

$$B = \left[A^{-1}\right]^T. \tag{1.34}$$

In general, it is not possible to find a canonical transformation that makes the Hamiltonian independent of $N - 1$ coordinates. Then, a goal in *perturbation theory* is to find a canonical transformation that turns $\mathcal{H}(\mathbf{x}, \mathbf{y})$ into

$$\mathcal{H}'(\mathbf{x}', \mathbf{y}') = \mathcal{H}'_0(x'_1, \mathbf{y}') + \epsilon \mathcal{H}'_1(\mathbf{x}', \mathbf{y}'). \tag{1.35}$$

In this case $\mathcal{H}'_0$ is called the *integrable approximation* and $\mathcal{H}'_1$ the *perturbation*. The latter can be neglected if one is interested in the dynamics up to a time $t < 1/\epsilon$. Depending on the goal in terms of accuracy, ϵ has to be sufficiently small.

After this broad and shallow introduction to Hamiltonian dynamics, let's turn to the planar three-body problem. The problem admits a Hamiltonian description, with Hamiltonian function:

$$\mathcal{H} = \sum_{j=1,2} \frac{\|\mathbf{p}_j\|^2}{2\mu_j} - \frac{G(M_\star + m_j)\mu_j}{\|\mathbf{r}_j\|} + \frac{\mathbf{p}_1 \cdot \mathbf{p}_2}{M_\star} - \frac{Gm_1 m_2}{\|\Delta\|}, \quad (1.36)$$

where $\mathbf{r}_j$ is the *heliocentric* position vector of planet j of reduced mass $\mu_j = m_j M_\star/(m_j + M_\star)$, $\mathbf{p}_j = m_j \mathbf{v}_j$, with $\mathbf{v}_j$ being the *barycentric* velocity vector (not a typo!: positions and velocities have to be taken in different reference frames if one wants a Hamiltonian description of the problem; Poincaré 1897) and $\Delta = \mathbf{r}_1 - \mathbf{r}_2$.

Using the canonical Delaunay variables:

$$\Lambda_j = \mu_j \sqrt{G(M_\star + m_j)a_j}, \quad \lambda_j = M_j + \varpi_j$$
$$\Gamma_j = \Lambda_j(1 - \sqrt{1 - e_j^2}), \quad \gamma_j = -\varpi_j$$

where a_j, e_j are the semi major axes and eccentricities, ϖ_j are the perihelion longitudes, M_j the mean anomalies and G is the gravitational constant, the Hamiltonian (Eq. 1.36) becomes:

$$\mathcal{H} = -G^2 \sum_{j=1,2} \frac{\mu_j^3 (M_\star + m_j)^2}{2\Lambda_j^2} + \mathcal{H}_1(\Lambda_{1,2}, \Gamma_{1,2}, \lambda_{1,2}, \gamma_{1,2}). \quad (1.37)$$

The first term in the r.h.s. of Eq. (1.37), denoted $\mathcal{K}$ hereafter, taken alone, is an integrable Hamiltonian, but the flow that it describes is trivially that of two uncoupled Keplerian motions: the actions $\Lambda_{1,2}$, $\Gamma_{1,2}$ are constant, the longitude of perihelia $\gamma_{1,2}$ are constants, and only the mean longitudes $\lambda_{1,2}$ move with constant frequency $G^2 \mu_j^3 (M_\star + m_j)^2 / \Lambda_j^3 = \sqrt{G(M_\star + m_j)/a_j^3}$. So, this kind of integrable approximation of the full Hamiltonian is not sufficient for our purposes and we need to find a better one.

To this end, we expand $\mathcal{H}_1$ in power series of $\sqrt{\Gamma_j} \sim \sqrt{\Lambda_j/2} e_j$ and in Fourier series of the angles λ_j, γ_j. The general form is therefore

$$\mathcal{H}_1 = \sum_{l_1,l_2,k_1,k_2,j_1,j_2} c_{l_1,l_2,k_1,k_2,j_1,j_2}(\Lambda_1, \Lambda_2) \Gamma_1^{j_1/2} \Gamma_2^{j_2/2} \cos(k_1\lambda_1 + k_2\lambda_2 + l_1\gamma_1, l_2\gamma_2)$$
$$+ s_{l_1,l_2,k_1,k_2,j_1,j_2}(\Lambda_1, \Lambda_2) \Gamma_1^{j_1/2} \Gamma_2^{j_2/2} \sin(k_1\lambda_1 + k_2\lambda_2 + l_1\gamma_1, l_2\gamma_2)$$
$$(1.38)$$

The so-called *D'Alembert rules* give us information on which terms of this series can have non-zero coefficients, namely:

- only the c coefficients can be non-zero, because the Hamiltonian must be invariant for a change of sign of all angles (measuring angles clockwise or counter-clockwise is arbitrary), so that the Fourier expansion can contain only cos terms;

- only the $c_{l_1,l_2,k_1,k_2,j_1,j_2}$ coefficients with $k_1 + k_2 - l_1 - l_2 = 0$ can be non-zero, because the Hamiltonian has to be invariant by a rotation of the reference frame, namely increasing all angles by an arbitrary phase δ_0 (remember that $\gamma = -\varpi$, so if all orbital angles are increased by δ_0, γ is decreased by δ_0);
- only the $c_{l_1,l_2,k_1,k_2,j_1,j_2}$ coefficients with $j_1 = |l_1| + 2n$ and $j_2 = |l_2| + 2i$ (with n and i non-negative integer numbers) can be non-zero. This is because the Hamiltonian is not singular for circular orbits (i.e. $\Gamma_1 = 0$ and/or $\Gamma_2 = 0$) so that it has become a polynomial expression in the canonical variables $p_1 = \sqrt{2\Gamma_1}\cos\gamma_1$, $q_1 = \sqrt{2\Gamma_1}\sin\gamma_1$, $p_2 = \sqrt{2\Gamma_2}\cos\gamma_2$, $q_2 = \sqrt{2\Gamma_2}\sin\gamma_2$.

If we are interested in two planets near a mean motion resonance where $P_2 \sim k/(k-1)P_1$, where P_1 and P_2 are the orbital periods and k is a positive integer number, the angle $k\lambda_2 - (k-1)\lambda_1$ will have an almost null time-derivative (as one can see remembering that $\dot\lambda = 2\pi/P$ and using the relationship between the orbital periods written above). Thus, it is a slow angle, whereas both λ_1 and λ_2 are fast angles, as it is their difference. To highlight this difference in timescales, let us define new angles:

$$\delta\lambda = \lambda_1 - \lambda_2, \quad \theta = k\lambda_2 - (k-1)\lambda_1. \tag{1.39}$$

Using the rule (1.34), this linear transformation of the angles can be made canonical by changing the actions as:

$$\Delta\lambda = k\Lambda_1 + (k-1)\Lambda_2, \quad \Theta = \Lambda_1 + \Lambda_2, \tag{1.40}$$

so that $(\Delta\lambda, \delta\lambda)$ and (Θ, θ) are pairs of canonical action-angle variables.

Since we are interested in the long-term evolution of the dynamics, we can average the Hamiltonian over $\delta\lambda$, which means that the averaged Hamiltonian will be independent of this angle and, consequently, $\Delta\lambda$ will be a constant of motion. Because the units of semi-major axis are arbitrary, one can always chose them so that $\Delta\lambda = 1$. In other words, changing the values of $\Delta\lambda$ does not change the dynamics; it simply changes the unit of measure of the semi major axes.

Using the D'Alembert rules described above, the function in Eq. (1.38) takes the form

$$\mathcal{H}_1 = \sum_{m,n,i>0,j>0} d_{m,n,i,j}(\Theta, \Delta\lambda)\Gamma_1^{|n-m|/2+j}\Gamma_2^{|m|/2+i}\cos[n(\theta + \gamma_1) + m(\gamma_2 - \gamma_1)], \tag{1.41}$$

where the coefficients $d_{m,n,i,j}$ come from the original coefficients $c_{l_1,l_2,k_1,k_2,j_1,j_2}$ through a simple index algebra that follows trivially from the redefinition of the angles (Eq. 1.39).

Because there are only two possible combinations of angles in the harmonics of Eq. (1.41), it is convenient to identify each of them with a single angle, namely:

$$\psi_1 = \theta + \gamma_1 \quad \delta\gamma = \gamma_2 - \gamma_1, \quad \gamma_2' = \gamma_2. \tag{1.42}$$

Again, using the rule (Eq. 1.34) this linear transformation of the angles can be made canonical by changing the actions as:

$$\Psi_1 = \Theta, \quad \Psi_2 = \Theta - \Gamma_1, \quad \mathcal{L} = \Gamma_1 + \Gamma_2 - \Theta. \tag{1.43}$$

Now, the Hamiltonian $\mathcal{K} + \mathcal{H}_1$ depends only on the angles ψ_1 and $\delta\gamma$, and $\mathcal{L}$ is a new constant of motion (related to the angular momentum of the system).

This Hamiltonian is still not integrable, because it depends on two angles. An integrable approximation (dependent on only one angle) can be obtained if one retains in Eq. (1.41) only the terms linear in the planets' eccentricities, i.e. proportional to $\sqrt{\Gamma_{1,2}}$, and doing some cumbersome change of variables (Sessin and Ferraz-Mello 1984; Batygin and Morbidelli 2013). In this case one can then trace global dynamical diagrams by plotting the level curves of the Hamiltonian (see for instance Fig. 3 in Batygin and Morbidelli 2013). However, even in the general non-integrable case one can look for the stable equilibrium points in $(\Psi_1, \psi_1, \Psi_2, \delta\gamma)$ as a function of $\mathcal{L}$ (recall that $\Delta\lambda$ can be fixed to unity). The locus of equilibrium points, once transformed back into the original orbital elements, describes a curve in $e_2, a_2/a_1$ or, equivalently, $e_1, a_2/a_1$, like that shown in Fig. 1.13. Note that on the curve $a_2/a_1 \to \infty$ as $e_1 \to 0$, $e_2 \to 0$. This feature comes from the fact that, from Hamilton's equations $\dot{\gamma}_{1,2} = \partial \mathcal{H}/\partial \Gamma_{1,2}$ applied to (1.41), one has $\dot{\gamma}_{1,2} \propto \Gamma_{1,2}^{-1/2}$, i.e. $\dot{\gamma}_{1,2} \to \infty$ as $\Gamma_{1,2} \propto e_{1,2}^2 \to 0$. Thus, to have the equilibrium $\dot{\psi}_1 = 0$ the value of $\dot{\theta}$ has to diverge, which means that a_2/a_1 has to diverge as well.

This feature of the curve of equilibrium points is the key to understand resonant capture. If the planets are far from resonance (i.e. a_2/a_1 is much larger than the resonant ratio assuming Keplerian motion; a 3:2 resonance being located at $a_2/a_1 = 1.3103$ in Fig. 1.13), the protoplanetary disk exerts damping forces on their eccentricities, so that the planet's orbits are basically circular. This means, from the shape of the curve of equilibrium points, that the planets will be on the equilibrium. As migration proceeds and a_2 approaches a_1 (this is the case if the outer planet migrates faster towards to the star, which happens in Type-I migration if it is more massive or if the inner planet is blocked at a planet trap), the ratio a_2/a_1 decreases. If this happens slowly compared to the libration period around the equilibrium point, the dynamical evolution has to react *adiabatically* (Neishtadt 1984). This means that the amplitude of libration around the equilibrium point has to be conserved. Because initially the planets have vanishing amplitude of libration (their eccentricities are basically null as those characterizing the equilibrium point), this means that the planets have to evolve from one equilibrium point to the other, i.e. they have to follow the curve traced in Fig. 1.13. As a_2/a_1 asymptotically approaches the Keplerian resonant value, the eccentricities of the two planets increase.

If the convergent migration is too fast, the adiabatic condition is broken. The amplitude of libration is not conserved. The planets can jump off resonance and continue to approach each other more closely. But, because the libration frequency of a resonance $P_2 \sim k/(k-1)P_1$ increases with k, eventually the planets will find a resonance with k large enough and libration frequencies fast enough that the adiabatic

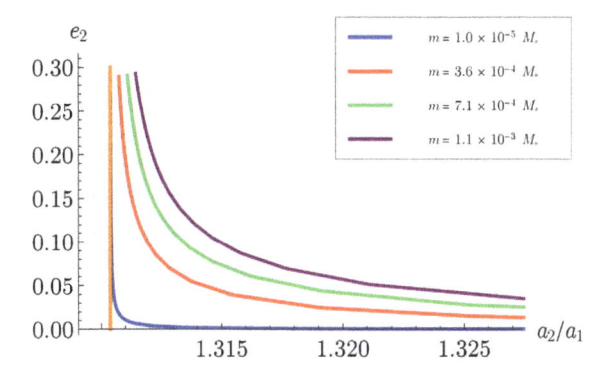

1 Planet Formation: Key Mechanisms and Global Models

Fig. 1.13 The locus of equilibrium points for the 3:2 mean motion resonance in the plane e_2 versus a_2/a_1, for different planetary masses (here assumed to be equal to each other) in different colors. The vertical orange line shows the location of the resonance in the Keplerian approximation (the same a_2/a_1 for any eccentricity because the orbital frequencies depends only on semi major axes). Adapted from Pichierri et al. (2018)

condition is satisfied. Then, they will be trapped in resonance. In essence, the faster is the convergent migration, the higher the index k of the resonance in which the planets will be trapped. But trapping always occurs, eventually. These arguments apply for each pair of neighboring planets in a multi-planet system. This is why the formation of configurations in resonant chains is a typical outcome of planet migration.

If the adiabatic condition is satisfied, is resonant trapping stable? Fig. 1.13 suggests that the eccentricities of the planets should grow indefinitely. But in reality the disk exerts eccentricity damping, so that the eccentricities grow until an equilibrium is established between the eccentricity damping from the disk and the resonant conversion of convergent migration into eccentricity pumping. A precise formula to compute this equilibrium in a variety of configurations (damping exerted on both planets or only on one, inner planet at a planet-trap or not etc.) can be found in the appendix of Pichierri et al. (2018). The equilibrium eccentricity is typically of order $(H/r)^2$, where (H/r) is the aspect ratio of the protoplanetary disk.

There is a complication. For the adiabatic principle to be applied, the dissipative forces have to act on the parameters of the Hamiltonian, not on the dynamical variables. This is the case for migration. The change in semi-major axis ratio changes the otherwise constant of motion $\mathcal{L}$, i.e. a parameter of the Hamiltonian. But the eccentricity damping affects $\Gamma_{1,2}$, i.e. dynamical variables. Then, the equilibrium point can become a focus, which means that the dynamical evolution spirals around it. The spiral can be inward if the focus is stable (which means that any initial amplitude of libration would shrink to 0) or outwards if the focus is unstable (which means that the amplitude of libration grows indefinitely, even if it is initially arbitrarily small). This unstable evolution, first pointed out by Goldreich and Schlichting (2014), is called *overstability*.

Figure 1.14 shows a summary of the most detailed investigation of the stability/overstability of a resonance in presence of eccentricity damping (Deck and Batygin 2015). If the eccentricity damping is proportional to the planetary masses (as in the case where both planets are embedded in the disk), for the 3:2 and higher-k resonances the resonant configuration is stable whenever $m_1 > m_2$. In the opposite case, the stability/overstability depends on the total mass ratio $\epsilon_p = (m_1 + m_2)/M_\star$. There is a limit value of the ratio m_1/m_2 below which the system is overstable, and this ratio decreases with increasing ϵ_p (Fig. 1.14a). For the 2:1 resonance the situation is qualitatively similar, but the planets can be overstable even if $m_1/m_2 > 1$ if ϵ_p is small enough (Fig. 1.14b). On the other hand, if there is no eccentricity damping on the inner planet ($\tau_{e_1} \to \infty$, which happens if the inner planet has been pushed into a disk's cavity), the resonant configuration is always stable for any m_1/m_2 ratio.

Even if resonant planets are not in the overstable regime, they may be unstable because of other processes. Due to their proximity and eccentric orbits they may approach each other too much over their resonant trajectories and be destabilized by a close-encounter (Pichierri et al. 2018). There can also be subtle secondary resonances between a combination of the libration frequencies and the synodic frequency $\dot{\lambda}_1 - \dot{\lambda}_2$. These secondary resonances cannot be described with the averaged Hamiltonian (Eq. 1.41) because the terms in $\lambda_1 - \lambda_2$ have been removed by the averaging procedure. But they can be studied following a more precise and convoluted approach. As the number of planets in a resonant chain increases, the number of libration frequencies increases as well and therefore a richer set of combined frequencies is possible. This explains, at least at the qualitative level, why long resonant chains are more fragile than short chains, as observed in numerical simulations (Matsumoto et al. 2012; Cossou et al. 2013). The long-term evolution of multi-planet resonant chains remains nevertheless an active area of research in celestial mechanics.

1.3 Global Models of Planet Formation

Building global models of planet formation is akin to putting together a puzzle. We have a vague picture of what the puzzle should look like (i.e., from exoplanet demographics) but observational biases cloud our view. Moreover, the puzzle pieces—the planet formation mechanisms—often change in number and in shape. The puzzle-builders must constantly have an eye both on the evolution of the big picture and on the set of viable puzzle pieces. One must not hesitate to discard a model when it no longer serves.

We will describe our current best global models for the origin of super-Earth systems (Sect. 1.3.1), giant planet systems (Sect. 1.3.2) and our own Solar System (Sect. 1.3.3). Then we will look at how water may be delivered to rocky planets (Sect. 1.3.4).

1 Planet Formation: Key Mechanisms and Global Models

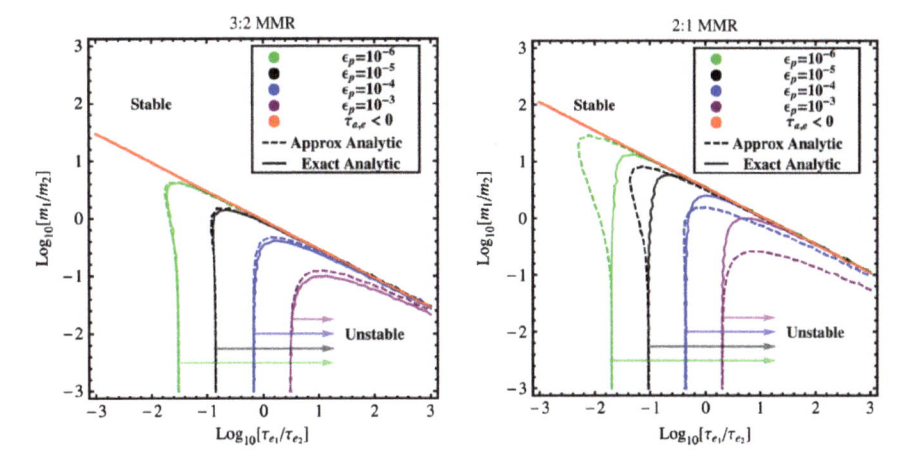

Fig. 1.14 Each plot shows the region of the parameter plane τ_{e_1}/τ_{e_2} and m_1/m_2, where $\tau_{e_{1,2}}$ is the damping timescale of the eccentricity of planets 1 and 2, respectively, with masses m_1 and m_2, for different planetary masses $\epsilon_p = (m_1 + m_2)/M_\star$. The left plot is for the 3:2 resonance, the right plot for the 2:1 resonance. For a planet of mass m at semi-major axis a the eccentricity damping timescale is $\tau_e \propto (H/a)^4/(m\Sigma\sqrt{a})$, where Σ is the surface density of the disk and H/a its aspect ratio. Thus, if the disk has a constant aspect ratio and $\Sigma \propto 1/\sqrt{a}$ the system should place on the dashed diagonal line (i.e. $m_1/m_2 = tau_{e_1}/tau_{e_2}$). This shows, for instance, that a 3:2 system is stable if $m_1 < m_2$. If instead the inner planet is in a low-density region (i.e. a disk's cavity), $\tau_{e_1} \to \infty$ and therefore the system is stable for all mass ratios. Adapted from Deck and Batygin (2015)

1.3.1 Origin of Close-in Super-Earths

The key properties of super-Earth systems that must be matched by any formation model can be very simply summarized as follows (we refer the reader to Chap. 3 for detailed reviews):

- A large fraction (roughly one third to one half) of stars have close-in super-Earths (with periods shorter than 100 days; Fressin et al. 2013; Mulders et al. 2018), but many (perhaps most) of them do not.
- Most super-Earth systems only have a single planet detected in transit, whereas a fraction of systems is found with many planets in transit (Lissauer et al. 2011; Fang and Margot 2012; Tremaine and Dong 2012).
- In multi-planet systems, pairs of neighboring super-Earths are rarely found to be in mean motion resonance (Lissauer et al. 2011; Fabrycky et al. 2014).
- The masses of super-Earths extend from Earth to Neptune, with a preference for a few $M_\oplus$ (Weiss et al. 2013; Marcy et al. 2014; Wolfgang et al. 2016; Chen and Kipping 2017).

Models for the formation of super-Earths were developed before they were even discovered (Raymond et al. 2008a). While certain models have been refined in recent years, only a single new model has been developed.

In-situ accretion is the most intuitive and simplest model for super-Earth formation, yet it has a fatal flaw. That model proposes that super-Earths simply accreted from local material very close to their stars in a similar fashion to the classical model of terrestrial planet formation in the Solar System (see Sect. 1.3.3.1 below). In-situ accretion was proposed in 2008 (Raymond et al. 2008a) and discarded because the masses implied in the innermost parts of disks seemed prohibitively large. In 2013 this idea was revisited by Chiang and Laughlin (2013), who used the population of known super-Earths to generate a "minimum-mass extrasolar nebula" representing a possible precursor disk that would have formed the population of super-Earths. The high masses inferred in inner disks conflicts with measurements, but those measurements are only of the outermost parts of disks (Williams and Cieza 2011). While it is possible to imagine that inner disks can pile up material, there is a simply-understood timescale problem. With very high densities in the inner disk, the growth timescale for super-Earths is extremely fast (Lissauer and Stevenson 2007; Raymond et al. 2007b; Raymond and Cossou 2014; Inamdar and Schlichting 2015; Schlichting 2014; Ogihara et al. 2015). In fact, the growth timescales are so fast and the requisite disks so massive that migration is simply unavoidable (Ogihara et al. 2015). Even aerodynamic drag is strong enough to cause rapid orbital drift (Inamdar and Schlichting 2015). Thus, we cannot consider the planets to have formed in-situ because they must have migrated and their final orbits cannot represent their starting ones. Nonetheless, it has been shown that if the right conditions were to arise, with the requisite amount of solid material close to the star, accretion should indeed produce planetary systems similar to the observed super-Earths (Hansen and Murray 2012, 2013; Dawson et al. 2015, 2016; Moriarty and Ballard 2016; Lee et al. 2014; Lee and Chiang 2016, 2017).

A number of the first processes to be explored for forming super-Earths relied on giant planets. For example, migrating giant planets can shepherd material interior to their orbits and stimulate the growth of super-Earths (Zhou et al. 2005; Fogg and Nelson 2005, 2007; Raymond et al. 2006b; Mandell et al. 2007). Moving secular resonances driven by giant planet interactions can do the same (Zhou et al. 2005). However, these models have been ruled out as the main formation pathways for super-Earths, because most super-Earth systems do not appear to have an associated giant planet (although testing the correlation between super-Earths and outer gas giants is an active area of study, e.g. Zhu and Wu 2018; Barbato et al. 2018; Bryan et al. 2019).

The migration model has proven quite successful model in reproducing the observed population of super-Earths (Terquem and Papaloizou 2007; Ogihara and Ida 2009; McNeil and Nelson 2010; Ida and Lin 2010; Cossou et al. 2014; Izidoro et al. 2017, 2019; Raymond et al. 2018b; Carrera et al. 2018). In that model, large planetary embryos grow throughout the disk and migrate inward, driven by the gaseous disk (Fig. 1.15 shows an example of a simulation of the migration model). It is natural to think that embryos would form first past the snow line, where pebble accretion is thought to be more efficient (see Sect. 1.2.3 and Morbidelli et al. 2015). However, it is also possible that in some disks large embryos can form very close to their stars. This might occur if inward-drifting pebbles are concentrated at a pressure bump, perhaps

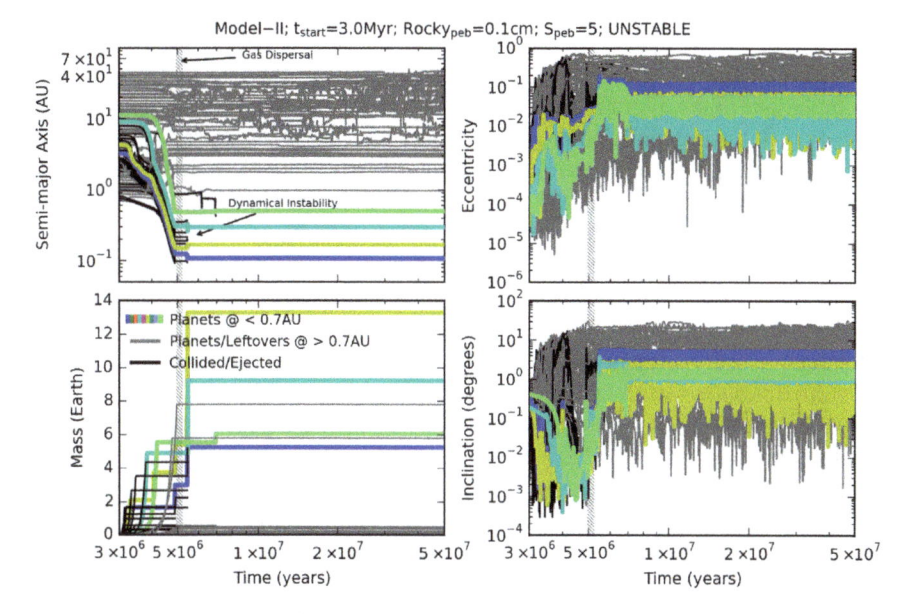

1 Planet Formation: Key Mechanisms and Global Models

Fig. 1.15 Evolution of a simulation of the *breaking the chains* migration model for the origin of close-in super-Earths. The panels show the evolution of a population of ∼lunar-mass planetary embryos that grow by accreting pebbles, migrate inward and form a resonant chain anchored at the inner edge of the disk. The resonant chain is destabilized shortly after the dissipation of the gaseous disk, leading to a late phase of scattering and collisions that spreads out the planets and increases their eccentricities and inclinations. Figure courtesy of A. Izidoro, from Izidoro et al. (2019)

associated with the inner edge of a dead zone[4] (Chatterjee et al. 2014, 2015). While the migration rate and direction depends on the disk model (Lyra et al. 2010; Kretke and Lin 2012; Bitsch et al. 2013, 2014), inward migration is generally favored. The inner edge of the disk creates a strong positive torque (Flock et al. 2019) that acts to trap any inward-migrating embryo (Masset et al. 2006). This leads to a convergence of bodies near the disk's inner edge. Convergent migration leads to resonant trapping (see Sect. 1.2.6) and tends to produce planets in long chains of resonances. Of course, the observed super-Earths are not found in resonance. However, most resonant chains become dynamically unstable as the gaseous disk dissipates. This leads to a phase of giant collisions between embryos that is quite similar to that simulated by studies that ignored the migration phase and invoked a large population of embryos in a dissipating gaseous disk (Hansen and Murray 2012, 2013; Dawson et al. 2015). The instability phase leads to scattering among embryos, breaks the resonant chains, and causes systems to spread out and become dynamically excited. A small fraction of resonant chains remain stable after the disk dissipates; these may represent iconic

---

[4] This idea forms the basis of a model of super-Earth formation sometimes called *inside-out planet formation* (Boley and Ford 2013; Chatterjee et al. 2014, 2015; Hu et al. 2016, 2018). That model invokes the direct formation of super-Earths from pebbles at pressure bumps in the inner disk.

systems such as Trappist-1 (Gillon et al. 2017; Luger et al. 2017) and Kepler-223 (Mills et al. 2016).

Simulations of the migration model have shown that the surviving systems quantitatively match the population of observed super-Earths as long as more than 90% of resonant chains become unstable (Izidoro et al. 2017, 2019). When run through a simulated observing pipeline, the significant dynamical excitation of the surviving systems implies large enough inclinations such that most viewing geometries can only see a single planet in transit. This solves the so-called Kepler dichotomy problem (Johansen et al. 2012; Fang and Margot 2012) and implies that all super-Earth systems are inherently multiple. The period ratio distribution of simulated systems matches observations, again taking observational biases into account (Izidoro et al. 2017, 2019; Mulders et al. 2019).

Yet questions remain. If super-Earth formation is as efficient as in simulations, why don't all stars have them? One possibility is that when an outer gas giant planet forms, it blocks the inward migration of large embryos and those instead become ice giants (Izidoro et al. 2015a, b). This implies an anti-correlation between the presence of outer gas giants and systems with many super-Earths, and it remains unclear if such an anti-correlation exists (Bryan et al. 2019; Zhu and Wu 2018; Barbato et al. 2018). The fraction of stars with gas giants appears to be far less than the fraction of stars without super-Earths, which makes it difficult to imagine gas giants being the main cause. Is it possible, instead, that outer ice giants or super-Earths are the culprit? Probably not. While their occurrence rate is high (Gould et al. 2010; Suzuki et al. 2016b), outer ice giants cannot efficiently block the inward migration of other planets and are generally too low-mass to block the pebble flux (Morbidelli et al. 2012; Lambrechts and Johansen 2014; Bitsch et al. 2018). Perhaps, instead, many disks are subdivided into radial zones (Johansen et al. 2009). Pebbles trapped within a given zone may not be able to drift past the zone boundary such that the pebble flux in certain regions of the disk would remain too low for planets to grow fast enough for long-range migration (Lambrechts et al. 2019a). Such a scenario would also be compatible with the ringed structures seen in many ALMA disks (Andrews et al. 2018).

The compositions of super-Earths may also provide a constraint for formation models. It naively seems that the migration model should produce very volatile-rich super-Earths because the main source of mass is beyond the snow line, where the efficiency of pebble accretion is higher (Morbidelli et al. 2015; Izidoro et al. 2019; Bitsch et al. 2019a, b). This need not be true in 100% of cases, as inward-migrating icy embryos can in some cases stimulate the growth of inner, purely rocky planets (Raymond et al. 2018b). However, most super-Earths in the migration model should be ice-rich (Izidoro et al. 2019). It is unclear whether this is consistent with the observed distribution of bulk densities of super-Earths. While it has been claimed that most super-Earths appear to be "rocky" (Owen and Wu 2013, 2017; Lopez 2017; Jin and Mordasini 2018), measurement uncertainties preclude any clear determination of the compositions of super-Earths (Dorn et al. 2015, 2018). In addition, "rocky" planets may in some context include water contents up to ∼20% by mass (Gupta and Schlichting 2019). For context, that is similar to the approximate water contents

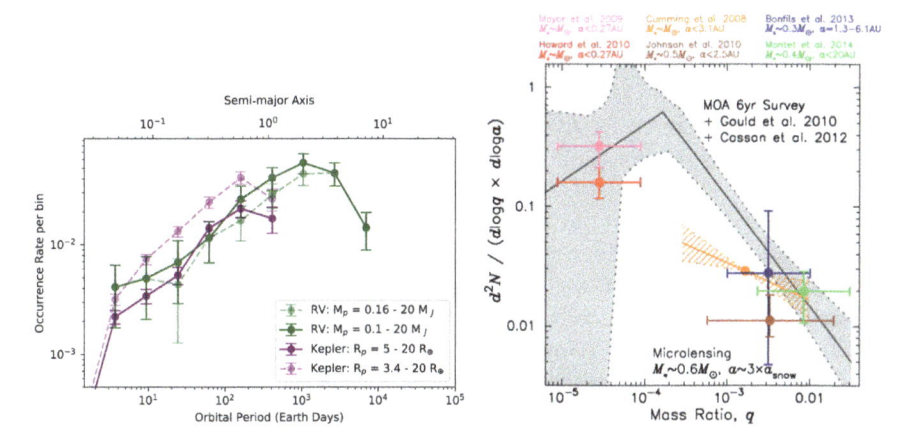

1 Planet Formation: Key Mechanisms and Global Models   47

Fig. 1.16 Mass and orbital radius distributions of giant exoplanets. *Left-hand panel:* The frequency of giant planets as a function of orbital distance from radial velocity surveys (green) and the Kepler transit statistics (purple). Figure courtesy of R. B. Fernandes, from Fernandes et al. (2019). *Right-hand panel:* The occurrence rate of giant planets as a function of planet-to-star mass ratio q. The different symbols correspond to different radial velocity studies whereas the black line and grey regions are the results from microlensing surveys. Figure courtesy of T. K. Suzuki, from Suzuki et al. (2016a)

of comet 67P (Pätzold et al. 2019), Pluto, and the most water-rich meteorites known (thought to originate from the outer Solar System; Alexander 2019a; 2019b). Finally, short-lived radionuclides such as Al-26 may efficiently dry out some super-Earths (Lichtenberg et al. 2019). The rocky versus icy nature of super-Earths remains an important outstanding issue.

The fact that super-Earths seem to ubiquitously have large radii consistent with atmospheres of a few percent H/He by mass (Weiss and Marcy 2014; Wolfgang et al. 2016; Fulton et al. 2017; Fulton and Petigura 2018) confirm that these planets formed during the gaseous disk phase and certainly by a process quite different than the formation of our own terrestrial planets. Formation models are starting to be coupled to models of atmospheric accretion and loss (Inamdar and Schlichting 2015; Ginzburg et al. 2016; Carrera et al. 2018). However, given the complexities in these processes (Lambrechts and Lega 2017) this remains an ongoing challenge.

1.3.2 Giant Exoplanets: Formation and Dynamics

Drawing from numerous radial velocity, transit and microlensing surveys for exoplanets, the essential constraints on giant planet formation (further reviewed in Chaps. 3 and 4) are:

- Gas giants exist around roughly 10% of Sun-like stars (Cumming et al. 2008; Mayor et al. 2011; Winn and Fabrycky 2015; Foreman-Mackey et al. 2016; Suzuki

et al. 2016b; Fernandes et al. 2019) and are more/less common around more/less massive stars (Lovis and Mayor 2007; Johnson et al. 2007; Clanton and Gaudi 2014, 2016).
- Most gas giants are on relatively wide orbits past 0.5–1 au (Butler et al. 2006; Udry et al. 2007; Mayor et al. 2011; Rowan et al. 2016; Wittenmyer et al. 2016, 2020).
- Gas giant exoplanets tend to have much higher eccentricities than Jupiter and Saturn, following a broad eccentricity distribution with a median of ∼0.25 (Butler et al. 2006; Udry and Santos 2007; Wright et al. 2009; Bonomo et al. 2017).
- There is a strong correlation between a star's metallicity and the probability that it hosts a gas giant planet (Gonzalez et al. 1997; Santos et al. 2001; Fischer and Valenti 2005), especially among hot Jupiters and gas giants with eccentric orbits (Dawson and Murray-Clay 2013).

There exist two categories of formation models for giant planets (Boley 2009; Helled et al. 2014). The first is a top-down, collapse scenario in which a localized instability in a protoplanetary disk can lead to the direct formation of one or more giant planets. The second is the bottom-up, core accretion scenario (which we argue below should really be called the *core-migration-accretion* scenario). In recent years the core accretion model has become the dominant one, yet it is plausible that some systems may be explained by the disk instability model.

Disk Instability

The disk instability scenario (Boss 1997, 1998, 2000; Mayer et al. 2002, 2007) invokes a localized region of the disk that becomes gravitationally unstable. This is usually quantified with the Toomre Q criterion (Toomre 1964):

$$\frac{c_s \, \Omega}{\pi \, G \, \Sigma} < Q_{\text{crit}} \approx 1, \tag{1.44}$$

where c_s is the sound speed, Ω is the local orbital frequency, G is the gravitational constant, and Σ is the local surface density of the disk. Simulations show that in disks that have $Q \lesssim 1$, fragmentation does indeed occur and Jupiter-mass clumps form quickly (Boss 1998; Mayer et al. 2007; Boley et al. 2010). To become true gas giants, these clumps must cool quickly to avoid being sheared apart by the Keplerian flow.

It is generally thought that only the outermost parts of massive disks are able to attain the criteria needed for instability (Meru and Bate 2011; Armitage 2011; Kimura and Tsuribe 2012). The very distant, massive planets such as those of the HR8799 system (Marois et al. 2008, 2010) may perhaps be explained by disk instability, as other mechanism struggle to form such massive planets so far away from their stars. However, it is not clear whether disk instability can produce Jupiter-like planets (although the giant planet orbiting the low-mass star GJ 3512 is a very good candidate; Morales et al. 2019). Planets or clumps that form very far out would also migrate inward rapidly (Baruteau et al. 2011) and it is unclear whether they would survive. In the *tidal downsizing model*, gravitational instability forms clumps in the outer parts of disks that migrate inward but are often disrupted to act as the seeds of much smaller planets (Nayakshin et al. 2015).

Core-*Migration*-Accretion

The bottom-up, core accretion scenario forms the basis of the current paradigm of giant planet formation. In its standard form the core-accretion model essentially represents a combination of processes summarized in Sect. 1.2.3 (growth of protoplanets) and 2.5 (gas accretion). As we shall see, another important process (orbital migration—described in Sect. 1.2.4) must be included.

Since its inception, the canonical picture has suffered setbacks related to core growth, migration, and gas accretion (see Sect. 1.2.3). Early models that invoked planetesimals as the building blocks of large cores could not explain the rapid growth of the $\sim 10\ M_\oplus$ cores needed to trigger rapid gas accretion (Levison and Stewart 2001; Thommes et al. 2003; Rafikov 2004; Chambers 2006a; Levison et al. 2010). Even in the most optimistic scenarios, planetesimal accretion was simply too slow. In recent years it has been shown that pebble accretion is far more efficient and can indeed—given that there is a sufficiently massive and drawn-out supply of pebbles (Johansen and Lambrechts 2017; Bitsch et al. 2019a)—explain the rapid growth of giant planet cores (Ormel and Klahr 2010; Lambrechts and Johansen 2012, 2014). It has also been shown that gas accretion is actually not halted, and only modestly slowed down, by the generation of a gap in the disk (D'Angelo et al. 2003; Uribe et al. 2013).

Migration remains a giant issue for the core accretion model, so much so that the model itself could plausibly be renamed the core-*migration*-accretion model. It is not so much a problem as an added dimension. The mass scale at which gas accretion becomes an important phenomenon is similar to that at which migration becomes important. Thus, planets that are undergoing gas accretion must necessarily be migrating at the same time. Growth tracks of planets must include both radial and mass growth.

Figure 1.17 shows the growth tracks of four different simulated planets. Starting from $\sim$lunar-mass cores, each planet's growth is initially determined simply by the flux of pebbles across its orbit (Lambrechts and Johansen 2014; Ida et al. 2016). Once each planet reaches several Earth masses, it starts to migrate inward. At the same time it starts to slowly accrete gas. Above the pebble isolation mass (defined in Sect. 1.2.3), pebble accretion is stopped but migration and gas accretion continue. The two inner planets in the simulation grew fast enough to undergo rapid gas accretion and become gas giants, whereas the two outer planets did not accrete enough gas before the disk dispersed and thus ended up as ice giants.

Figure 1.17 leads to a naive-sounding but surprisingly profound question: *Why is Jupiter at 5 au?* In the model from Fig. 1.17, any core that started within roughly 10 au ended up as a hot Jupiter. Likewise, in order to finish at 5 au a core needed to start at 15–20 au (see also Bitsch et al. 2015a; Ndugu et al. 2018). How, then did our own Jupiter avoid this fate? There are currently four possible solutions. First, perhaps Jupiter's core simply did form at 15–20 au (Bitsch et al. 2015b). While this is hard to rule out, it seems unlikely because most material that originated interior to Jupiter's orbit would have remained interior to Jupiter's final orbit (Raymond and

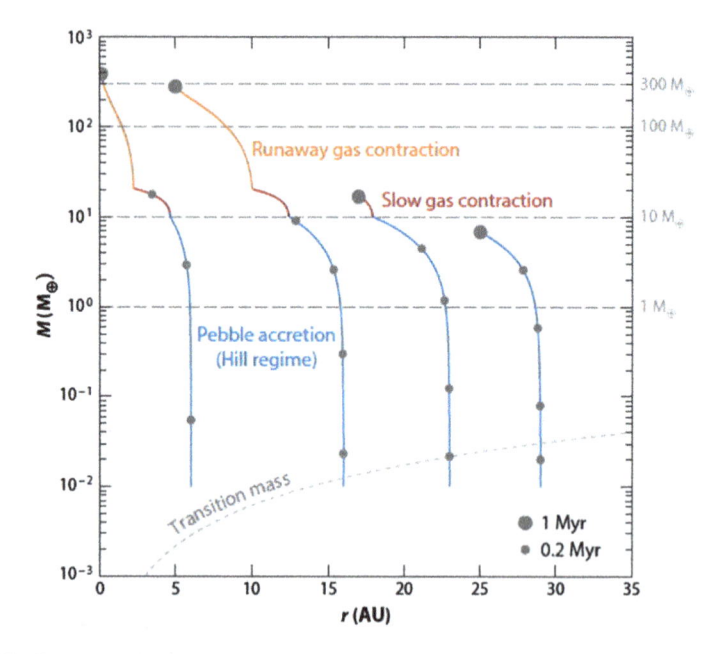

Fig. 1.17 Growth tracks for four different types of planets from simulations that include pebble accretion and migration and gas accretion. The leftmost planet grows fast enough to migrate a long distance and become a hot Jupiter. The second planet from the left starts as a core at ∼15 au and ends up as a Jupiter-like planet at 5 au. The two planets on the right are ice giant analogs that never undergo rapid gas accretion. In this case the pebble isolation mass—the mass above which the pebble flux is blocked and pebble accretion shut off (Lambrechts et al. 2014; Bitsch et al. 2018)— was fixed at 10 $M_\oplus$. Each small dot denotes a time interval of 0.2 Myr along the growth tracks and each large dot a time interval of 1 Myr. Figure courtesy of A. Johansen, from Johansen and Lambrechts (2017)

Izidoro 2017a; Pirani et al. 2019). This would therefore require that only a few Earth masses of material formed in the entire Solar System interior to 15-20 au. Second, perhaps Jupiter's migration was much slower than that shown in Fig. 1.17. That could be the case if the disk's viscosity was much lower than generally assumed, which is consistent with new disk models (see Sect. 1.2.1). Third, perhaps Jupiter's migration was halted because the inner gas disk was evaporated away by energetic radiation from the active young Sun (Alexander and Pascucci 2012). Indeed, disks are thought to disperse by being photo-evaporated by their central stars (Alexander et al. 2014), and this may be a consequence. Fourth, perhaps Jupiter's migration was held back by Saturn. Hydrodynamical simulations show that, while a Jupiter-mass planet on its own migrates inward, the Jupiter-Saturn pair can avoid rapid inward migration and sometimes even migrate outward (Masset and Snellgrove 2001; Morbidelli et al. 2007; Pierens and Nelson 2008; Zhang and Zhou 2010; Pierens and Raymond 2011; Pierens et al. 2014). This forms the basis of the Grand Tack model of Solar System formation, which we will discuss in detail in Sect. 1.3.3.2. However, that model's potential fatal flaw is that avoiding inward migration requires that Jupiter and Saturn maintain a specific mass ratio of roughly 3-to-1 (Masset and Snellgrove 2001) and it is uncertain whether that ratio can be maintained in the face of gas accretion.

Despite these uncertainties, the core accretion model can match a number of features of the known exoplanet population (Ida et al. 2004; Mordasini et al. 2009; Ndugu et al. 2018), including the observed giant planet-metallicity correlation (Fischer and Valenti 2005) and the much higher abundance of Neptune-mass compared with Jupiter-mass planets (Butler et al. 2006; Udry et al. 2007; Suzuki et al. 2016b).

Unstable Giant Planet Systems and the Exoplanet Eccentricity Distribution

Gas giant exoplanets are often found on non-circular orbits. This is surprising because, as we saw in Sect. 1.2, planets form in disks and so on orbits similar to the disks' streamlines, which are all circular and coplanar. The broad eccentricity of giant planets was for years a subject of broad speculation (for a historical discussion and a comprehensive list of proposed mechanisms, see Ford and Rasio, 2008). The most likely culprit is a mechanism often referred to as *planet-planet scattering* (Rasio and Ford 1996; Weidenschilling and Marzari 1996; Lin and Ida 1997). The planet-planet scattering model proposes that the giant exoplanets formed in systems with many planets and that those that we see are the survivors of dynamical instabilities. Instabilities lead to orbit crossing, followed by a phase of gravitational scattering that usually concludes when one or more planets are ejected from the system entirely, typically leaving the surviving planets with eccentric orbits (Marzari and Weidenschilling 2002; Adams and Laughlin 2003; Montesinos et al. 2005; Chatterjee et al. 2008; Jurić and Tremaine 2008; Nagasawa et al. 2008; Raymond et al. 2008b, 2009a, 2010, 2011, 2012; Beaugé and Nesvorný 2012).

Whereas instabilities among systems of rocky planets tend to lead to collisions, among systems of giant planets they lead to scattering and ejection. This can be understood by simply considering the conditions required for a planet to be able to give a strong enough gravitational kick to eject another object. The Safronov number Θ is simply the ratio of the escape speed from a planet's surface to the escape speed from the star at that orbital radius. It is defined as:

$$\Theta^2 = \left(\frac{G m_p}{R_p}\right)\left(\frac{a_p}{G M_\star}\right) = \frac{m_p}{M_\star}\frac{a_p}{R_p}, \tag{1.45}$$

where m_p and $M_\star$ are the planet and star mass, respectively, a_p is the planet's orbital radius, and R_p its physical radius. When $\Theta < 1$ collisions are (statistically-speaking) favored over ejection, and when $\Theta > 1$ ejection is favored. Scattering and ejection are therefore favored for massive planets far from their stars. It is interesting to note that Θ is defined in the same way as the gravitational focusing factor F_G from Eq. (1.17) in Sect. 1.2.2, which described the gravitational cross section of a growing protoplanet such that $F_G = 1 + (V_{esc}/V_{rel})^2$. Here, in the context of scattering, the dynamics are the same but in reverse. While the gravitational focusing factor F_G describes from how far away can a planet accrete another object, Θ is a measure of how far a planet can launch another object.

The planet-planet scattering model matches the observed eccentricity distribution of giant exoplanets (Adams and Laughlin 2003; Jurić and Tremaine 2008; Chatterjee et al. 2008; Raymond et al. 2009a, 2010). The only requirement is that a large fraction of systems have undergone instabilities. In simulations performed to date, at least 75%—and probably more likely 90–95%—of giant planets systems must have undergone planet-planet scattering in their past (Jurić and Tremaine 2008; Raymond et al. 2010, 2011). Scattering can even explain the highest-eccentricity giant exoplanets (Carrera et al. 2019).

There is an anti-correlation between giant exoplanet masses and eccentricities. More massive planets are observed to have higher eccentricities (Jones et al. 2006; Ribas and Miralda-Escudé 2007; Wright et al. 2008). This is the opposite of what one would expect from planet-planet scattering. Given that scattering is a process of equipartition of energy, one would naturally expect the low-mass planets to end up with high eccentricities and the high-mass planets with low-eccentricities. This is indeed what is seen in simulations of unstable systems starting with a dispersion of different planet masses. The solution to this conundrum may be quite simply that massive planets form in systems with many, roughly equal-mass planets (Raymond et al. 2010; Ida et al. 2013). In that case, the most massive planets do indeed end up on the most eccentric orbits, as observed.

A *Breaking the Chains* Scenario for Giant Planets

The evolution of giant planet systems may well parallel that of super-Earths. As we saw in Sect. 1.3.1, the prevailing model for super-Earth formation—called *breaking the chains* (Izidoro et al. 2017)—involves migration into resonant chains followed by instability. Could this same evolutionary pathway apply to giant planet systems (Fig. 1.18)?

The answer appears to be yes. Giant planets may indeed follow a *breaking the chains*-style evolution that is similar, but not identical, to that of super-Earth systems.

Survival of resonances after the gaseous disk phase may be somewhat more common for giant planet systems. Only a few dozen giant planets are known to be found in resonance (e.g., the GJ876 resonant chain of giant planets; Rivera et al. 2010). Yet it is possible that many more resonant pairs of giant planets are hiding in plain site. The radial velocity signatures of resonant planets can mimic those of a single eccentric planet (Anglada-Escudé et al. 2010) and it is possible that up to 25% of the current sample of eccentric planets are actually pairs of resonant planets (Boisvert et al. 2018). The PDS 70 protoplanetary disk may host a pair of young giant planets in resonance (Bae et al. 2019), and other disk signatures may require multi-resonant planets to explain them (Dodson-Robinson and Salyk 2011). When multiple giant planets form within a given disk it may thus slow and limit migration. This contrasts with the case of super-Earths, which migrate all the way to the disk's inner edge.

Instabilities appear to be ubiquitous among both super-Earths and giant exoplanets. As discussed above, the eccentricities of giant exoplanets are easily matched if most systems are survivors of instability. Instabilities may be triggered by the dispersal of the gaseous disk (Matsumura et al. 2010), chaotic diffusion within the giant planets' orbits (Marzari and Weidenschilling 2002; Batygin and Morbidelli 2015), or external perturbations, e.g. from wide binary stars (Kaib et al. 2013).

Connection with Rocky Planets and Debris Disks

When giant planets go unstable they affect their entire system. Given their large masses and the high eccentricities they reach during the scattering phase, giant planets can wreak havoc on their inner and outer systems. Giant planet scattering systematically disrupts inner rocky planet systems (Veras and Armitage 2005, 2006; Matsumura et al. 2013; Carrera et al. 2016) or their building blocks (Raymond et al. 2011, 2012), usually by driving inner bodies onto such eccentric orbits that they collide with their host stars (see Fig. 1.19). Scattering also tends to destroy outer planetesimals disks by ejecting planetesimals into interstellar space (Raymond et al. 2011, 2012, 2013; Marzari 2014). Outer planetesimal belts—when they survive and contain enough mass to self-excite to a moderate degree—evolve collisionally to produce cold dust observed as debris disks (Wyatt 2008; Krivov 2010; Matthews et al. 2014; Hughes et al. 2018).

Raymond et al. (2011, 2012) proposed that, by influencing both the inner and outer parts of their systems, giant planets induce a correlation between rocky inner planets and debris disks. Hints of a correlation have been found (Wyatt 2012) but more data are needed. It has also been suggested that giant planets—especially those on very eccentric orbits—should anti-correlate with the presence of debris disks. While there may be a tendency for debris disks to be less dense in systems with eccentric giant planets (Bryden et al. 2009), no strong correlation or anti-correlation between giant planets and debris disks has been observed to date (Moro-Martín et al. 2007, 2015).

Nonetheless, connections between planetary system architecture and the presence and characteristics of dust remains an active area of study.

1.3.3 Solar System Formation

The standard timeline of Solar System formation proceeds as follows. Time zero is generally taken as the formation of Calcium- and Aluminum-rich Inclusions, or CAIs. CAIs are the oldest known inclusions in chondritic meteorites aged to be 4.568 Gyr old (Amelin et al. 2002; Bouvier and Wadhwa 2010; Connelly et al. 2012). Radioactive ages of iron meteorites suggest that their parent bodies several hundred to a thousand-km scale planetary embryos—were fully formed within 1 Myr after CAIs (Halliday and Kleine 2006; Kruijer et al. 2014; Schiller et al. 2015). There are two isotopically distinct types of meteorites: the so-called carbonaceous and non-carbonaceous groups (Warren 2011). There is a broad distribution in the ages of both types of meteorites, whose overlap indicate that these very different types of meteorites accreted simultaneously (Amelin et al. 2002; Warren 2011; Kruijer et al. 2017). Chondrules—the primitive, mm-scale building blocks of chondritic meteorites—are found at the size scale at which objects drift rapidly through the disk due to aerodynamic drag (Weidenschilling 1977b; Lambrechts and Johansen

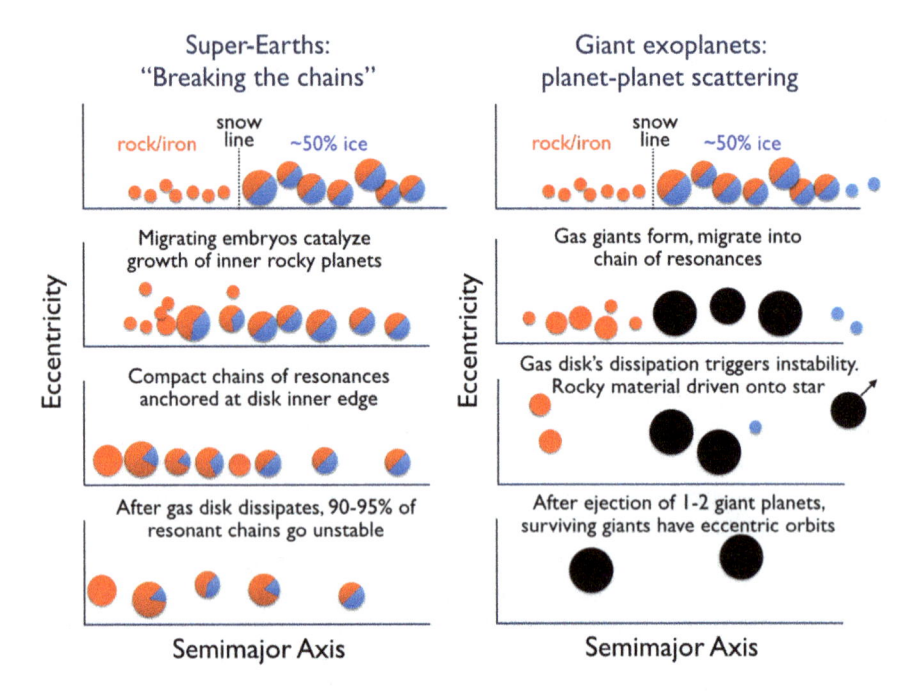

Fig. 1.18 Cartoon comparison of the evolution of super-Earth and giant planets systems. The same general processes could very well causing both populations to follow *breaking the chains*-like pathways (Raymond et al. 2018d)

2012; Johansen et al. 2015). The separation of the two different types of chondritic meteorites is interpreted as indicating that two isotopically-distinct reservoirs were kept spatially segregated, perhaps by the rapid growth of a large planetary embryo to the pebble isolation mass (Budde et al. 2016; Kruijer et al. 2017). This may thus constrain the timing of the growth of Jupiter's core to have been very rapid, reaching the critical mass to block the pebble flux (of $\sim 10-20\ M_\oplus$; Bitsch et al., 2018) within 1–1.5 Myr after CAIs (Kruijer et al. 2017).

The gaseous disk probably lasted for roughly 5 Myr. Gaseous disks around other stars are observed to dissipate on a few Myr timescale (see Fig. 1.2). The oldest chondritic meteorites are the CB chondrites, which formed roughly 5 Myr after CAIs albeit perhaps in the absence of gas (Kita et al. 2005; Krot et al. 2005; Johnson et al. 2015). Given that gas is thought to be needed for planetesimal formation (Johansen et al. 2014), this implies 5 Myr as an upper limit on the gas disk lifetime (at least in some regions).

Hafnium-Tungsten measurements of Martian meteorites indicate that Mars' formation was basically finished within 5–10 Myr (Nimmo and Kleine 2007; Dauphas and Pourmand 2011), meaning that it grew very little after the disk had dispersed. In contrast, Hf/W measurements indicate that Earth's last differentiation event—generally thought to have been the Moon-forming impact—did not take place until $\sim$100 Myr after CAIs (Touboul et al. 2007; Kleine et al. 2009). However, uncertain-

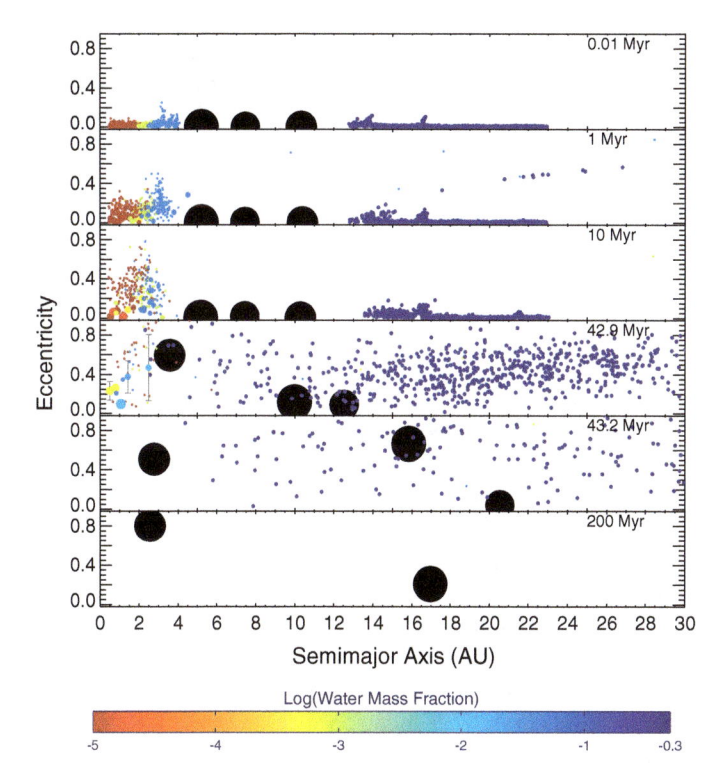

1 Planet Formation: Key Mechanisms and Global Models

Fig. 1.19 Evolution of a simulation in which a system of gas giant went unstable, resulting in the destruction of both an inner system of growing terrestrial planets (that were driven into collisions with the central star) and an outer disk of ice-rich planetesimals (that was ejected into interstellar space; see Raymond et al. (2018a,c). From Raymond et al. (2012)

ties in the degree of equilibration of the Hf/W isotopic system during giant impacts make it hard to determine a more accurate timeframe (Fischer and Nimmo 2018).

Highly-siderophile ("iron-loving") elements should in principle be sequestered in the core during differentiation events. Thus, any highly-siderophile elements in the terrestrial mantle and crust should in principle have been delivered *after* the Moon-forming impact (Kimura et al. 1974). Simulations show that there is a clear anti-correlation between the timing of the last giant impact on Earth and the mass in planetesimals delivered after the last giant impact (Jacobson et al. 2014). The total amount of highly-siderophile elements can therefore constrain the timing of the impact. Assuming a chondritic composition, Earth accreted the last ∼0.5% of its mass as a veneer after the Moon-forming impact (Morbidelli and Wood 2015). This implies that the Moon-forming impact took place roughly 100 Myr after CAIs (Jacobson et al. 2014), consistent with the Hf/W chronometer.

The Solar System's giant planets are thought to have undergone an instability. The so-called *Nice model* showed that the giant planets' current orbital configuration (Tsiganis et al. 2005; Nesvorný and Morbidelli 2012)—as well as the orbital properties of Jupiter's Trojans (Morbidelli et al. 2005; Nesvorný et al. 2013), the irregular satellites (Nesvorný et al. 2014), and the Kuiper belt (Levison et al. 2008; Nesvorný 2015)—could be explained by a dynamical instability in the giant planets' orbits (Nesvorný 2018b). The instability is generally thought to have been triggered by interactions between the giant planets and a remnant planetesimal disk, essentially the primordial Kuiper belt (Gomes et al. 2005; Morbidelli et al. 2007; Levison et al. 2011; Quarles and Kaib 2019), although a self-driven instability is also possible (Raymond et al. 2010; de Sousa et al. 2020). The instability was first proposed to explain the so-called terminal lunar cataclysm, i.e. an abrupt increase in the flux of projectiles hitting the Moon roughly 500 Myr after CAIs (Tera et al. 1974; Gomes et al. 2005; Morbidelli et al. 2012). However, newer analyses suggest that no terminal lunar cataclysm ever took place (Chapman et al. 2007; Boehnke and Harrison 2016; Zellner 2017; Morbidelli et al. 2018; Hartmann 2019), and that the instability may have taken place anytime in the first $\sim$100 Myr after CAIs (Morbidelli et al. 2018; Nesvorný et al. 2018; Mojzsis et al. 2019).

Over the past decade, global models of Solar System formation have been revolutionized (see Fig. 1.20). Decades-old models that assumed local growth of the planets (i.e., the so-called *classical model*; Weidenschilli 1977a; Wetherill 1978, 1992; Chambers and Wetherill 1998; Chambers 2001; Raymond et al. 2004, 2009b; 2014; O'Brien et al. 2006) have been supplanted with models that explain the distribution of the planets and small body belts by invoking processes such as long-range migration of the giant planets (the *Grand Tack* model; Walsh et al. 2011; Raymond and Morbidelli 2014; Brasser et al. 2016), non-uniform planetesimal formation within the disk (the *Low-mass Asteroid belt* model; Hansen 2009; Izidoro et al. 2015c; Drążkowska et al. 2016; Raymond and Izidoro 2017b) and an early instability among the giant planets' orbits (the *Early Instability* model; Clement et al., 2018, 2019a,b). Pebble accretion has been proposed to play an important role during the terrestrial planets' growth (Levison et al. 2015), the importance of which may be constrained by isotopic measurements of different types of meteorites as well as Earth samples (Warren 2011; Kruijer et al. 2017; Schiller et al. 2018; Budde et al. 2019).

Here we first describe the classical model and then the other competing models. We then compare the predictions of different models.

1.3.3.1 The Classical Model

The classical model of terrestrial planet formation makes two dramatically simplifying assumptions. First, it assumes that the planets formed roughly in place. This implies that one can reconstruct the approximate mass distribution of the protoplanetary disk by simply spreading the planets' masses out in concentric annuli (the so-called 'minimum-mass solar nebula' model; Weidenschilling 1977a; Hayashi 1981; Davis 2005). Second, it assumes that giant- and terrestrial planet formation can be treated separately. Thus, one can study the influence of the giant planets' orbits on terrestrial planet formation after the gas disk had dispersed without the need to consider their earlier effects.

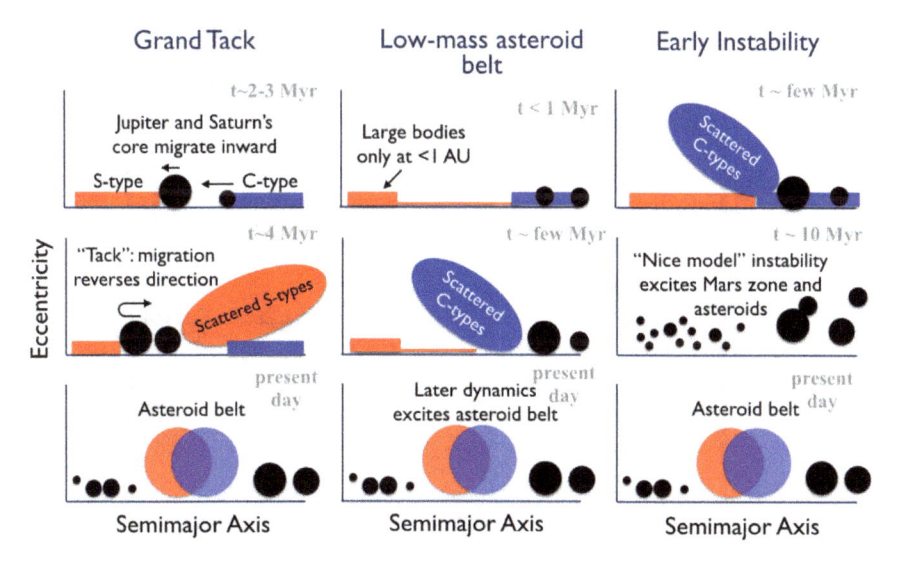

1 Planet Formation: Key Mechanisms and Global Models 57

Fig. 1.20 Cartoon comparison between three plausible models for Solar System formation described in Sects. 3.3.2, 3.3.3, and 3.3.4. From Raymond et al. (2018d)

Figure 1.21 shows a simulation of the classical model of terrestrial planet formation (from Raymond et al. 2006a). In this case Jupiter is included on a supposed pre-instability orbital configuration, on a near-circular orbit. The terrestrial disk is initially composed of roughly 2000 planetary embryos with masses between that of Ceres and the Moon. Accretion is driven from the inside-out by gravitational self-stirring and from the outside-in by secular and resonant forcing from Jupiter. There is a long chaotic phase of growth that involves many giant impacts between planetary embryos (see e.g. Agnor et al. 1999; Kokubo and Ida 2007; Quintana et al. 2016). The planets grow on a timescale of 10–100 Myr. Remnant planetesimals are cleared out on a longer timeframe. There is sufficient radial mixing during growth for water-rich material from past 2.5 au to have been delivered to the growing Earth (see extensive discussion of the origin of Earth's water in Sect. 1.3.4).

The classical model has a fatal flaw: it systematically produces Mars analogs that are far more massive than the real Mars (Wetherill 1991). This can be seen in the simulation from Fig. 1.21, which formed passable Venus and Earth analog, but a Mars analog roughly as massive as Earth. The true problem is not Mars' absolute mass but the large Earth/Mars mass ratio. The classical model tends to produce systems in which neighboring planets have comparable masses rather than the ∼9:1 Earth/Mars ratio (Raymond et al. 2006a; O'Brien et al. 2006; Raymond et al. 2009b; Morishima et al. 2008, 2010; Fischer and Ciesla 2014; Kaib and Cowan 2015). Classical model simulations tend to have two other problems related to the asteroid belt: large embryos are often stranded in the belt (Raymond et al. 2009b), and they do not match the belt's eccentricity and inclination distributions (Izidoro et al. 2015c).

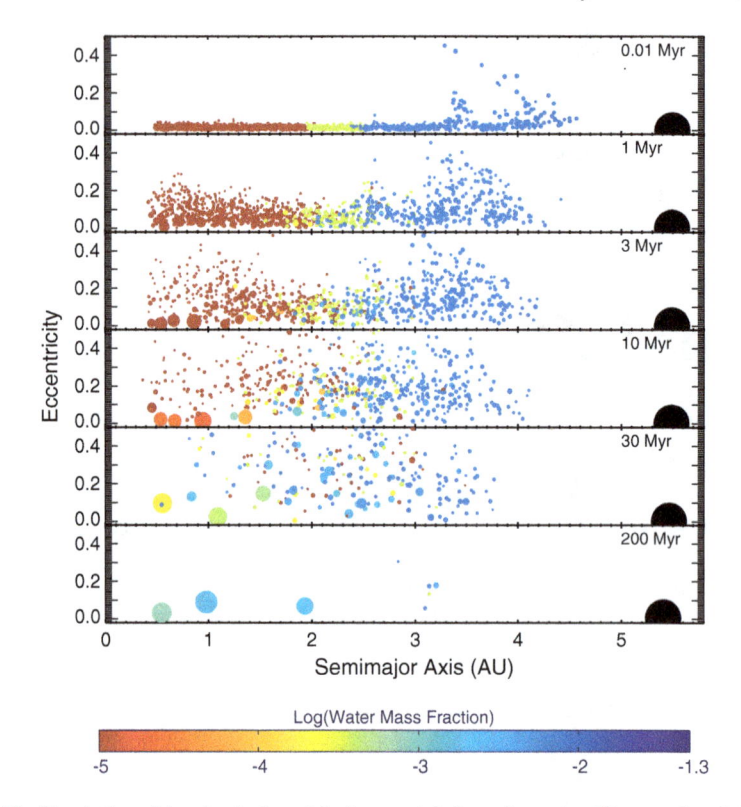

Fig. 1.21 Simulation of the classical model of terrestrial planet formation (Raymond et al. 2006a). The size of each planetary embryo scales with its mass m as $m^{1/3}$. Colors represent the water contents, initially calibrated to match the water contents of current orbital radii of different populations of primitive asteroids (Gradie and Tedesco 1982; Raymond et al. 2004). Jupiter is the large black circle

The building blocks of the terrestrial planets were roughly Mars-mass and thus the inner Solar System may only have contained roughly two dozen embryos (Morbidelli et al. 2012). One may wonder how often a small Mars arises because, by chance, it avoided any late giant impacts. Fischer and Ciesla (2014) showed that this happens in a few percent of simulations. However, when one takes the asteroid belt constraints into consideration the success rate of the classical model in matching the inner Solar System drops by orders of magnitude (Izidoro et al. 2015c).

The 'small Mars' problem—first pointed out by Wetherill (1991) and re-emphasized by Raymond et al. (2009b)—is thus the Achilles heel of the classical model. It prompted the development of alternate models, whose goal was to explain how two neighboring planets could have such different masses.

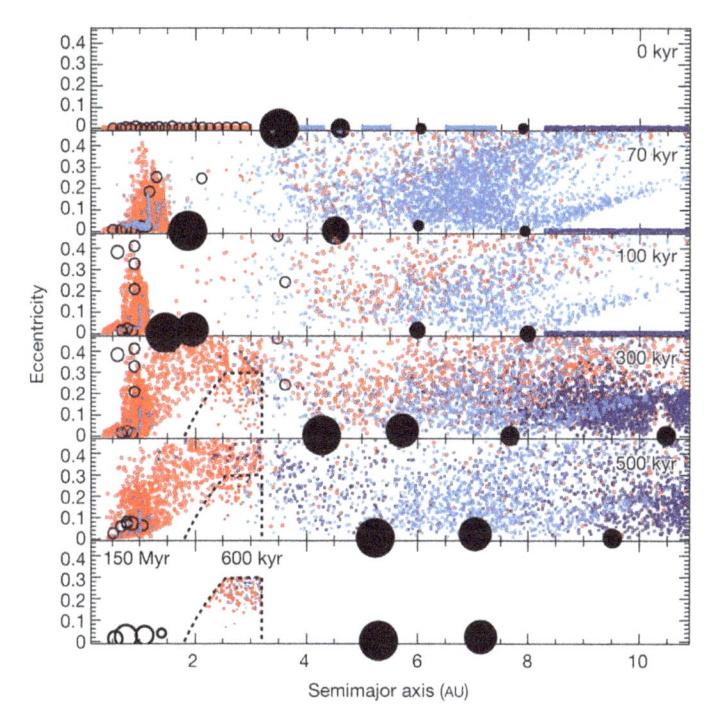

1 Planet Formation: Key Mechanisms and Global Models 59

Fig. 1.22 Evolution of the Grand Tack model (Walsh et al. 2011). The giant planets' migration (black symbols) sculpts the distribution of rocky bodies in the inner Solar System. Here Jupiter migrated inward, then Saturn migrated inward, and then the two planets migrated back out. The red dots indicate presumably dry, S-type planetesimals that formed interior to Jupiter's orbit, whereas the blue dots represent planetesimals originally between and beyond the giant planets' orbits. Rocky planetary embryos are shown as empty circles. The bottom three panels include dashed contours of the present-day main asteroid belt

1.3.3.2 The Grand Tack Model

The Grand Tack model (Walsh et al. 2011) proposed that Jupiter's migration during the gaseous disk phase was responsible for depleting Mars' feeding zone. The Grand Tack is based on hydrodynamical simulations of orbital migration (see Sect. 2.4 and 2.5 for more details). Simulations show the giant planets carve annular gaps in the disk (Lin and Papaloizou 1986; Ward 1997; Crida et al. 2006). A single planet usually migrates inward on a timescale that is related to the disk's viscous timescale (D'Angelo et al. 2003; Dürmann and Kley 2015). Multiple planets often migrate convergently and become trapped in mean-motion resonances (Snellgrove et al. 2001; Lee and Peale 2002), often sharing a common gap. When the inner planet is two to four times more massive than the outer one and both planets are in the gap-opening regime, gas piles up in the inner disk and a toque imbalance causes the two planets to migrate *outward* (Masset and Snellgrove 2001; Morbidelli et al.

2007; Pierens and Nelson 2008; Zhang and Zhou 2010; Pierens and Raymond 2011; Pierens et al. 2014).

The Grand Tack model proposes that Jupiter formed first and migrated inward. Saturn formed more slowly, migrated inward and caught up to Jupiter. The two planets became locked in resonance and migrated outward together until either the disk dissipated or certain conditions slowed their migration (e.g., if the disk was flared such that Saturn dropped below the gap-opening limit in the outer disk).

The evolution of the Grand Tack model is shown in Fig. 1.22. If Jupiter's inward migration reached 1.5–2 au then the inner disk would have been truncated at $\sim$1 au, depleting Mars' feeding zone but not Earth's (Walsh et al. 2011; Brasser et al. 2016). The terrestrial planets in simulations provide a good match to the real ones (Walsh et al. 2011; Jacobson and Morbidelli 2014; Brasser et al. 2016).

The asteroid belt is severely depleted by Jupiter's migration but is not completely emptied (Walsh et al. 2011, 2012). The S-types were scattered outward by Jupiter's inward migration, then back inward by Jupiter's outward migration, with an efficiency of implantation of $\sim$0.1%. The C-types were scattered inward by Jupiter and Saturn's outward migration. The final belt provides a reasonable match to the real one—particularly when evolved over Gyr timescales (Deienno et al. 2016). It is worth noting, however, that the initial conditions in Fig. 1.22 neglect additional (possibly much more important) phases of implantation that took place during Jupiter and Saturn's rapid gas accretion (Raymond and Izidoro 2017a).

The Grand Tack's potential Achilles heel is the mechanism of outward migration itself (see discussion in Raymond and Morbidelli 2014). The most stringent constraint on outward migration is that it requires a relatively limited range of mass ratios between Jupiter and Saturn (roughly between a ratio of 2:1 and 4:1; Masset and Snellgrove 2001). It remains to be seen whether long-range outward migration remains viable when gas accretion is consistently taken into account, as the gas giants' mass ratio should be continuously changing during this phase. There may also be geochemical constraints related to the speed of accretion in the Grand Tack model (Zube et al. 2019).

1.3.3.3 The Low-Mass Asteroid Belt Model

The Low-Mass Asteroid belt model makes the assumption that Mars' small mass is a consequence of a primordial mass deficit between Earth and Jupiter's present-day orbits. Gas disks are generally expected to have smooth radial distributions, but this is not the case for dust. Dust drifts within disks and ALMA images show that dust rings are common in protoplanetary disks (ALMA Partnership 2015; Andrews et al. 2018). Given that planetesimal formation via the streaming instability is sensitive to the local conditions (Simon et al. 2016; Yang et al. 2017), dust rings may be expected to produce rings of planetesimals. One model of dust drift and coagulation, combined with the conditions needed to trigger the streaming instability, showed that planetesimals may indeed form rings (Drążkowska et al. 2016).

Fig. 1.23 Distribution of simulated terrestrial planets formed assuming that all of the terrestrial planets' mass was initially found in a narrow annulus between the orbits of Venus and Earth (denoted by the dashed vertical lines). The open circles show simulated planets and the solid ones are the real planets. From Hansen (2009)

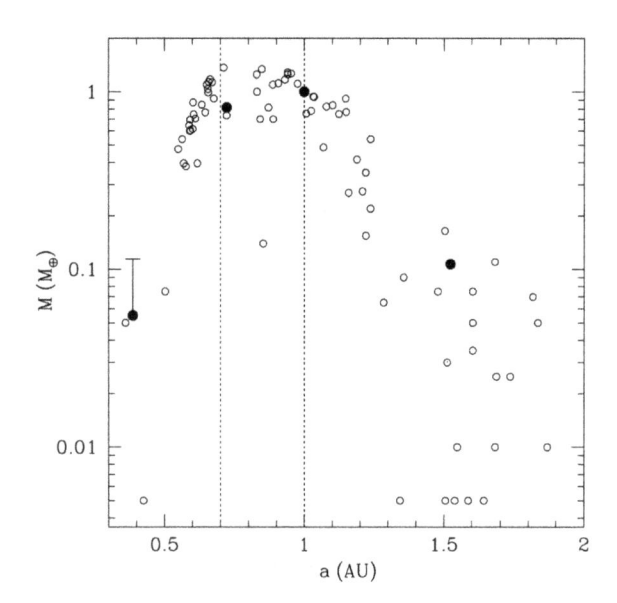

Since the 1990s it was known that if an edge existed in the initial distribution of planetary embryos or planetesimals, any planet that formed beyond that edge would be much less massive than the planets that formed within the main disk (Wetherill 1992, 1996; Chambers and Wetherill 1998; Agnor et al. 1999; Chambers 2001). However, it was Hansen (2009) who proposed that such an edge might really have existed in the early Solar System. Indeed, he showed that the terrestrial planets can be matched if they formed from a narrow annulus of rocky material between the orbits of Venus and Earth (see Fig. 1.23). In this model the large Earth/Mars mass ratio is a simple consequence of the depletion of material past Earth's orbit, and the small Mercury/Venus mass ratio is a consequence of the analogous depletion interior to Venus' orbit.

The most extreme incarnation of the Low-mass asteroid belt model proposes that no planetesimals formed between the orbits of Earth and Jupiter (Hansen 2009; Raymond and Izidoro 2017b), but the model is also consistent with planetesimal forming in the belt, just at a reduced efficiency. Yet, even an empty primordial asteroid belt would have been dynamically re-filled with objects originating across the Solar System. Rapid gas accretion onto Jupiter and later Saturn destabilizes the orbits of nearby planetesimals, many of which are gravitationally scattered in all directions. Under the action of gas drag, a fraction of planetesimals are trapped on stable, low-eccentricity orbits, preferentially in the asteroid belt (Raymond and Izidoro 2017a), and some are scattered past the asteroid belt to the terrestrial planet region itself (see Fig. 1.27 and Sect. 1.3.4). On a longer timescale, the growing terrestrial planets scatter remnant planetesimals outward, a small fraction of which are trapped in the main belt, preferentially in the inner parts (Raymond and Izidoro 2017b).

The main uncertainty in the Low-mass asteroid belt are the initial conditions. When disk models, dust growth and drift, disk observations and interpretation (including studies of ALMA-detected disks), and meteorite constraints (including the broad age distribution of non-carbonaceous chondrites, which indicate many different planetesimal formation events; Kruijer et al. 2017) are all accounted for, will it be a reasonable assumption that the terrestrial planets could have formed from a narrow ring of planetesimals?

1.3.3.4 The Early Instability Model

It has been demonstrated numerically that dynamical instabilities among giant planets cause severe damage to the small bodies in their same systems (see Fig. 1.19). The Solar System's giant planets are thought to have undergone an instability, albeit a much less violent one than the typical instability incurred by giant exoplanet systems. Nonetheless, the Solar System's instability is violent enough that, if they were fully-formed at the time, the terrestrial planets would have had a very low probability of survival (Kaib and Chambers 2016). Yet the timing of the instability is uncertain, and it could have happened anytime in the first 100 million years of Solar System evolution (Morbidelli et al. 2018; Mojzsis et al. 2019). One may then wonder whether a very early instability could have played a role in shaping the distribution of the terrestrial planets.

The Early Instability model, conceived and developed by Clement et al. (2018, 2019a, b), is built on the premise that the giant planets' instability took place within roughly 10 Myr of the dissipation of the gaseous disk. The evolution of one realization of the model is shown in Fig. 1.24. The early evolution of the Early Instability model is identical to that in the classical model. The giant planets' dynamical instability—triggered after 10 Myr in the simulation from Fig. 1.24—strongly excites the orbits of inner Solar System objects extending through the asteroid belt in to Mars' feeding zone (Deienno et al. 2018; Clement et al. 2019b). The belt is strongly depleted and dynamically excited, and Mars' growth is effectively stunted. In a fraction of simulations no Mars forms at all (Clement et al. 2018)! The growth of Earth and Venus are largely unperturbed and qualitatively similar to the classical model.

The instability itself is stochastic in nature. Matching the instability statistically requires a large number of numerical realizations, which produce a spectrum of Solar Systems with different properties. Such simulations have more constraints than many because they include all of the planets and not just the terrestrial planets. One remarkable feature of the Early Instability model is that systems that provide the best match to the outer Solar System are the same ones that provide the best match to the terrestrial planets (Clement et al. 2018, 2019a, b).

The main uncertainty in the Early Instability model is simply the timing of the instability. Mars' formation was largely complete within 5–10 Myr (Nimmo and Kleine 2007; Dauphas and Pourmand 2011), shortly after the disappearance of the gaseous disk. To affect terrestrial planet formation, the giant planets must therefore have gone unstable within perhaps 5 Myr of the disk's dispersal. This would also

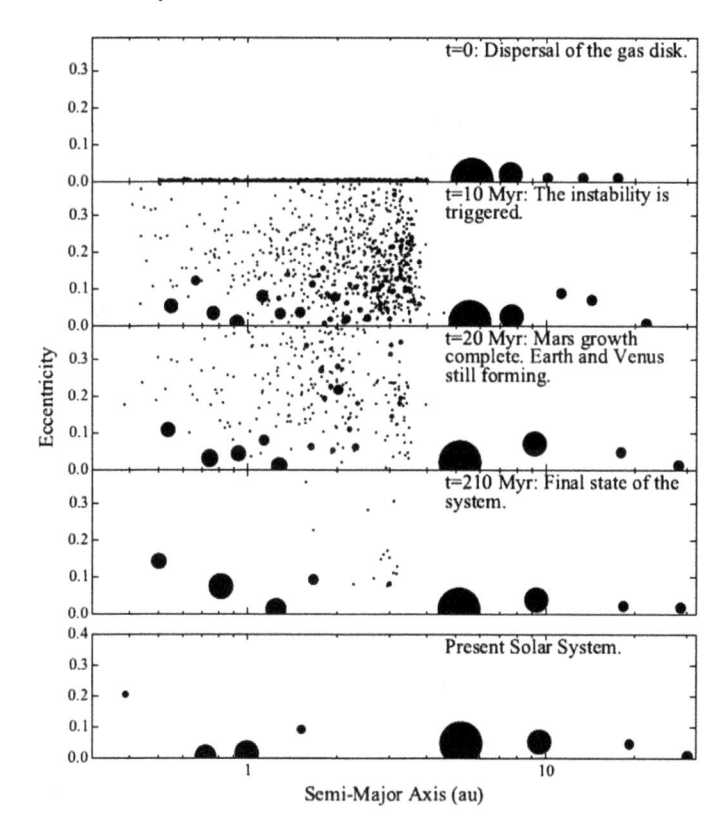

1 Planet Formation: Key Mechanisms and Global Models

Fig. 1.24 Evolution of the Early Instability model. In this case the giant planets' instability was triggered 10 Myr after the dissipation of the disk. The instability strongly excited the orbits of growing planetesimals and planetary embryos in the asteroid belt and Mars' feeding zone. Mars' growth was stunted, but Earth and Venus continued to accrete for ~100 Myr. The final system has a large Earth/Mars mass ratio and an overall planetary system architecture similar to the actual Solar System (shown at bottom for comparison). Figure courtesy of M. S. Clement, from Clement et al. (2018)

imply a cometary bombardment in the inner Solar System that was coincident with the late phases of terrestrial planet growth. While such a bombardment would deliver only a very small amount of mass to Earth (Morbidelli et al. 2000; Gomes et al. 2005), it would provide a large component of Earth's noble gases (Marty et al. 2016). The Xe isotopic compositions of the mantle and atmosphere are different (Mukhopadhyay 2012; Caracausi et al. 2016), and it has been suggested that a comet-delivered component contributed to the atmospheric Xe budget but not to the mantle Xe (Marty et al. 2017). It remains to be understood whether or not this implies that the bulk of Earth accreted with little cometary influx, which would constrain the timing of cometary delivery and presumably of the instability itself.

1.3.3.5 Other Models

It is worth noting that the three scenarios outlined above as alternatives to the classical model are not the only ones that have been proposed. For example, it was suggested that sweeping secular resonances during the dissipation of the gaseous disk could have depleted and excited the asteroid belt and generated an edge in the terrestrial planets' mass distribution (Nagasawa et al. 2005; Thommes et al. 2008; Bromley and Kenyon 2017). The main uncertainty in that model is whether a non-zero eccentricity of Jupiter can realistically be maintained during the gas dissipation phase.

It has also been suggested that pebble accretion may have played a role in terrestrial planet formation (Levison et al. 2015; Chambers et al. 2016). Isotopic analyses of meteorites may help to constrain the degree to which carbonaceous pebbles from the outer Solar System contributed to the growth of the terrestrial planets (Schiller et al. 2018; Budde et al. 2019). In general, the formation of Earth-mass planets within the disk lifetime poses a problem because, given that the gas disk is required for pebble accretion, such planets should grow fast enough to migrate inward and become super-Earths (Lambrechts et al. 2019a).

1.3.4 Origin of Water on Earth and Rocky Exoplanets

We now turn our attention to the origin of planets' water. Cosmochemical tracers such as isotopic ratios can be used to constrain the potential sources of Earth's water (see, e.g., Morbidelli et al. 2000; Marty 2012). While bulk density measurements of solid exoplanets can in principle be used to trace water contents (Valencia et al. 2007; Sotin et al. 2007; Fortney et al. 2007; Seager et al. 2007), in practice it is extremely challenging (Adams et al. 2008; Selsis et al. 2007; Dorn et al. 2015). There have been a few recent reviews on the origins of Earth's water in dynamical and cosmochemical context (O'Brien et al. 2018; McCubbin and Barnes 2019; Meech and Raymond 2019).

Earth is our benchmark for water. While it appears blue from space, all of Earth's oceans only add up to $\sim 2.5 \times 10^{-4} \, M_\oplus$ of water. This is generally referred to as one "ocean" of water. The water budget of Earth's interior is quite uncertain. Different studies infer different quantities of water trapped in hydrated silicates, with overall budgets between roughly one and ten oceans (Hirschmann 2006; Mottl et al. 2007; Marty 2012; Halliday 2013). Earth's core may be very dry (Badro et al. 2014) or may contain fifty or more oceans of water (Nomura et al. 2014). While papers often quote Earth as being roughly one part in a thousand water by mass, it is important to be aware of the uncertainty.

Figure 1.25 shows the D/H ratios for a number of Solar System objects (for a compilation of references to the D/H measurements, see Morbidelli et al. 2000; Marty and Yokochi 2006; Alexander et al. 2012, 2018). The Sun and gas-dominated planets have D/H ratios roughly six times lower than Earth's. This is interpreted as the isotopic composition of the gaseous protoplanetary disk. Carbonaceous chondrite

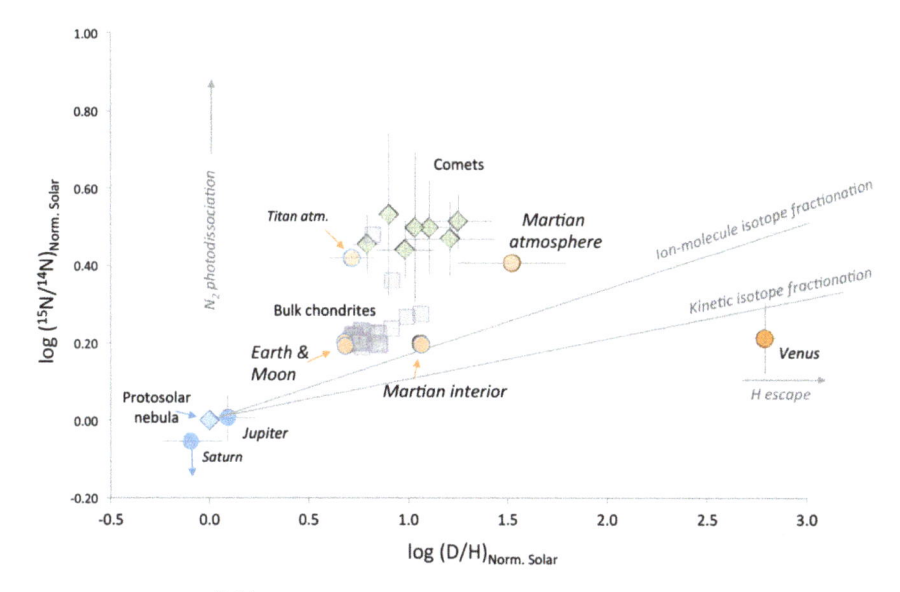

1 Planet Formation: Key Mechanisms and Global Models 65

Fig. 1.25 Measured $^{15/14}$N isotope ratios vs. D/H ratios of Solar System objects. Arrows indicate how different processes would affect an object's evolution in this isotope space. Figure courtesy of B. Marty, from Marty et al. (2016)

meteorites have similar D/H values to Earth. Most measured comets have higher D/H values than Earth, although two recently-measured Jupiter-family comets (Hartogh et al. 2011; Lis et al. 2013) and one Oort cloud comet (Biver et al. 2016) have Earth-like D/H ratios. In contrast, ESA's Rosetta mission measured the D/H ratio of comet 67P/C-G to be more than three times the terrestrial value. A recent result (Sosa and Fernández 2011; Lis et al. 2019) found that very active comets tended to have Earth-like D/H whereas less active ones have higher D/H. One interpretation is that comets' original water was Earth-like and loss processes during outgassing have fractionated the surviving D/H. If true, that would mean render the D/H ratio useless as a discriminant between comets and carbonaceous sources of water. One must then resort to dynamical constraints or perhaps to other isotopic systems.

Figure 1.25 naively would suggest carbonaceous chondrite-like objects as the source of Earth's water and Nitrogen. Yet, even if Earth's water were delivered by carbonaceous objects, that is only a part of the story. A complete model must explain the full evolution of our planetary system and, in the context of Earth's water, ask: where did the water-bearing objects themselves originate?

We can break down the various models for water delivery into six rough scenarios that we outline below. Two of these models invoke local sources of water whereas the others propose that Earth's water was delivered from farther out in the disk. For a comprehensive review of these models, see Meech and Raymond (2019).

Adsorption of Water Vapor Onto Silicate Grains

In a simple picture of the structure of disks, water should exist as a solid past the snow line and as a vapor closer-in. If water vapor was indeed present where silicate grains were coalescing to form the terrestrial planets, then "in-gassing" may have attached hydrogen molecules to silicate grains (Stimpfl et al. 2006; Muralidharan et al. 2008; Asaduzzaman et al. 2014; Sharp 2017; D'Angelo et al. 2019). This process is called adsorption. The mechanism can in principle have seeded Earth with a few oceans of water, albeit without taking any water loss processes into account, such as ^{26}Al-driven heating (Grimm and McSween 1993; Monteux et al. 2018) or impact-related losses (Genda and Abe 2005; Svetsov 2007). Yet at face value it should have seeded Earth with nebular water, which has a D/H ratio six times smaller than Earth's. It also cannot explain the abundance of other volatiles such as C, N and the noble gases, which would then require an alternate source containing little hydrogen. Finally, it begs the question of why the Enstatite chondrites—meteorites that provide the closest chemical match to Earth's bulk composition (Dauphas 2017)—appear to have formed without any water.

Oxidation of H-Rich Primordial Envelope

As planetary embryos grow, they gravitationally accrete H-rich envelopes if they are more massive than a few tenths of Earth's mass (Ikoma et al. 2000; Lambrechts and Lega 2017). This hydrogen could have chemically reacted with Earth's surface magma ocean and generated water by hydrating silicates (Ikoma and Genda 2006). Given that the D/H ratio of nebular gas was many times smaller than Earth's water, this model, like the previous one, predicts that Earth's initial D/H ratio was small. In this case, however, it is possible to envision that Earth's D/H changed in time. The loss of a thick hydrogen envelope would certainly entail mass fractionation, and for a certain range of parameters Earth's final D/H ratio can be matched even though its water was acquired from gas (Genda and Ikoma 2008). However, the collateral effects of this presumed atmospheric loss have not been quantified, and appear to be at odds with other constraints, such as the abundance of ^{129}Xe from the decay of ^{129}I (Marty et al. 2016). In addition, it seems quite a coincidence for Earth to match carbonaceous chondrites in multiple isotopic systems as a result of such loss processes, which would affect different molecules differently.

Pebble "Snow"

Planets beyond the snow line accrete water as a solid. It is simply a building block. Yet the snow line is a moving target. As the disk evolves and cools, the snow line generally moves inward (Sasselov and Lecar 2000; Lecar et al. 2006; Martin and Livio 2012). A planet on a static orbit would see the snow line sweep past it. Such a planet would start off in the rocky part of the disk but, once the snow line passed interior to its orbit, would find itself outside the snow line, in the presumably icy part of the disk (see Fig. 1.26).

As the snow line sweeps inward, new ice does not come from the condensation of water vapor. This is because the speed at which gas moves through the disk is far faster than the motion of the snow line itself (Morbidelli et al. 2016). Thus, the gas just interior to the snow line is dry. Rather, ice at the snow line comes from pebbles and dust that drift inward through the disk. The source of these drifting pebbles is thought to move outward as an analogous wave of dust coagulation and growth sweeps radially out through the disk (Birnstiel et al. 2012; Lambrechts et al. 2014; Birnstiel et al. 2016; Ida et al. 2016). If anything blocks the inward flow of pebbles—such as a growing giant planet core (Lambrechts and Johansen 2014; Bitsch et al. 2018)—the snow line will continue to move inward but it will not bring any ice along with it (Morbidelli et al. 2016). The source of water will also drop if the outward-sweeping growth front producing pebbles reaches the outer edge of the disk (Ida et al. 2019).

In the Solar System there is evidence that the pebble-sized building blocks carbonaceous and noncarbonaceous were segregated as of 1–1.5 Myr after CAIs (Budde et al. 2016; Kruijer et al. 2017). This would suggest that carbonaceous pebbles did not deliver Earth's water, as their widespread presence in the inner Solar System would presumably have produced a category of meteorites intermediate in isotopic composition between the carbonaceous and noncarbonaceous. The Enstatite chondrites, which formed near the end of the disk's lifetime and are the closest chemical match to Earth, are dry (Alexander et al. 2018).

Nonetheless, pebble snow may be a key mechanism in delivering water to rocky exoplanets (see below).

Wide Feeding Zone

A planet's feeding zone is simply the radial distribution of its constituents. In the classical model, the terrestrial planets formed from a broad disk of planetary embryos extending from Venus' orbit out to Jupiter's. For example, in the simulation from Fig. 1.21, each of the three surviving terrestrial planets accreted material from a broad swath of the disk. Given that the constituent planetesimals should have a radial gradient in composition based on the local temperature (Dodson-Robinson et al. 2009), this implies that volatiles are naturally incorporated into growing planets if the planets' feeding zones extend out far enough (Morbidelli et al. 2000; Raymond et al. 2004, 2007a).

The planets' broad feeding zones—in concert with the isotopic match between Earth's water and that of carbonaceous chondrites—led to the scenario that Earth's water was delivered by primordial carbonaceous planetesimals or planetary embryos (Morbidelli et al. 2000; Raymond et al. 2004, 2007a). However, this idea was built on the classical model. As the classical model crumbled under the weight of the small Mars problem (see Sect. 1.3.3) the idea that Earth's water was a result of its broad feeding zone no longer made sense. The Early Instability model (Clement et al. 2018, 2019a,b) starts from essentially the same initial conditions as the classical model and presents a viable solution to the small Mars problem all the while delivering water to Earth by the same mechanism. However, like the classical model (Raymond et al. 2009b), the Early Instability may only adequately deliver water to Earth in simulations that fail to match Mars' small mass (Clement and Rubie, personal communication). The Early Instability model acknowledges that in order

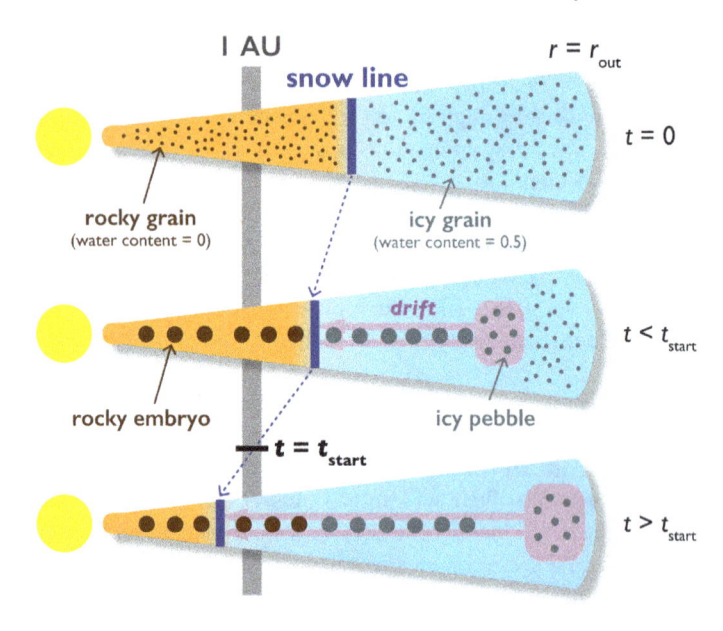

Fig. 1.26 The *pebble snow* model (Oka et al. 2011; Sato et al. 2016; Ida et al. 2019). Rocky planetary embryos grow from rocky grains interior to the snow line. As the disk cools, the snow line sweeps inward and ice-rich pebbles from the outer parts of the disk deliver water to the terrestrial planet region. Figure courtesy of T. Sato, from Sato et al. (2016)

to understand in the first place the existence of water-rich asteroids intermixed with water-poor asteroids in the asteroid belt, which may have been the result of the growth and/or migration of the giant planets. That leads us to the next scenario.

External Pollution

The orbits of leftover planetesimals are strongly perturbed by the growth and migration of the giant planets. The phase of rapid gas accretion is particularly dramatic. The mass of the giant planets can increase from $\sim 10-20$ $M_\oplus$ up to hundreds of Earth masses on a $\sim 10^5$ year timescale (Pollack et al. 1996; Ida et al. 2004). This rapid gas accretion destabilizes the orbits of any nearby planetesimals that managed to avoid being accreted. Many planetesimals undergo close gravitational encounters with the growing planet and are scattered in all directions. Under the action of gas drag, many planetesimals are trapped interior to the giant planet's orbit (Raymond and Izidoro 2017a; Ronnet et al. 2018). This happens when a planetesimal on an eccentric orbit that crosses the giant planet's orbit at apocenter undergoes sufficient orbital energy loss due to gas drag to drop its apocenter away from the giant planet, releasing it from the gas giant's dynamical clutches.

Figure 1.27 shows this process in action. Jupiter's growth triggers a pulse of planetesimal scattering, and a second is triggered when Saturn forms. The outcomes of these pulses can vary because the disk itself changes in time as it both slowly dissipates and, more importantly, is sculpted by the giant planets.

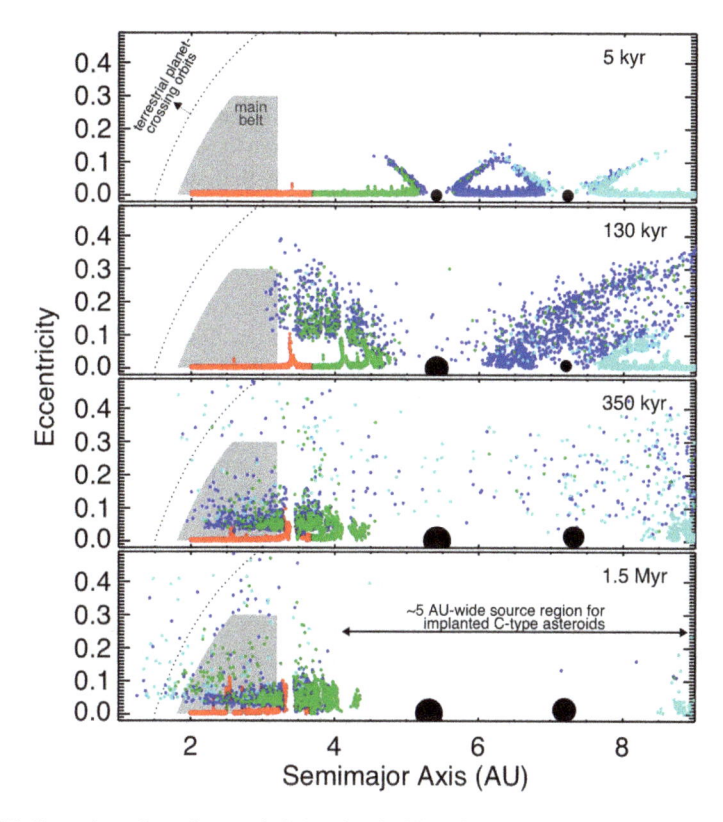

1 Planet Formation: Key Mechanisms and Global Models

Fig. 1.27 Scattering of ten thousand planetesimals driven by the growth of Jupiter (from 100 – 200 kyr) and Saturn (from 300 – 400 kyr). Planetesimals from the Jupiter-Saturn region end up polluting the inner Solar System *en masse*. Some are trapped in the outer parts of the main asteroid belt (shaded) and some are scattered past the belt to the terrestrial planet region (above the dashed line). And even so, this example shows the minimum expected impact of the giant planets' growth, as it assumed the giant planets to have formed on low-eccentricity, non-migrating orbits in 3:2 resonance. Planetesimal colors correspond simply to each object's starting location. There is an underlying gaseous disk in the simulation, whose structure and overall density evolve in time in a consistent way. In this example, planetesimals were assumed to be 100 km in diameter for the gas drag calculation. From Raymond and Izidoro (2017a)

The relative number of planetesimals that are trapped in the main asteroid belt (providing an orbital match to the C-types) vs. those that are scattered *past* the asteroid belt to the terrestrial planets (to deliver water) varies as a function of the strength of gas drag (Raymond and Izidoro 2017a). When gas drag is strong—i.e., for a massive inner gas disk or smaller planetesimals—scattered planetesimals are rapidly decoupled from Jupiter and are generally trapped in the outer parts of the main belt. When gas drag is weak—for a low-density inner disk or large planetesimals—it takes a large number of orbits for gas drag to decouple planetesimals' orbits from Jupiter. Planetesimals are more frequently scattered farther inward to pollute the terrestrial planet-forming region and to deliver water.

A similar mechanism is driven by giant planet migration. In the Grand Tack model, Jupiter and Saturn's outward migration scatters a large number of planetesimals inward (Walsh et al. 2011, 2012; see Fig. 1.22). Given that the giant planets are migrating away, scattered planetesimals are more easily decoupled from the giant planets. The distribution of scattered planetesimals that survive on orbits interior to the giant planets' have a much weaker size dependence than for those that were scattered during the giant planets' growth and for which gas drag played a central role. The dependence on the gas surface density is also much weaker.

Both migration and growth can thus generate populations of terrestrial planet-crossing planetesimals that originated beyond Jupiter. Given their distant origins these planetesimals tend to have high-eccentricity orbits, but simulations show that they are nonetheless accreted by the terrestrial planets on a geochemically consistent timescale and in sufficient quantity to easily deliver Earth's water budget (O'Brien et al. 2014). In addition, they naturally match the compositions of carbonaceous meteorites.

In the wide feeding zone model, the C-type asteroids represented the distant, mother source of Earth's water. In the pollution model, Earth's water was delivered from the same parent population as the objects that were implanted into the asteroid belt as C-types. Thus, Earth's water and C-type asteroids are brother and sister.

At present, the pollution scenario is the leading model to explain the origin of Earth's water. It matches the amount and chemical signature of Earth's water, and naturally fits within different models of Solar System formation.

Inward Migration

Earth-mass planets migrate relatively rapidly in protplanetary disks (see Sects. 1.2.2 and 1.3.1). Given that large embryos are thought to form fastest past the snow line (Lambrechts and Johansen 2014; Morbidelli et al. 2015), this implies that many inward-migrating planets or cores should be water-rich (Bitsch et al. 2019b). Indeed, the migration model for super-Earth formation predicts that most should be water-rich (Raymond et al. 2018b; Izidoro et al. 2019). Inward-migrating gas giants can shepherd material interior to their orbits and trigger the formation of very water-rich terrestrial planets (Raymond et al. 2006b; Mandell et al. 2007).

A number of lines of evidence point to the terrestrial planets having formed from ∼Mars-mass planetary embryos (Morbidelli et al. 2012). This is below the mass threshold for large-scale migration (Ward 1997), and so we do not think that Earth accreted any water from inward-migrating embryos.

However, migration may prove a central water delivery mechanism in other systems (see below).

Extrapolation to Exoplanets

Let us now very simply extrapolate these six water delivery (or water production) mechanisms to exoplanet systems. Most of these mechanisms can account for Earth's water budget with perhaps an order of magnitude variation.

We propose that the water contents of rocky planets are likely to have a bimodal distribution. Migration is the key player. If embryos grew large enough and underwent long-range migration then they must invariably pollute the terrestrial planet-forming region (Izidoro et al. 2014). This should lead to elevated water contents of ∼10% by mass. However, if migration was not involved, then the other mechanisms discussed can still deliver an Earth-like or higher water budget. Gas giants are rare enough around Sun-like and low-mass stars that they probably do not play the central role in water delivery in general terms (although they likely dominate in their own systems). Pebble snow can in principle deliver tens to hundreds of oceans, but requires a clear path between the outward-sweeping icy pebble front and the rocky planet zone that is unencumbered by pressure bumps, including those produced by growing planets (which are themselves growing by consuming icy pebbles).

If migration is indeed the key, then one might predict an observational marker between the action of migration and the planets' water and other volatile contents. Such an observational test is beyond current capabilities but may be imaginable in the coming decade or two.

1.4 The Future of Planet Formation Studies

We conclude this broad-sweeping review with our vision for the future of planet formation. We cannot pretend to have a coherent view of all of the theoretical and empirical challenges that will push the field forward. Nonetheless, we proceed by highlighting three action items: a key bottleneck for planet formation models, one particularly promising path forward, and a call for connection.

A Key Bottleneck for Planet Formation

We consider the central bottleneck in planet formation to be understanding the underlying structure of protoplanetary disks (see Sect. 2.1 and discussion in Morbidelli and Raymond, 2016). Disks are the birthplace of planets. Their underlying structure controls how dust grows and drifts, where dust or pebbles pile up to become sufficiently concentrated to form planetesimals, as well as how fast and in what direction planets migrate. Simple viscous disk models do not match observations and have been supplanted by models that include effects such as ambipolar diffusion and wind-driven angular momentum loss. ALMA observations of disks (Andrews et al. 2018) provide ever more stringent constraints on such models, yet to date there is no underlying model that matches the population of disk observations. Of course, interactions with other stars during the embedded cluster phase leads to a diversity of disk properties, but with statistically described distributions (Bate 2018). We encourage future work to develop comprehensive, trustworthy disk models.

Coupling Dynamical and Chemical Models

One key path forward for understanding the early evolution of the Solar System is coupling dynamical and accretion models with cosmochemical constraints. Some constraints are already being used by current models. For example, the D/H ratio— especially when combined with the $^{15/14}$N ratio (Marty and Yokochi 2006; Marty et al. 2016)—appears to be a powerful tracer of Earth's water. Isotope systems such as Hf/W already provide strong constraints on the timing of planetary accretion. Some formation models also incorporate simple chemical models (Rubie et al. 2015). Yet, there remain many connections to be made between dynamical models and cosmochemistry.

A Call to Connect Different Constraints and Models

We conclude this chapter with a call to create as many connections as possible. This is a large variety of disciplines linked together to create the field of planet formation. These include observations of protoplanetary disks and debris disks, exoplanet studies, meteorite analysis, planetary surfaces (e.g., crater modeling), orbital dynamics, gasdynamics, small body studies, and a variety of different types of numerical modeling. We encourage the reader to strive to make connections between their specialty and others. Many of the most interesting dynamical models come from connections between subdisciplines. For example, the idea of the Solar System's instability (the *Nice model* Tsiganis et al., 2005; Gomes et al., 2005; Nesvorný, 2018b) was born from a dynamical model to explain the now-defunct terminal lunar cataclysm. The *Grand Tack* model of Solar System formation (Walsh et al. 2011) was inspired by numerical studies of planet migration designed to explain the origin of hot Jupiters (Lin et al. 1996). Giant exoplanets orbiting stars with wide companions have more eccentric orbits than giant exoplanets around single stars; this is a result of Oort cloud comet-like oscillations in the orbits of wide binary stars (Kaib et al. 2013).

Connecting the dots in new ways is essential to moving the field of planet formation forward.

Acknowledgements We are grateful to the Agence Nationale pour la Recherche, who sponsored our research from 2014-2018 (project *MOJO*).

References

Abod, C.P., Simon, J.B., Li, R., et al.: Astrophys. J. **883**, 192 (2019)
Absil, O., Defrère, D., Coudé du Foresto, V., et al.: Astron. Astrophys. **555**, A104 (2019)
Adams, E.R., Seager, S., Elkins-Tanton, L.: Astrophys. J. **673**, 1160 (2008)
Adams, F.C., Hollenbach, D., Laughlin, G., Gorti, U.: Astrophys. J. **611**, 360 (2004)
Adams, F.C., Laughlin, G.: Icarus **163**, 290 (2003)
Agnor, C.B., Canup, R.M., Levison, H.F.: Icarus **142**, 219 (1999)
Alexander, C.M.O.: Geochemica Cosmochemica Acta **254**, 246 (2019a)
Alexander, C.M.O.: Geochemica Cosmochemica Acta **254**, 277 (2019b)
Alexander, C.M.O., Bowden, R., Fogel, M.L., et al.: Science **337**, 721 (2012)
Alexander, C.M.O., McKeegan, K.D., Altwegg, K.: Space Sci. Rev. **214**, 36 (2018)
Alexander, R., Pascucci, I., Andrews, S., Armitage, P., Cieza, L.: Protostars Planets VI, p. 475 (2014)
Alexander, R.D., Armitage, P.J.: Astrophysi. J. Lett. **639**, L83 (2006)
Alexander, R.D., Pascucci, I.: Monthly Not. Roy. Astronom. Soc. **422**, L82 (2012)
ALMA Partnership, Brogan, C.L., Pérez, L.M., Hunter, T.R., et al.: Astrophys. J. Lett. **808**, L3 (2015)

Amelin, Y., Krot, A.N., Hutcheon, I.D., Ulyanov, A.A.: Science **297**, 1678 (2002)
Andrews, S.M., Huang, J., Pérez, L.M., et al.: Astrophys. J. Lett. **869**, L41 (2018)
Andrews, S.M., Williams, J.P.: Astrophys. J. **631**, 1134 (2005)
Andrews, S.M., Williams, J.P.: Astrophys. J. **671**, 1800 (2007a)
Andrews, S.M., Williams, J.P.: Astrophys. J. **659**, 705 (2007b)
Andrews, S.M., Wilner, D.J., Hughes, A.M., et al.: Astrophys. J. **700**, 1502 (2009)
Andrews, S.M., Wilner, D.J., Hughes, A.M., et al.: Astrophys. J. **723**, 1241 (2010)
Andrews, S.M., Wilner, D.J., Zhu, Z., et al.: Astrophys. J. Lett. **820**, L40 (2016)
Anglada-Escudé, G., López-Morales, M., Chambers, J.E.: Astrophys. J. **709**, 168 (2010)
Anglada-Escudé, G., Tuomi, M., Gerlach, E., et al.: Astron. Astrophys. **556**, A126 (2013)
Armitage, P.J.: Annu. Rev. Astron. Astrophys. **49**, 195 (2011)
Armitage, P.J., Eisner, J.A., Simon, J.B.: Astrophys. J. Lett. **828**, L2 (2016)
Asaduzzaman, A.M., Zega, T.J., Laref, S., et al.: Earth Planet. Sci. Lett. **408**, 355 (2014)
Badro, J., Cote, A., Brodholt, J.: PNAS **132**, 94 (2014)
Bae, J., Zhu, Z., Baruteau, C., et al.: Astrophys. J. Lett. **884**, L41 (2019)
Bai, X.N.: Astrophys. J. **821**, 80 (2016)
Bai, X.N.: Astrophys. J. **845**, 75 (2017)
Bai, X.N., Stone, J.M.: Astrophys. J. **769**, 76 (2013)
Baillié, K., Charnoz, S., Pantin, E.: Astron. Astrophys. **577**, A65 (2015)
Balbus, S.A., Hawley, J.F.: Revi. Mod. Phys. **70**, 1 (1998)
Barbato, D., Sozzetti, A., Desidera, S., et al.: Astron. Astrophys. **615**, A175 (2018)
Barenfeld, S.A., Carpenter, J.M., Sargent, A.I., et al.: Astrophys. J. **851**, 85 (2017)
Baruteau, C., Meru, F., Paardekooper, S.J.: Monthly Not. Roy. Astron. Soc. **416**, 1971 (2011)
Batalha, N.M., Rowe, J.F., Bryson, S.T., et al.: Astrophys. J. Supple. Ser. **204**, 24 (2013)
Bate, M.R.: Monthly Not. Roy. Astron. Soc. **475**, 5618 (2018)
Batygin, K., Morbidelli, A.: Astron. Astrophys. **556**, A28 (2013)
Batygin, K., Morbidelli, A., Holman, M.J.: Astrophys. J. **799**, 120 (2015)
Beaugé, C., Nesvorný, D.: Astrophys. J. **751**, 119 (2012)
Bell, K.R., Lin, D.N.C.: Astrophys. J. **427**, 987 (1994)
Benítez-Llambay, P., Masset, F., Koenigsberger, G., Szulágyi, J.: Nature **520**, 63 (2015)
Benítez-Llambay, P., Pessah, M.E.: Astrophys. J. Lett. **855**, L28 (2018)
Béthune, W., Lesur, G., Ferreira, J.: Astron. Astrophys. **600**, A75 (2017)
Birnstiel, T., Fang, M., Johansen, A.: Space Sci. Rev. **205**, 41 (2016)
Birnstiel, T., Klahr, H., Ercolano, B.: Astron. Astrophys. **539**, A148 (2012)
Bitsch, B., Crida, A., Morbidelli, A., et al.: Astron. Astrophys. **549**, A124 (2013)
Bitsch, B., Morbidelli, A., Lega, E., Crida, A.: Astron. Astrophys. **564**, A135 (2014)
Bitsch, B., Johansen, A., Lambrechts, M., Morbidelli, A.: Astron. Astrophys. **575**, A28 (2015a)
Bitsch, B., Lambrechts, M., Johansen, A.: Astron. Astrophys. **582**, A112 (2015b)
Bitsch, B., Morbidelli, A., Johansen, A., et al.: Astron. Astrophys. **612**, A30 (2018)
Bitsch, B., Izidoro, A., Johansen, A., et al.: Astron. Astrophys. **623**, A88 (2019a)
Bitsch, B., Raymond, S.N., Izidoro, A.: Astron. Astrophys. **624**, A109 (2019b)
Biver, N., Moreno, R., Bockelée-Morvan, D., et al.: Astron. Astrophys. **589**, A78 (2016)
Blum, J., Wurm, G.: Annu. Rev. Astron. Astrophys. **46**, 21 (2008)
Boehnke, P., Harrison, T.M.: Proc. Natl. Acad. Sci. **113**, 10802 (2016)
Boisvert, J.H., Nelson, B.E., Steffen, J.H.: Monthly Not. Roy. Astron. Soc. **480**, 2846 (2018)
Boley, A.C.: Astrophys. J. Lett. **695**, L53 (2009)
Boley, A.C., Ford, E.B.: arXiv:1306.0566 (2013)
Boley, A.C., Hayfield, T., Mayer, L., Durisen, R.H.: Icarus **207**, 509 (2010)
Bonomo, A.S., Desidera, S., Benatti, S., et al.: Astron. Astrophys. **602**, A107 (2010)
Boss, A.P.: Science **276**, 1836 (1997)
Boss, A.P.: Astrophys. J. **503**, 923 (1998)
Boss, A.P.: Astrophys. J. Lett. **536**, L101 (2000)
Bouvier, A., Wadhwa, M.: Nat. Geosci. **3**, 637 (2010)

Bowler, B.P., Nielsen, E.L.: Occurrence rates from direct imaging surveys. In: Deeg, H., Belmonte, J. (eds.) Handbook of Exoplanets. Springer, Cham (2018)
Brasser, R., Matsumura, S., Ida, S., et al.: Astrophys. J. **821**, 75 (2016)
Briceño, C., Vivas, A.K., Calvet, N., et al.: Science **291**, 93 (2001)
Bromley, B.C., Kenyon, S.J.: Astron. J. **153**, 216 (2017)
Bryan, M.L., Knutson, H.A., Lee, E.J., et al.: Astron. J. **157**, 52 (2019)
Bryden, G., Beichman, C.A., Carpenter, J.M., et al.: Astrophys. J. **705**, 1226 (2009)
Bryden, G., Beichman, C.A., Trilling, D.E., et al.: Astrophys. J. **636**, 1098 (2006)
Budde, G., Burkhardt, C., Brennecka, G.A., et al.: Earth Planet. Sci. Lett. **454**, 293 (2016)
Budde, G., Burkhardt, C., Kleine, T.: Nat. Astron. **3**, 736 (2019)
Butler, R.P., Wright, J.T., Marcy, G.W., et al.: Astrophys. J. **646**, 505 (2006)
Campante, T.L., Barclay, T., Swift, J.J., et al.: Astrophys. J. **799**, 170 (2015)
Caracausi, A., Avice, G., Burnard, P.G., et al.: Nature **533**, 82 (2016)
Carpenter, J.M., Bouwman, J., Mamajek, E.E., et al.: Astrophys. J. Suppl. Ser. **181**, 197 (2009)
Carrera, D., Johansen, A., Davies, M.B.: Astron. Astrophys. **579**, A43 (2015)
Carrera, D., Davies, M.B., Johansen, A.: Monthly Not. Roy. Astron. Soc. **463**, 3226 (2016)
Carrera, D., Gorti, U., Johansen, A., Davies, M.B.: Astrophys. J. **839**, 16 (2017)
Carrera, D., Ford, E.B., Izidoro, A., et al.: Astrophys. J. **866**, 104 (2018)
Carrera, D., Raymond, S.N., Davies, M.B.: Astron. Astrophys. **629**, L7 (2019)
Cassan, A., Kubas, D., Beaulieu, J.P.: Nature **481**, 167 (2012)
Chambers, J.E.: Icarus **152**, 205 (2001)
Chambers, J.E.: Icarus **180**, 496 (2006a)
Chambers, J.E.: Astrophys. J. **879**, 98 (2006b)
Chambers, J.E., Wetherill, G.W.: Icarus **136**, 304 (1998)
Chambers, K.C., Magnier, E.A., Metcalfe, N.: arxiv:1612.05560
Chandrasekhar, S.: Astrophys. J. **97**, 255 (1943)
Chang, S.H., Gu, P.G., Bodenheimer, P.H.: Astrophys. J. **708**, 1692 (2010)
Chapman, C.R., Cohen, B.A., Grinspoon, D.H.: Icarus **189**, 233 (2007)
Chatterjee, S., Ford, E.B., Matsumura, S., Rasio, F.A.: Astrophys. J. **686**, 580 (2008)
Chatterjee, S., Tan, J.C.: Astrophys. J. **780**, 53 (2014)
Chatterjee, S., Tan, J.C.: Astrophys. J. Lett. **798**, L32 (2015)
Chen, J., Kipping, D.: Astrophys. J. **834**, 17 (2017)
Chiang, E., Laughlin, G.: Monthly Not. Roy. Astron. Soc. **431**, 3444 (2013)
Chiang, E., Youdin, A.N.: Annu. Rev. Earth Planet. Sci. **38**, 493 (2010)
Chiang, E.I., Goldreich, P.: Astrophys. J. **490**, 368 (1997)
Clanton, C., Gaudi, B.S.: Astrophys. J. **791**, 91 (2014)
Clanton, C., Gaudi, B.S.: Astrophys. J. **819**, 125 (2016)
Clement, M.S., Kaib, N.A., Raymond, S.N., Walsh, K.J.: Icarus **311**, 340 (2018)
Clement, M.S., Kaib, N.A., Raymond, S.N., et al.: Icarus **321**, 778 (2019a)
Clement, M.S., Raymond, S.N., Kaib, N.A.: Astron. J. **157**, 38 (2019b)
Coleman, G.A.L., Papaloizou, J.C.B., Nelson, R.P.: Monthly Not. Roy. Astron. Soc. **470**, 3206 (2017)
Connelly, J.N., Amelin, Y., Krot, A.N., Bizzarro, M.: Astrophys. J. Lett. **675**, L121 (2008)
Connelly, J.N., Bizzarro, M., Krot, A.N., et al.: Science **338**, 651 (2012)
Cossou, C., Raymond, S.N., Pierens, A.: Astron. Astrophys. **553**, L2 (2013)
Cossou, C., Raymond, S.N., Hersant, F., Pierens, A.: Astron. Astrophys. **569**, A56 (2014)
Cresswell, P., Dirksen, G., Kley, W., Nelson, R.P.: Astron. Astrophys. **473**, 329 (2007)
Crida, A., Morbidelli, A., Masset, F.: Icarus **181**, 587 (2006)
Cumming, A., Butler, R.P., Marcy, G.W.: Proc. Astron. Soc. Pac. **120**, 531 (2008)
D'Angelo, G., Kley, W., Henning, T.: Astrophys. J. **586**, 540 (2003)
D'Angelo, M., Cazaux, S., Kamp, I., et al.: Astron. Astrophys. **622**, A208 (2019)
Dauphas, N.: Nature **541**, 521 (2017)
Dauphas, N., Pourmand, A.: Nature **473**, 489 (2011)

Davis, S.S.: Astrophys. J. Lett. **627**, L153 (2005)
Dawson, R.I., Murray-Clay, R.A.: Astrophys. J. Lett. **767**, L24 (2013)
Dawson, R.I., Chiang, E., Lee, E.J.: Monthly Not. Roy. Astronom. Soc. **453**, 1471 (2015)
Dawson, R.I., Lee, E.J., Chiang, E.: Astrophys. J. **822**, 54 (2016)
de Sousa, R., Morbidelli, A., Raymond, S., et al.: arXiv:1912.10879 (2020)
Deck, K.M., Batygin, K.: Astrophys. J. **810**, 119 (2015)
Deienno, R., Gomes, R.S., Walsh, K.J., et al.: Icarus **272**, 114 (2016)
Deienno, R., Izidoro, A., Morbidelli, A., et al.: Astrophys. J. **864**, 50 (2018)
DeMeo, F.E., Carry, B.: Icarus **226**, 723 (2013)
DeMeo, F.E., Carry, B.: Nature **505**, 629 (2014)
Demory, B.O., Gillon, M., Deming, D., et al.: Astron. Astrophys. **533**, A114 (2011)
Desch, S.J., Turner, N.J.: Astrophys. J. **811**, 156 (2015)
Dodson-Robinson, S.E., Salyk, C.: Astrophys. J. **738**, 131 (2011)
Dodson-Robinson, S.E., Willacy, K., Bodenheimer, P., et al.: Icarus **200**, 672 (2009)
Dong, S., Zhu, Z.: Astrophys. J. **778**, 53 (2013)
Dorn, C., Khan, A., Heng, K., et al.: Astron. Astrophys. **577**, A83 (2015)
Dorn, C., Mosegaard, K., Grimm, S.L., Alibert, Y.: Astrophys. J. **865**, 20 (2018)
Drążkowska, J., Alibert, Y., Moore, B.: Astron. Astrophys., **594**, A105 (2016)
Drążkowska, J., Alibert, Y.: Astron. Astrophys., **608**, A92 (2017)
Drążkowska, J., Dullemond, C.P.: Astron. Astrophys. **614**, A62 (2018)
Duffell, P.C., Haiman, Z., MacFadyen, A.I., et al.: Astrophys. J. Lett. **792**, L10 (2014)
Dullemond, C.P., Birnstiel, T., Huang, J., et al.: Astrophys. J. Lett. **869**, L46 (2018)
Dürmann, C., Kley, W.: Astron. Astrophys. **574**, A52 (2015)
Eisner, J.A., Carpenter, J.M.: Astrophys. J. **598**, 1341 (2003)
Eklund, H., Masset, F.S.: Monthly Not. Roy. Astron. Soc. **469**, 206 (2017)
Ertel, S., Absil, O., Defrère, D., et al.: Astron. Astrophys. **570**, A128 (2014)
Fabrycky, D.C., Lissauer, J.J., Ragozzine, D., et al.: Astrophys. J. **790**, 146 (2014)
Fang, J., Margot, J.L.: Astrophys. J. **761**, 92 (2012)
Fernandes, R.B., Mulders, G.D., Pascucci, I., et al.: Astrophys. J. **874**, 81 (2019)
Fischer, D.A., Valenti, J.: Astrophys. J. **622**, 1102 (2005)
Fischer, D.A., Marcy, G.W., Butler, R.P., et al.: Astrophys. J. **675**, 790 (2008)
Fischer, R.A., Ciesla, F.J.: Earth Planet. Sci. Lett. **392**, 28 (2014)
Fischer, R.A., Nimmo, F.: Earth Planet. Sci. Lett. **499**, 257 (2018)
Flock, M., Fromang, S., Turner, N.J., Benisty, M.: Astrophys. J. **835**, 230 (2017)
Flock, M., Turner, N.J., Mulders, G.D., et al.: Astron. Astrophys. **630**, A147 (2019)
Fogg, M.J., Nelson, R.P.: Astron. Astrophys. **441**, 791 (2005)
Fogg, M.J., Nelson, R.P.: Astron. Astrophys. **461**, 1195 (2007)
Ford, E.B., Lystad, V., Rasio, F.A.: Nature **434**, 873 (2005)
Ford, E.B., Rasio, F.A.: Astrophys. J. **686**, 621 (2008)
Foreman-Mackey, D., Morton, T.D., Hogg, D.W., et al.: Astron. J. **152**, 206 (2016)
Fortney, J.J., Marley, M.S., Barnes, J.W.: Astrophys. J. **659**, 1661 (2007)
Fressin, F., Torres, G., Charbonneau, D., et al.: Astrophys. J. **766**, 81 (2013)
Fulton, B.J., Petigura, E.A.: Astron. J. **156**, 6 (2018)
Fulton, B.J., Petigura, E.A., Howard, A.W., et al.: Astron. J. **154**, 109 (2017)
Genda, H., Abe, Y.: Nature **433**, 842 (2005)
Genda, H., Ikoma, M.: Icarus **194**, 42 (2008)
Gillon, M., Triaud, A.H.M.J., Demory, B.O., et al.: Nature **542**, 456 (2017)
Ginzburg, S., Schlichting, H.E., Sari, R.: Astrophys. J. **825**, 29 (2016)
Goldreich, P., Schlichting, H.E.: Astron. J. **147**, 32 (2014)
Goldreich, P., Tremaine, S.: Astrophys. J. **233**, 857 (1979)
Goldreich, P., Tremaine, S.: Astrophys. J. **241**, 425 (1980)
Gomes, R.S., Morbidelli, A., Levison, H.F.: Icarus **170**, 492 (2004)
Gomes, R., Levison, H.F., Tsiganis, K., Morbidelli, A.: Nature **435**, 466 (2005)

Gonzalez, G.: Monthly Not. Roy. Astron. Soc. **285**, 403 (1997)
Gould, A., Dong, S., Gaudi, B.S., et al.: Astrophys. J. **720**, 1073 (2010)
Gradie, J., Tedesco, E.: Science **216**, 1405 (1982)
Greaves, J.S., Rice, W.K.M.: Monthly Not. Roy. Astron. Soc. **412**, L88 (2011)
Greenberg, R., Hartmann, W.K., Chapman, C.R., et al.: Icarus **35**, 1 (1978)
Gressel, O., Turner, N.J., Nelson, R.P., McNally, C.P.: Astrophys. J. **801**, 84 (2015)
Grimm, R.E., McSween, H.Y.: Science **259**, 653 (1993)
Gupta, A., Schlichting, H.E.: Monthly Not. Roy. Astron. Soc. **487**, 24 (2019)
Hahn, J.M.: The Dynamics of Planetary Systems and Astrophysical Disks. Wiley (2009)
Haisch, K.E., Jr., Lada, E.A., Lada, C.J.: Astrophys. J. Lett. **553**, L153 (2001)
Halliday, A.N.: Geochemica Cosmochemica Acta **105**, 146 (2013)
Halliday, A.N., Kleine, T.: In: Lauretta, D.S., McSween, H.Y. Jr. (eds.) Meteorites and the Early Solar System II, p. 775. University of Arizona Press, Tucson (2006)
Hansen, B.M.S.: Astrophys. J. **703**, 1131 (2009)
Hansen, B.M.S., Murray, N.: Astrophys. J. **751**, 158 (2012)
Hansen, B.M.S., Murray, N.: Astrophys. J. **775**, 53 (2013)
Hartmann, L., Calvet, N., Gullbring, E., D'Alessio, P.: Astrophys. J. **495**, 385 (1998)
Hartmann, W.K.: In: Lunar and Planetary Science Conference, Lunar and Planetary Science Conference, p. 1064 (2019)
Hartogh, P., Lis, D.C., Bockelée-Morvan, D., et al.: Nature **478**, 218 (2011)
Hasegawa, Y., Pudritz, R.E.: Astrophys. J. **760**, 117 (2012)
Hayashi, C.: Prog. Theor. Phys. Suppl. **70**, 35 (1981)
Helled, R., Bodenheimer, P., Podolak, M., et al.: In: Protostars and Planets VI, p. 643 (2014)
Henon, M., Heiles, C.: Astron. J. **69**, 73 (1964)
Hillenbrand, L.A., Carpenter, J.M., Kim, J.S., et al.: Astrophys. J. **677**, 630 (2008)
Hirschmann, M.M.: Annu. Rev. Earth Planet. Sci. **34**, 629 (2006)
Hollenbach, D., Johnstone, D., Lizano, S., Shu, F.: Astrophys. J. **428**, 654 (1994)
Howard, A.W., Marcy, G.W., Johnson, J.A., et al.: Science **330**, 653 (2010)
Howard, A.W., Marcy, G.W., Bryson, S.T., et al.: Astrophys. J. Suppl. Ser. **201**, 15 (2012)
Hu, X., Tan, J.C., Zhu, Z., et al.: Astrophys. J. **857**, 20 (2018)
Hu, X., Zhu, Z., Tan, J.C., Chatterjee, S.: Astrophys. J. **816**, 19 (2016)
Hubickyj, O., Bodenheimer, P., Lissauer, J.J.: Icarus **179**, 415 (2005)
Hughes, A.M., Duchêne, G., Matthews, B.C.: Annu. Revi. Astron. Astrophys. **56**, 541 (2018)
Ida, S., Lin, D.N.C.: Astrophys. J. **604**, 388 (2004)
Ida, S., Lin, D.N.C.: Astrophys. J. **673**, 487 (2008)
Ida, S., Lin, D.N.C.: Astrophys. J. **719**, 810 (2010)
Ida, S., Lin, D.N.C., Nagasawa, M.: Astrophys. J. **775**, 42 (2013)
Ida, S., Guillot, T., Morbidelli, A.: Astron. Astrophys. **591**, A72 (2016)
Ida, S., Guillot, T.: Astron. Astrophys. **596**, L3 (2016)
Ida, S., Yamamura, T., Okuzumi, S.: Astron. Astrophys. **624**, A28 (2019)
Ikoma, M., Genda, H.: Astrophys. J. **648**, 696 (2006)
Ikoma, M., Nakazawa, K., Emori, H.: Astrophys. J. **537**, 1013 (2000)
Inamdar, N.K., Schlichting, H.E.: Monthly Not. Roy. Astron. Soc. **448**, 1751 (2015)
Izidoro, A., Bitsch, B., Raymond, S.N., et al.: arXiv:1902.08772 (2019)
Izidoro, A., Morbidelli, A., Raymond, S.N.: Astrophys. J. **794**, 11 (2014)
Izidoro, A., Raymond, S.N., Morbidelli, A., et al.: Astrophys. J. Lett. **800**, L22 (2015a)
Izidoro, A., Morbidelli, A., Raymond, S.N., et al.: Astron. Astrophys. **582**, A99 (2015b)
Izidoro, A., Raymond, S.N., Morbidelli, A., Winter, O.C.: Monthly Not. Roy. Astron. Soc. **453**, 3619 (2015c)
Izidoro, A., Ogihara, M., Raymond, S.N., et al.: Monthly Not. Roy. Astron. Soc. **470**, 1750 (2017)
Jacobson, S.A., Morbidelli, A.: Philos. Trans. Roy. Soc. Lond. Ser. A **372**, 0174 (2014)
Jacobson, S.A., Morbidelli, A., Raymond, S.N., et al.: Nature, **508**, 84
Jacquet, E., Balbus, S., Latter, H.: Monthly Not. Roy. Astron. Soc. **415**, 3591 (2011)

Jin, S., Mordasini, C.: Astrophys. J. **853**, 163 (2018)
Johansen, A., Youdin, A., Klahr, H.: Astrophys. J. **697**, 1269 (2009)
Johansen, A., Davies, M.B., Church, R.P., Holmelin, V.: Astrophys. J. **758**, 39 (2012)
Johansen, A., Blum, J., Tanaka, H., et al.: Protostars and Planets VI, p. 547 (2014)
Johansen, A., Mac Low, M.M., Lacerda, P., Bizzarro, M.: Sci. Adv. **1**, 1500109 (2015)
Johansen, A., Lambrechts, M.: Ann. Rev. Earth Planet. Sci. **45**, 359 (2017)
Johnson, J.A., Butler, R.P., Marcy, G.W., et al.: Astrophys. J. **670**, 833 (2007)
Johnson, B.C., Walsh, K.J., Minton, D.A., et al.: Sci. Adv. **2**, e1601658 (2016)
Jones, H.R.A., Butler, R.P., Tinney, C.G., et al.: Monthly Not. Roy. Astron. Soc. **369**, 249 (2006)
Jurić, M., Tremaine, S.: Astrophys. J. **686**, 603 (2008)
Kaib, N.A., Chambers, J.E.: Monthly Not. Roy. Astron. Soc. **455**, 3561 (2016)
Kaib, N.A., Cowan, N.B.: Icarus **252**, 161 (2015)
Kaib, N.A., Raymond, S.N., Duncan, M.: Nature **493**, 381 (2013)
Kanagawa, K.D., Tanaka, H., Muto, T., et al.: Monthly Not. Roy. Astron. Soc. **448**, 994 (2015)
Kanagawa, K.D., Tanaka, H., Szuszkiewicz, E.: Astrophys. J. **861**, 140 (2018)
Kataoka, A., Tanaka, H., Okuzumi, S., Wada, K.: Astron. Astrophys. **557**, L4 (2013)
Kerridge, J.F.: Geochemica Cosmochemica Acta **49**, 1707 (1985)
Kimura, K., Lewis, R.S., Anders, E.: Geochemica Cosmochemica Acta **38**, 683 (1974)
Kimura, S.S., Tsuribe, T.: Proc. Astron. Soci. Jpn. **64**, 116 (2012)
Kita, N.T., Huss, G.R., Tachibana, S., et al.: Astron. Soc. Pac. Conf. Ser. **341**, 558 (2005)
Kleine, T., Touboul, M., Bourdon, B., et al.: Geochemica Cosmochemica Acta **73**, 5150 (2009)
Kokubo, E., Ida, S.: Icarus **123**, 180 (1996)
Kokubo, E., Ida, S.: Icarus **131**, 171 (1998)
Kokubo, E., Ida, S.: Icarus **143**, 15 (2000)
Kokubo, E., Ida, S.: Astrophys. J. **671**, 2082 (2007)
Kral, Q., Krivov, A.V., Defrère, D., et al.: Astron. Rev. **13**, 69 (2017)
Krasinsky, G.A., Pitjeva, E.V., Vasilyev, M.V., Yagudina, E.I.: Icarus **158**, 98 (2002)
Kretke, K.A., Lin, D.N.C.: Astrophys. J. **755**, 74 (2012)
Krivov, A.V.: Res. Astron. Astrophys. **10**, 383 (2010)
Krot, A.N., Amelin, Y., Cassen, P., Meibom, A.: Nature **436**, 989 (2005)
Kruijer, T.S., Burkhardt, C., Budde, C., Kleine, T.: PNAS **114**, 6712 (2017)
Kruijer, T.S., Touboul, M., Fischer-Gödde, M., et al.: Science **344**, 1150 (2014)
Kuchynka, P., Folkner, W.M.: Icarus **222**, 243 (2013)
Lambrechts, M., Johansen, A.: Astron. Astrophys. **544**, A32 (2012)
Lambrechts, M., Johansen, A.: Astron. Astrophys. **572**, A107 (2014)
Lambrechts, M., Johansen, A., Morbidelli, A.: Astron. Astrophys. **572**, A35 (2014)
Lambrechts, M., Lega, E.: Astron. Astrophys. **606**, A146 (2017)
Lambrechts, M., Morbidelli, A., Jacobson, S.A., et al.: Astron. Astrophys. **627**, A83 (2019a)
Lambrechts, M., Lega, E., Nelson, R.P., Crida, A., Morbidelli, A.: Astron. Astrophys. **630**, A82 (2019b)
Laplace, P.S.: 1822, Mécanique céleste
Laskar, J.: Astron. Astrophys. **317**, L75 (1997)
Laws, C., Gonzalez, G., Walker, K.M., et al.: Astron. J. **125**, 2664 (2003)
Lecar, M., Podolak, M., Sasselov, D., Chiang, E.: Astrophys. J. **640**, 1115 (2006)
Lee, M.H., Peale, S.J.: Astrophys. J. **567**, 596 (2002)
Lee, E.J., Chiang, E., Ormel, C.W.: Astrophys. J. **797**, 95 (2014)
Lee, E.J., Chiang, E.: Astrophys. J. **817**, 90 (2016)
Lee, E.J., Chiang, E.: Astrophys. J. **842**, 40 (2017)
Lega, E., Crida, A., Bitsch, B., Morbidelli, A.: Monthly Not. Roy. Astron. Soc. **440**, 683 (2014)
Lestrade, J.F., Matthews, B.C., Sibthorpe, B., et al.: Astron. Astrophys. **548**, A86 (2012)
Levison, H.F., Stewart, G.R.: Icarus **153**, 224 (2001)
Levison, H.F., Morbidelli, A., Vanlaerhoven, C., et al.: Icarus **196**, 258 (2008)
Levison, H.F., Thommes, E., Duncan, M.J.: Astron. J. **139**, 1297 (2010)

Levison, H.F., Morbidelli, A., Tsiganis, K., Nesvorný, D., Gomes, R.: Astron. J. **142**, 152 (2011)
Levison, H.F., Kretke, K.A., Walsh, K.J., Bottke, W.F.: Proc. Natl. Acad. Sci. **112**, 14180 (2015)
Lichtenberg, T., Golabek, G.J., Burn, R., et al.: Nat. Astron. **3**, 307 (2019)
Lin, D.N.C., Bodenheimer, P., Richardson, D.C.: Nature **380**, 606 (2019)
Lin, D.N.C., Ida, S.: Astrophys. J. **477**, 781 (1997)
Lin, D.N.C., Papaloizou, J.: Monthly Not. Roy. Astronom. Soc. **186**, 799 (1979)
Lin, D.N.C., Papaloizou, J.: Astrophys. J. **309**, 846 (1986)
Lis, D.C., Biver, N., Bockelée-Morvan, D., et al.: Astrophys. J. Lett. **774**, L3 (2013)
Lis, D.C., Bockelée-Morvan, D., Güsten, R., et al.: Astron. Astrophys. **625**, L5 (2019)
Lissauer, J.J.: Icarus **69**, 249 (1987)
Lissauer, J.J.: Annu. Rev. Astron. Astrophys. **31**, 129 (1993)
Lissauer, J.J., Hubickyj, O., D'Angelo, G., Bodenheimer, P.: Icarus **199**, 338 (2009)
Lissauer, J.J., Ragozzine, D., Fabrycky, D.C., et al.: Astrophys. J. Suppl. Ser. **197**, 8 (2011)
Lissauer, J.J., Stevenson, D.J. Protostars and Planets V, p. 591 (2007)
Lodders, K.: Astrophys. J. **591**, 1220 (2003)
Looney, L.W., Mundy, L.G., Welch, W.J.: Astrophys. J. **592**, 255 (2003)
Lopez, E.D.: Monthly Not. Roy. Astron. Soc. **472**, 245 (2017)
Lovis, C., Mayor, M.: Astron. Astrophys. **472**, 657 (2007)
Luger, R., Sestovic, M., Kruse, E., et al.: Nat. Astron. **1**, 0129 (2017)
Lykawka, P.S., Ito, T.: Astrophys. J. **883**, 130 (2019)
Lynden-Bell, D., Pringle, J.E.: Monthly Not. Roy. Astron. Soc. **168**, 603 (1974)
Lyra, W., Paardekooper, S.J., Mac Low, M.M.: Astrophys. J. Lett. **715**, L68 (2010)
Mamajek, E.E.: Am. Inst. Phys. Conf. Ser. **1158**, 3 (2009)
Manara, C.F., Morbidelli, A., Guillot, T.: Astron. Astrophys. **618**, L3 (2018)
Mandell, A.M., Raymond, S.N., Sigurdsson, S.: Astrophys. J. **660**, 823 (2007)
Marcy, G.W., Butler, R.P., Vogt, S.S., et al.: Astrophys. J. **555**, 418 (2011)
Marcy, G.W., Isaacson, H., Howard, A.W., et al.: Astrophys. J. Suppl. Ser. **210**, 20 (2014)
Marois, C., Macintosh, B., Barman, T., et al.: Science **322**, 1348 (2008)
Marois, C., Zuckerman, B., Konopacky, Q.M., Macintosh, B., Barman, T.: Nature **468**, 1080 (2010)
Marshall, J.P., Moro-Martín, A., Eiroa, C., et al.: Astron. Astrophys. **565**, A15 (2014)
Martin, R.G., Livio, M.: Monthly Not. Roy. Astron. Soc. **425**, L6 (2012)
Martin, R.G., Livio, M.: Astrophys. J. **810**, 105 (2015)
Marty, B.: Earth Planet. Sci. Lett. **313**, 56 (2012)
Marty, B., Altwegg, K., Balsiger, H., et al.: Science **356**, 1069 (2017)
Marty, B., Avice, G., Sano, Y., et al.: Earth Planet. Sci. Lett. **441**, 91 (2016)
Marty, B., Yokochi, R.: Rev. Mineral Geophys. **62**, 421 (2006)
Marzari, F.: Monthly Not. Roy. Astron. Soc. **444**, 1419 (2014)
Marzari, F., Weidenschilling, S.J.: Icarus **156**, 570 (2002)
Masset, F.S.: Monthly Not. Roy. Astron. Soc. **472**, 4204 (2017)
Masset, F.S., Snellgrove, M.: Monthly Not. Roy. Astron. Soc. **320**, L55 (2001)
Masset, F.S., Morbidelli, A., Crida, A., et al.: Astrophys. J. **642**, 478 (2006)
Matsumoto, Y., Nagasawa, M., Ida, S.: Icarus **221**, 624 (2012)
Matsumura, S., Ida, S., Nagasawa, M.: Astrophys. J. **767**, 129 (2013)
Matsumura, S., Thommes, E.W., Chatterjee, S., et al.: Astrophys. J. **714**, 194 (2010)
Matthews, B.C., Krivov, A.V., Wyatt, M.C., et al.: In: Beuther, H., Klessen, R.S., Dullemond, C.P., Henning, T. (eds.) Protostars and Planets VI, p. 521 (2014)
Mayer, L., Lufkin, G., Quinn, T., Wadsley, J.: Astrophys. J. Lett. **661**, L77 (2007)
Mayer, L., Quinn, T., Wadsley, J., Stadel, J.: Science **298**, 1756 (2002)
Mayor, M., Marmier, M., Lovis, C., et al.: arXiv:1109.2497 (2011)
McCubbin, F.M., Barnes, J.J.: Earth Planet. Sci. Lett. **526**, 115771 (2019)
McNally, C.P., Nelson, R.P., Paardekooper, S.J., Benítez-Llambay, P.: Monthly Not. Roy. Astron. Soc. **484**, 728 (2019)
McNeil, D.S., Nelson, R.P.: Monthly Not. Roy. Astron. Soc. **401**, 1691 (2010)

Meech, K., Raymond, S.N.: arXiv:1912.04361 (2019)
Meru, F., Bate, M.R.: Monthly Not. Roy. Astron. Soc. **410**, 559 (2011)
Meyer, M.R., Carpenter, J.M., Mamajek, E.E., et al.: Astrophys. J. Lett. **673**, L181 (2008)
Millholland, S., Wang, S., Laughlin, G.: Astrophys. J. Lett. **849**, L33 (2017)
Mills, S.M., Fabrycky, D.C., Migaszewski, C., et al.: Nature **533**, 509 (2016)
Moeckel, N., Throop, H.B.: Astrophys. J. **707**, 268 (2009)
Mojzsis, S.J., Brasser, R., Kelly, N.M., et al.: Astrophys. J. **881**, 44 (2019)
Montesinos, B., Eiroa, C., Krivov, A.V., et al.: Astron. Astrophys. **593**, A51 (2016)
Monteux, J., Golabek, G.J., Rubie, D.C., Tobie, G., Young, E.D.: Space Sci. Rev. **214**, 39 (2018)
Moorhead, A.V., Adams, F.C.: Icarus **178**, 517 (2005)
Morales, J.C., Mustill, A.J., Ribas, I., et al.: Science **365**, 1441 (2019)
Morbidelli, A., Bitsch, B., Crida, A., et al.: Icarus **267**, 368 (2016)
Morbidelli, A., Chambers, J., Lunine, J.I., et al.: Meteor. Planet. Sci. **35**, 1309 (2000)
Morbidelli, A., Levison, H.F., Tsiganis, K., Gomes, R.: Nature **435**, 462 (2005)
Morbidelli, A., Tsiganis, K., Crida, A., et al.: Astron. J. **134**, 1790 (2007)
Morbidelli, A., Nesvorny, D.: Astron. Astrophys. **546**, A18 (2012)
Morbidelli, A., Lambrechts, M., Jacobson, S., Bitsch, B.: Icarus **258**, 418 (2015)
Morbidelli, A., Nesvorny, D., Laurenz, V., et al.: Icarus **305**, 262 (2018)
Morbidelli, A., Raymond, S.N.: J. Geophys. Res. (Planets) **121**, 1962 (2016)
Morbidelli, A., Wood, B.J.: Washington DC Am. Geophys. Union Geophys. Monogr. Ser. **212**, 71 (2015)
Mordasini, C., Alibert, Y., Benz, W.: Astron. Astrophys. **501**, 1139 (2009)
Mori, S., Bai, X.N., Okuzumi, S.: Astrophys. J. **872**, 98 (2019)
Moriarty, J., Ballard, S.: Astrophys. J. **832**, 34 (2016)
Morishima, R., Schmidt, M.W., Stadel, J., Moore, B.: Astrophys. J. **685**, 1247 (2008)
Morishima, R., Stadel, J., Moore, B.: Icarus **207**, 517 (2010)
Moro-Martín, A., Carpenter, J.M., Meyer, M.R., et al.: Astrophys. J. **658**, 1312 (2007)
Moro-Martín, A., Marshall, J.P., Kennedy, G., et al.: Astrophys. J. **801**, 143 (2015)
Mottl, M., Glazer, B., Kaiser, R., Meech, K.: Chemie der Erde / Geochemistry **67**, 253 (2007)
Mukhopadhyay, S.: Nature **486**, 101 (2012)
Mulders, G.D., Mordasini, C., Pascucci, I., Ciesla, F.J., Emsenhuber, A., Apai, D.: Astrophys. J., **887**, 2 (2019)
Mulders, G.D., Pascucci, I., Apai, D., Ciesla, F.J.: Astron. J. **156**, 24 (2018)
Mundy, L.G., Looney, L.W., Welch, W.J.: Protostars and Planets IV, p. 355 (2000)
Muralidharan, K., Deymier, P., Stimpfl, M., et al.: Icarus **198**, 400 (2008)
Nagasawa, M., Lin, D.N.C., Thommes, E.: Astrophys. J. **635**, 578 (2005)
Nagasawa, M., Ida, S., Bessho, T.: Astrophys. J. **678**, 498 (2008)
Nayakshin, S.: Monthly Not. Roy. Astron. Soc. **454**, 64 (2015)
Ndugu, N., Bitsch, B., Jurua, E.: Monthly Not. Roy. Astron. Soc. **474**, 886 (2018)
Neishtadt, A.: J. App. Math. Mech. **48**, 133 (1984)
Nesvorný, D.: Astron. J. **150**, 73 (2015)
Nesvorný, D., Morbidelli, A.: Astron. J. **144**, 117 (2012)
Nesvorný, D., Youdin, A.N., Richardson, D.C.: Astron. J. **140**, 785 (2010)
Nesvorný, D., Vokrouhlicky, D., Morbidelli, A.: Astrophys. J. **768**, 45 (2013)
Nesvorný, D., Vokrouhlický, D., Deienno, R.: Astrophys. J. **784**, 22 (2014)
Nesvorný, D., Vokrouhlický, D., Bottke, W.F., Levison, H.F.: Nat. Astron. **2**, 878 (2018)
Nesvorný, D.: Annu. Rev. Astron. Astrophys. **56**, 137 (2018)
Nesvorný, D., Li, R., Youdin, A.N., et al.: Nat. Astron. **3**, 808 (2019)
Nimmo, F., Kleine, T.: Icarus **191**, 497 (2007)
Noll, K.S., Grundy, W.M., Chiang, E.I., et al.: In: Barucci, M.A., Boehnhardt, H., Cruikshank, D.P., Morbidelli, A. (eds.) The Solar System Beyond Neptune, vol. 592, p. 345. University of Arizona Press, Tucson
Nomura, R., Hirose, K., Uesegi, K., et al.: Science **343**, 522 (2014)

Nyquist, L.E., Kleine, T., Shih, C.Y., Reese, Y.D.: Geochemica Cosmochemica Acta **73**, 5115 (2009)
O'Brien, D.P., Morbidelli, A., Levison, H.F.: Icarus **184**, 39 (2006)
O'Brien, D.P., Walsh, K.J., Morbidelli, A., et al.: Icarus **239**, 74 (2014)
O'Brien, D.P., Izidoro, A., Jacobson, S.A., et al.: Space Sci. Rev. **214**, 47 (2018)
Ogihara, M., Ida, S.: Astrophys. J. **699**, 824 (2009)
Ogihara, M., Morbidelli, A., Guillot, T.: Astron. Astrophys. **578**, A36 (2015)
Ogihara, M., Kokubo, E., Suzuki, T.K., Morbidelli, A.: Astron. Astrophys. **612**, L5 (2018a)
Ogihara, M., Kokubo, E., Suzuki, T.K., Morbidelli, A.: Astron. Astrophys. **615**, A63 (2018b)
Oka, A., Nakamoto, T., Ida, S.: Astrophys. J. **738**, 141 (2011)
Okuzumi, S., Tanaka, H., Kobayashi, H., Wada, K.: Astrophys. J. **752**, 106 (2012)
Ormel, C.W., Klahr, H.H.: Astron. Astrophys. **520**, A43 (2010)
Owen, J.E., Ercolano, B., Clarke, C.J.: Monthly Not. Roy. Astron. Soc. **412**, 13 (2011)
Owen, J.E., Wu, Y.: Astrophys. J. **775**, 105 (2013)
Owen, J.E., Wu, Y.: Astrophys. J. **847**, 29 (2017)
Paardekooper, S.J.: Monthly Not. Roy. Astron. Soc. **444**, 2031 (2014)
Paardekooper, S.J., Baruteau, C., Crida, A., Kley, W.: Monthly Not. Roy. Astron. Soc. **401**, 1950 (2010)
Paardekooper, S.J., Baruteau, C., Kley, W.: Monthly Not. Roy. Astron. Soc. **410**, 293 (2011)
Paardekooper, S.J., Mellema, G.: Astron. Astrophys. **459**, L17 (2006)
Pätzold, M., Andert, T.P., Hahn, M., et al.: Monthly Not. Roy. Astron. Soc. **483**, 2337 (2019)
Petigura, E.A., Howard, A.W., Marcy, G.W.: Proc. Natl. Acad. Sci. **110**, 19273 (2013)
Pfalzner, S., Steinhausen, M., Menten, K.: Astrophys. J. Lett. **793**, L34 (2014)
Pichierri, G., Morbidelli, A., Crida, A.: Celest. Mech. Dyn. Astron. **130**, 54 (2018)
Pierens, A.: Monthly Not. Roy. Astron. Soc. **454**, 2003 (2015)
Pierens, A., Nelson, R.P.: Astron. Astrophys. **482**, 333 (2008)
Pierens, A., Raymond, S.N.: Astron. Astrophys. **533**, A131 (2011)
Pierens, A., Raymond, S.N.: Monthly Not. Roy. Astron. Soc. **462**, 4130 (2016)
Pierens, A., Raymond, S.N., Nesvorny, D., Morbidelli, A.: Astrophys. J. Lett. **795**, L11 (2014)
Pinte, C., Dent, W.R.F., Ménard, F., et al.: Astrophys. J. **816**, 25 (2016)
Pirani, S., Johansen, A., Bitsch, B., et al.: Astron. Astrophys. **623**, A169 (2019)
Piso, A.M.A., Youdin, A.N.: Astrophys. J. **786**, 21 (2014)
Poincaré, H.: Les methodes nouvelles de la mecanique celeste (1892)
Poincaré, H.: Bulletin Astronomique. Serie I **14**, 53 (1897)
Pollack, J.B., Hubickyj, O., Bodenheimer, P., et al.: Icarus **124**, 62 (1996)
Quarles, B., Kaib, N.: Astron. J. **157**, 67 (2019)
Quintana, E.V., Barclay, T., Borucki, W.J., et al.: Astrophys. J. **821**, 126 (2016)
Quintana, E.V., Barclay, T., Raymond, S.N., et al.: Science **344**, 277 (2014)
Rafikov, R.R.: Astron. J. **128**, 1348 (2004)
Rasio, F.A., Ford, E.B.: Science **274**, 954 (1996)
Raymond, S.N., Quinn, T., Lunine, J.I.: Icarus **168**, 1 (2004)
Raymond, S.N., Quinn, T., Lunine, J.I.: Icarus **183**, 265 (2006a)
Raymond, S.N., Mandell, A.M., Sigurdsson, S.: Science **313**, 1413 (2006b)
Raymond, S.N., Quinn, T., Lunine, J.I.: Astrobiology **7**, 66 (2007a)
Raymond, S.N., Scalo, J., Meadows, V.S.: Astrophys. J. **669**, 606 (2007b)
Raymond, S.N., Barnes, R., Mandell, A.M.: Monthly Not. Roy. Astron. Soc. **384**, 663 (2008a)
Raymond, S.N., Barnes, R., Armitage, P.J., Gorelick, N.: Astrophys. J. Lett. **687**, L107 (2008b)
Raymond, S.N., Barnes, R., Veras, D., et al.: Astrophys. J. Lett. **696**, L98 (2009a)
Raymond, S.N., O'Brien, D.P., Morbidelli, A., Kaib, N.A.: Icarus **203**, 644 (2009b)
Raymond, S.N., Armitage, P.J., Gorelick, N.: Astrophys. J. **711**, 772 (2010)
Raymond, S.N., Armitage, P.J., Moro-Martín, A., et al.: Astron. Astrophys. **530**, A62 (2011)
Raymond, S.N., Armitage, P.J., Moro-Martín, A., et al.: Astron. Astrophys. **541**, A11 (2012)
Raymond, S.N., Schlichting, H.E., Hersant, F., Selsis, F.: Icarus **226**, 671 (2013)

Raymond, S.N., Kokubo, E., Morbidelli, A., et al.: Protostars and Planets VI, p. 595 (2014)
Raymond, S.N., Cossou, C.: Monthly Not. Roy. Astron. Soc. **440**, L11 (2014)
Raymond, S.N., Morbidelli, A.: In: Complex Planetary Systems, Proceedings of the International Astronomical Union, IAU Symposium, vol. 310, p. 194 (2014)
Raymond, S.N., Izidoro, A.: Icarus **297**, 134 (2017a)
Raymond, S.N., Izidoro, A.: Sci. Adv. **3**, e1701138 (2017b)
Raymond, S.N., Armitage, P.J., Veras, D., et al.: Monthly Not. Roy. Astron. Soc. **476**, 3031 (2018a)
Raymond, S.N., Boulet, T., Izidoro, A., et al.: Monthly Not. Roy. Astron. Soc. **479**, L81 (2018b)
Raymond, S.N., Armitage, P.J., Veras, D.: Astrophys. J. Lett. **856**, L7 (2018c)
Raymond, S.N., Izidoro, A., Morbidelli, A.: arXiv:1812.01033 (2018d)
Ribas, I., Miralda-Escudé, J.: Astron. Astrophys. **464**, 779 (2007)
Rivera, E.J., Laughlin, G., Butler, R.P., et al.: Astrophys. J. **719**, 890 (2010)
Robert, F., Merlivat, L., Javoy, M.: Meteoritics **12**, 349 (1977)
Robert, C.M.T., Crida, A., Lega, E.: Astron. Astrophys. **617**, A98
Rogers, L.A.: Astrophys. J. **801**, 41 (2015)
Ronnet, T., Mousis, O., Vernazza, P., et al.: Astron. J. **155**, 224 (2018)
Rowan, D., Meschiari, S., Laughlin, G., et al.: Astrophys. J. **817**, 104 (2016)
Rowe, J.F., Bryson, S.T., Marcy, G.W., et al.: Astrophys. J. **784**, 45 (2014)
Rubie, D.C., Jacobson, S.A., Morbidelli, A., et al.: Icarus **248**, 89 (2015)
Safronov, V.S.: Evolution of the Protoplanetary Cloud and Formation of the Earth and Planets. Keter Publishing House (1972)
Santos, N.C., Israelian, G., Mayor, M.: Astron. Astrophys. **373**, 1019 (2001)
Sasselov, D.D., Lecar, M.: Astrophys. J. **528**, 995 (2000)
Sato, T., Okuzumi, S., Ida, S.: Astron. Astrophys. **589**, A15 (2016)
Schäfer, U., Yang, C.C., Johansen, A.: Astron. Astrophys. **597**, A69 (2017)
Schiller, M., Bizzarro, M., Fernandes, V.A.: Nature **555**, 507 (2018)
Schiller, M., Connelly, J.N., Glad, A.C., et al.: Earth Planet. Sci. Lett. **420**, 45 (2015)
Schlichting, H.E.: Astrophys. J. Lett. **795**, L15 (2014)
Schoonenberg, D., Ormel, C.W.: Astron. Astrophys. **602**, A21 (2017)
Seager, S., Kuchner, M., Hier-Majumder, C.A., Militzer, B.: Astrophys. J. **669**, 1279 (2007)
Selsis, F., Chazelas, B., Bordé, P., et al.: Icarus **191**, 453 (2007)
Sessin, W., Ferraz-Mello, S.: Celest. Mech. **32**, 307 (1984)
Shakura, N.I., Sunyaev, R.A.: Astron. Astrophys. **24**, 337 (1973)
Sharp, Z.D.: Chem. Geol. **448**, 137 (2017)
Simon, J.B., Armitage, P.J., Li, R., Youdin, A.N.: Astrophys. J. **822**, 55 (2016)
Simon, J.B., Armitage, P.J., Youdin, A.N., Li, R.: Astrophys. J. Lett. **847**, L12 (2017)
Sinukoff, E., Howard, A.W., Petigura, E.A., et al.: Astron. J. **153**, 70 (2017)
Snellgrove, M.D., Papaloizou, J.C.B., Nelson, R.P.: Astron. Astrophys. **374**, 1092 (2001)
Sosa, A., Fernández, J.A.: Monthly Not. Roy. Astron. Soc. **416**, 767 (2011)
Sotin, C., Grasset, O., Mocquet, A.: Icarus **191**, 337 (2007)
Stimpfl, M., Walker, A.M., Drake, M.J., et al.: J. Crystal Growth **294**, 83 (2006)
Stone, J.M., Ostriker, E.C., Gammie, C.F.: Astrophys. J. Lett. **508**, L99 (1998)
Su, K.Y.L., Rieke, G.H., Stansberry, J.A., et al.: Astrophys. J. **653**, 675 (2006)
Suzuki, T.K., Ogihara, M., Morbidelli, A., et al.: Astron. Astrophys. **596**, A74 (2016a)
Suzuki, D., Bennett, D.P., Sumi, T., et al.: Astrophys. J. **833**, 145 (2016b)
Svetsov, V.V.: Solar Syst. Res. **41**, 28 (2007)
Tanaka, H., Takeuchi, T., Ward, W.R.: Astrophys. J. **565**, 1257 (2002)
Tanaka, H., Ward, W.R.: Astrophys. J. **602**, 388 (2004)
Tera, F., Papanastassiou, D.A., Wasserburg, G.J.: Earth Planet. Sci. Lett. **22**, 1 (1974)
Terquem, C., Papaloizou, J.C.B.: Astrophys. J. **654**, 1110 (2007)
Thommes, E.W., Duncan, M.J., Levison, H.F.: Icarus **161**, 431 (2003)
Thommes, E.W., Matsumura, S., Rasio, F.A.: Science **321**, 814 (2008)
Throop, H.B., Bally, J.: Astron. J. **135**, 2380 (2008)

Tominaga, R.T., Takahashi, S.Z., Inutsuka, S.I.: Astrophys. J. **881**, 53 (2019)
Toomre, A.: Astrophys. J. **139**, 1217 (1964)
Touboul, M., Kleine, T., Bourdon, B., et al.: Nature **450**, 1206 (2007)
Tremaine, S., Dong, S.: Astron. J. **143**, 94 (2012)
Trilling, D.E., Bryden, G., Beichman, C.A., et al.: Astrophys. J. **674**, 1086 (2008)
Tsiganis, K., Gomes, R., Morbidelli, A., Levison, H.F.: Nature **435**, 459 (2005)
Turner, N.J., Fromang, S., Gammie, C., et al.: Protostars and Planets VI, p. 411 (2014)
Udry, S., Bonfils, X., Delfosse, X., et al.: Astron. Astrophys. **469**, L43 (2007)
Udry, S., Santos, N.C.: Annu. Rev. Astron. Astrophys. **45**, 397 (2007)
Uribe, A.L., Klahr, H., Henning, T.: Astrophys. J. **769**, 97 (2013)
Valencia, D., Sasselov, D.D., O'Connell, R.J.: Astrophys. J. **665**, 1413 (2007)
Varnière, P., Quillen, A.C., Frank, A.: Astrophys. J. **612**, 1152 (2004)
Veras, D., Armitage, P.J.: Astrophys. J. Lett. **620**, L111 (2005)
Veras, D., Armitage, P.J.: Astrophys. J. **645**, 1509 (2006)
Villeneuve, J., Chaussidon, M., Libourel, G.: Science **325**, 985 (2009)
Walsh, K.J., Morbidelli, A., Raymond, S.N., et al.: Nature **475**, 206 (2011)
Walsh, K.J., Morbidelli, A., Raymond, S.N., et al.: Meteor. Planet. Sci. **47**, 1941 (2012)
Ward, W.R.: Icarus **67**, 164 (1986)
Ward, W.R.: Icarus **126**, 261 (1997)
Warren, P.H.: Earth Planet. Sci. Lett. **311**, 93 (2011)
Weber, P., Benítez-Llambay, P., Gressel, O., et al.: Astrophys. J. **854**, 153 (2018)
Weidenschilling, S.J.: Astrophys. Space Sci. **51**, 153 (1977a)
Weidenschilling, S.J.: Monthly Not. Roy. Astron. Soc. **180**, 57 (1977b)
Weidenschilling, S.J., Marzari, F.: Nature **384**, 619 (1996)
Weiss, L.M., Marcy, G.W.: Astrophys. J. Lett. **783**, L6 (2014)
Weiss, L.M., Marcy, G.W., Petigura, E.A., et al.: Astron. J. **155**, 48 (2018)
Weiss, L.M., Marcy, G.W., Rowe, J.F., et al.: Astrophys. J. **768**, 14 (2013)
Wetherill, G.W.: In: Gehrels, T. (ed.) IAU Colloq. 52: Protostars and Planets, p. 565 (1978)
Wetherill, G.W.: In: Lunar and Planetary Institute Science Conference Abstracts, Lunar and Planetary Inst. Technical Report, vol. 22, p. 1495 (1991)
Wetherill, G.W.: Icarus **100**, 307 (1992)
Wetherill, G.W.: Icarus **119**, 219 (1996)
Wetherill, G.W., Stewart, G.R.: Icarus **77**, 330 (1989)
Wetherill, G.W., Stewart, G.R.: Icarus **106**, 190 (1993)
Williams, J.P., Cieza, L.A.: Annu. Rev. Astron. Astrophys. **49**, 67 (2011)
Windmark, F., Birnstiel, T., Güttler, C., et al.: Astrono. Astrophy. **540**, A73 (2012)
Winn, J.N., Fabrycky, D.C.: Annu. Rev. Astron. Astrophys. **53**, 409
Wittenmyer, R.A., Butler, R.P., Tinney, C.G., et al.: Astrophys. J. **819**, 28 (2016)
Wittenmyer, R.A., Wang, S., Horner, J., et al.: Monthly Not. Roy. Astron. Soc. **492**, 377 (2020)
Wolfgang, A., Rogers, L.A., Ford, E.B.: Astrophys. J. **825**, 19 (2016)
Wright, J.T., Marcy, G.W., Butler, R.P., et al.: Astrophys. J. Lett. **683**, L63 (2008)
Wright, J.T., Upadhyay, S., Marcy, G.W., et al.: Astrophys. J. **693**, 1084 (2009)
Wright, J.T., Fakhouri, O., Marcy, G.W., et al.: Proc. Astron. Soc. Pac. **123**, 412 (2011)
Wyatt, M.C.: Annu. Rev. Astron. Astrophys. **46**, 339 (2008)
Wyatt, M.C., Kennedy, G., Sibthorpe, B., et al.: Monthly Not. Roy. Astron. Soc. **424**, 1206 (2012)
Yang, C.C., Johansen, A., Carrera, D.: Astron. Astrophys. **606**, A80 (2017)
Youdin, A.N., Goodman, J.: Astrophys. J. **620**, 459 (2005)
Zellner, N.E.B.: Origins Life Evol. Biosphere **47**, 261 (2017)
Zhang, H., Zhou, J.L.: Astrophys. J. **714**, 532 (2010)
Zhang, Q.: Astrophys. J. Lett. **852**, L13 (2018)
Zhou, J.L., Aarseth, S.J., Lin, D.N.C., Nagasawa, M.: Astrophys. J. Lett. **631**, L85 (2005)
Zhu, W., Wu, Y.: Astrophys. J. **156**, 92 (2018)
Zube, N.G., Nimmo, F., Fischer, R.A., Jacobson, S.A.: Earth Planet. Sci. Lett. **522**, 210 (2019)

Part II
Star-Planet Interactions

Chapter 2
The Role of Interactions Between Stars and Their Planets

A. F. Lanza

Abstract Stars interact with their close-by planets by tides, radiation, and magnetic fields. This chapter briefly introduces such interactions focusing on the basic physical processes and considering their consequences for exoplanet demography, in particular, for the distributions of the orbital semi-major axis as well as of radius and mass of planets. The rotation of planet-hosting stars, as well as their magnetic activity, are considered because they play a relevant role in the interactions.

Keywords Planet-star interactions, stars · Activity, stars · Late-type, stars · Rotation, stars · Coronae

2.1 Introduction

Stars interact with their planets through different mechanisms and those interactions affect the physical parameters of the planets along their evolution having an impact on their statistical distributions. Interactions are important especially for close-by planets, that is, those preferentially detected by the planet-search techniques based on transits or radial-velocity measurements. For example, tides raised inside a star by a close-by massive planet produce a slow variation of the orbital semi-major axis with the planet eventually coming closer and closer to its host. On the other hand, tides raised by the star inside a close planet can synchronize its rotation in a time interval much shorter than the main-sequence lifetime of the star and this deeply affects the atmospheric dynamics of the planet and its evolution. In addition to tides, stars produce strong irradiation of the atmospheres of close-by planets that can produce their evaporation leaving a bare core with a significantly smaller radius than that of the initial planet. Magnetic fields and stellar winds can affect planets up to large distances as we see in the solar system. Therefore, a brief description of the main processes responsible for star-planet interaction is in order when considering

---

A. F. Lanza (✉)
INAF-Osservatorio Astrofisico di Catania, Via S. Sofia, 78 – 95123 Catania, Italy
e-mail: antonino.lanza@inaf.it

© Springer Nature Switzerland AG 2022
K. Biazzo et al. (eds.), *Demographics of Exoplanetary Systems*,
Astrophysics and Space Science Library 466,
https://doi.org/10.1007/978-3-030-88124-5_2

the demographics of extrasolar planets and their evolution, given that most of the known extrasolar planets orbit very close to their host stars.

In this chapter, we shall focus on the basic mechanisms of star-planet interactions, considering the roles of the different fields involved, that is

- gravity, responsible for tides;
- radiation, powering atmospheric evaporation and possibly radius inflation;
- magnetic fields, that permeate the atmospheres and winds of late-type stars and can provide us with hints on the internal fluid dynamics of planetary interiors.

We shall limit ourselves to the discussion of a few relevant processes and interactions considering their impact on the planet parameters, but without introducing a detailed and complete presentation of the subject, always giving more emphasis to the simplest physical models of the processes to enlighten the most important effects. In this way, we hope to provide the basic physical knowledge that will allow the reader to approach more advanced reviews or research articles in the field.

2.2 Tides

2.2.1 A Simple Description of Tides

When we treat the gravitational interaction between two bodies by assuming that both are point masses, we have the classic *two-body problem*. Tides arise when we consider that at least one of the two bodies has a finite extension, while the other is still treated as a point mass for simplicity. We denote as A the body of finite extension and as B the point mass orbiting around it. The mass of body A is M, while its mean radius is R; the mass of body B is $\mathcal{M}$ and their relative orbit is assumed to be circular, while the spin of body A is aligned with the orbital angular momentum.

The gravitational force of B depends on the distance of each point of A from B, thus it is not uniform across the extension of body A and produces a stress on body A. If body A is fluid as in the case of a star or a gaseous or liquid planet, its shape will adjust to the tidal potential originated by body B, producing an ellipsoidal deformation at the first order as we shall see when we consider a detailed quantitative formulation of the problem (see Sect. 2.2.2).

Limiting ourselves for the moment to a simplified treatment of the deformation of the fluid body A, we can describe it as a tidal bulge with a radius amplitude δR and consider the case of a small deformation, that is, $\delta R \ll R$ (cf. Zahn 2008). To first order, we can consider the relative deformation $\delta R/R$ to be proportional to the ratio of the difference of the gravitational acceleration between the two hemispheres of A produced by body B to the mean acceleration produced at its surface by body A itself, that is

$$\frac{\delta R}{R} \approx \frac{G\mathcal{M}R/d^3}{GM/R^2} = \frac{\mathcal{M}}{M}\left(\frac{R}{d}\right)^3, \qquad (2.1)$$

where G is the gravitation constant and d the separation between A and B. Assuming that body A is nearly homogeneous, the mass of the tidal bulge δM can be estimated as:

$$\delta M \approx \frac{\delta R}{R} M. \qquad (2.2)$$

If body A consisted of an ideal fluid without any internal dissipation, the tidal bulge would instantaneously adjust itself and point along the line joining the centers of A and B. However, in the case of a real fluid, there is always some dissipation that moves the tidal bulge ahead or behind the line OB joining the centers of the two bodies depending on whether body A is rotating faster or slower than the orbital motion of the system (cf. Fig. 2.1). This can be understood by describing dissipation as a friction that carries the tidal bulge ahead of the line OB when the rotation of A is faster than the orbital motion; conversely, the tidal bulge lags behind when rotation is slower than the orbital motion. From a geometrical point of view, we can describe this situation by specifying the angle α between the line joining the centers of the two bodies and the axis of symmetry of the bulge.

When there is dissipation inside body A, $\alpha \neq 0$ and there is a torque acting on body A that modifies its rotation by transferring angular momentum between its spin and the orbital motion. The tidal torque acting on body A is:

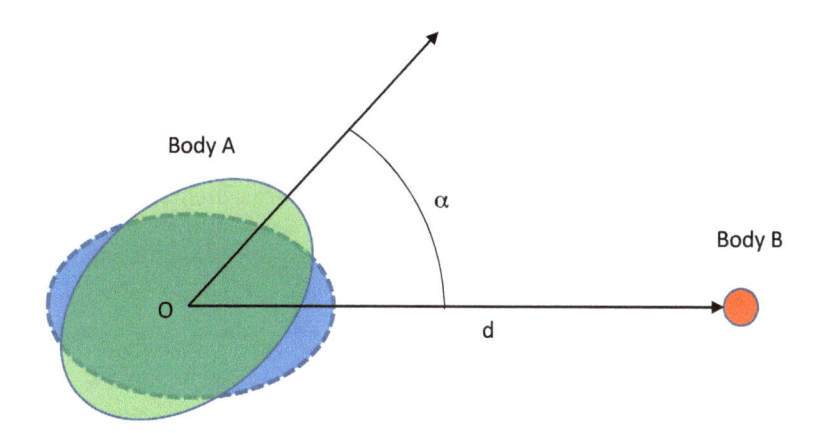

Fig. 2.1 Illustration of the tidal deformation of fluid body A by the tidal potential produced by body B. The blue ellipsoid with a dashed contour indicates the section of the tidally deformed body A when there is no dissipation and A, being made of an ideal fluid, can immediately adjust to the tidal potential. The green ellipsoid, conversely, indicates the section of body A in the case of a finite dissipation and a rotation faster than the orbital motion ($\Omega_s > \Omega_0$). In this case, the tidal lag angle α is positive because the tidal bulge is carried ahead of the line OB joining the centers of A and B by a faster rotation and tidal dissipation. Note that the amplitude of α has been greatly exaggerated to make the difference between the two cases (without and with dissipation inside body A) immediately evident. The separation between the two bodies is indicated with d, while O is the barycenter of body A

$$\Gamma \approx (f_2 - f_1) R \sin \alpha \approx -\delta M \left(\frac{GMR}{d^3} \right) R \sin \alpha = -\frac{GM^2}{R} \left(\frac{R}{d} \right)^6 \sin \alpha, \quad (2.3)$$

where f_2 and f_1 are the gravitational forces acting on the hemispheres of the bulge that are further or closer to body B, respectively (see Fig. 2.2), and we substituted Eqs. (2.1) and (2.2) into Eq. (2.3) to express δM in terms of the other system parameters.

To compute Γ, we need an expression for the tidal lag angle α. In the case of weak friction, α will depend linearly on the difference between the spin angular velocity Ω_s of body A, assumed to rotate rigidly, and the orbital angular velocity (or mean motion) Ω_0, and on the inverse of some characteristic time t_{diss} that measures tidal dissipation inside body A. In other words, in the *weak friction approximation*, we can write:

$$\alpha = \frac{(\Omega_s - \Omega_0)}{t_{\text{diss}}} \frac{R^3}{GM}, \quad (2.4)$$

where we have introduced the square of the free-fall frequency of body A, GM/R^3, to make α non-dimensional. Note that when $\Omega_s < \Omega_0$ and the tidal bulge lags behind the line joining the centers of the two bodies, Eqs. (2.3) and (2.4) lead to $\Gamma > 0$ which tends to accelerate the rotation of A transferring angular momentum from the orbit to bring body rotation into synchronism with the orbital motion, thus reducing the dissipation. The opposite happens when $\Omega_s > \Omega_0$ and the bulge is ahead of the line joining the centers of the two bodies. In this case, angular momentum is transferred from the rotation of A to the orbital motion, again in an attempt to drive the system towards the synchronous state and reduce dissipation.

If we denote with q the ratio of the masses of the two bodies, $q = \mathcal{M}/M$, and substitute Eq. (2.4) into Eq. (2.3) assuming $\alpha \ll 1$ in the weak-friction approximation, we have:

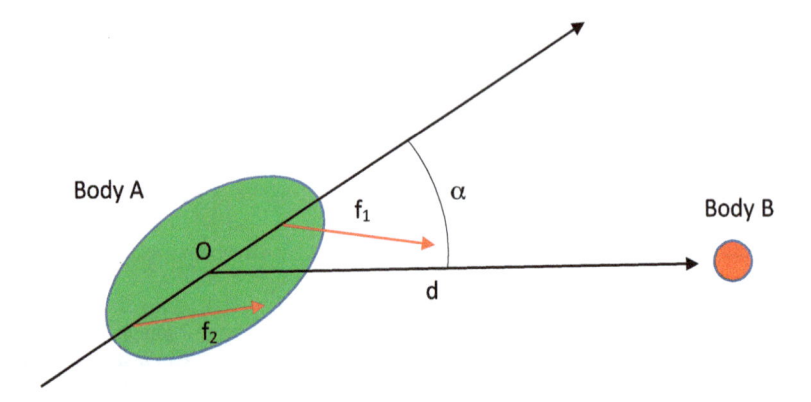

Fig. 2.2 Illustration of the gravitational forces f_1 and f_2 produced by the attraction of body B on the tidal bulge of body A. The lag angle α has been greatly exaggerated to better illustrate the effect. As in Fig. 2.1, body A is assumed to rotate faster than the orbital motion ($\Omega_s > \Omega_0$), so that $\alpha > 0$, while O is the barycenter of body A and d the separation between the two bodies

$$\Gamma = -\frac{(\Omega_s - \Omega_0)}{t_{\text{diss}}} q^2 M R^2 \left(\frac{R}{d}\right)^6. \tag{2.5}$$

Note that the tidal interaction decreases rapidly with the separation between the two bodies as a consequence of the factor $(R/d)^6$.

2.2.2 A More Detailed Theory of Tides

The simplified model introduced in Sect. 2.2.1 is useful to understand the fundamental effects of tides, but it is a rough simplification valid only in the case of a circular orbit and when the spin axis of body A is aligned with the orbital angular momentum. Moreover, it describes the tidal bulge as an ellipsoidal deformation, an assumption that has not been motivated yet. Therefore, a more detailed model is required and it will be introduced in the present section.

We consider a reference frame with the origin at the barycenter O of body A and the polar axis $\hat{z}$ directed along the spin axis of A assumed to rotate rigidly in an inertial space with an angular velocity Ω_s (see Fig. 2.3). A spherical polar coordinate system (r, θ, ϕ) is adopted with r being the distance from the origin O, θ the colatitude measured from the polar axis $\hat{z}$, and ϕ the azimuthal angle measured

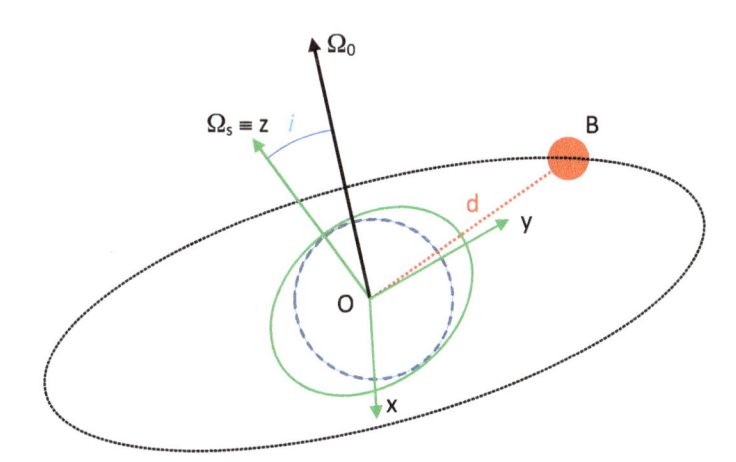

Fig. 2.3 Reference frame adopted to study tides in this section. The origin O is at the barycenter of the extended body A, the polar axis $\hat{z}$ is directed along the spin axis of A ($\hat{z} \equiv \hat{\Omega}_s$), while the $\hat{x}$ and $\hat{y}$ axes are in the equatorial plane of body A, which is normal to its spin axis. The section of body A, deformed by the tidal potential produced by body B, is indicated by the solid green line, while a section of uniform (polar) radius is indicated by the dashed blue circle. The orbital angular momentum is indicated by Ω_0 and forms the obliquity angle i with the spin axis of body A. The relative orbit of body B (in red) is indicated by the black dotted line and lies in the plane perpendicular to the vector Ω_0. The instantaneous separation between the centers of the two bodies is indicated by d

from some fixed axis in the inertial space. The gravitational potential of body B can be developed into a series of spherical harmonics in the reference frame of body A because it is a solution of the Laplace equation outside B. The zero-order term of this series is a constant that can be dropped without any loss of generality; the first-order term in $1/d$, where d is the separation between the two bodies, is responsible for the Keplerian (orbital) motion of the two bodies and is not relevant for the tidal interaction; the higher-order terms in $1/d$ are the relevant ones for our analysis and represent the *tidal potential* Ψ that verifies the Laplace equation outside body B: $\nabla^2 \Psi = 0$. If we further develop the time dependence of the potential into a Fourier series of the orbital frequency Ω_0, we obtain in a non-rotating frame (see Sect. 2.1 in Ogilvie 2014 for details):

$$\Psi = \Re \left\{ \sum_{l=2}^{\infty} \sum_{m=0}^{l} \sum_{n=-\infty}^{\infty} \frac{GM}{a} A_{l,m,n}(e,i) \left(\frac{r}{a}\right)^l Y_l^m(\theta,\phi) \exp(-jn\Omega_0 t) \right\}, \quad (2.6)$$

where $\Re\{x\}$ indicates the real part of the complex quantity x, l is the degree and m the azimuthal order of the spherical harmonic Y_l^m in the series development of the potential, n the order of the term in the Fourier series with respect to the time t, a the semi-major axis, e the eccentricity, and i the obliquity of the relative orbit in the assumed reference frame,[1] $A_{l,m,n}(e,i)$ a coefficient that depends on the eccentricity and the obliquity of the orbit, and $j = \sqrt{-1}$.

If the radius R of the extended body A is much smaller than the orbit semi-major axis a, only the terms with $l = 2$ are important in the development of the tidal potential. The amplitudes (but not the phases) of these *quadrupolar components* of the tidal potential are given in Table 2.1, correct to the first order in eccentricity and obliquity. For a circular and coplanar orbit ($e = 0$, $i = 0$) the static and the asynchronous tides are the only relevant components and they produce an ellipsoidal deformation of the fluid body A, thus justifying our assumption in Sect. 2.2.1. Note that for a circular orbit $n = m$ because of the rotational symmetry with respect to the azimuthal coordinate, while $l - m$ must be even for $i = 0$ because the spherical harmonics Y_l^m must be symmetric about the equatorial plane $\theta = \pi/2$.

It is useful to consider a reference frame rotating with body A (the *reference frame of the fluid* in tidal theory). In such a reference frame, the terms of the series of the tidal potential will contain time frequencies of the form

$$\hat{\omega}(n,m) = n\Omega_0 - m\Omega_s \quad (2.7)$$

that can be defined as the tidal frequencies in the frame rotating with the fluid. Specifically, for a given m and n, $\hat{\omega}(n,m)$ represents the frequency of the periodic variation of the tidal potential term with the same m and n as seen by an observer moving

---

[1] Note that the obliquity i is the angle between the spin axis of body A and the orbital angular momentum of body B. It is different from the inclination of the orbital plane I, that is, the angle between the orbital angular momentum and the line of sight.

2 The Role of Interactions Between Stars … 91

Table 2.1 Quadrupolar components of the tidal potential

l	m	n	$|A|$	Description
2	0	0	$\sqrt{\pi/5}$	Static tide
2	2	2	$\sqrt{6\pi/5}$	Asynchronous tide
2	0	1	$3e\sqrt{\pi/5}$	Eccentricity tide
2	2	1	$\frac{1}{2}e\sqrt{6\pi/5}$	Eccentricity tide
2	2	3	$\frac{7}{2}e\sqrt{6\pi/5}$	Eccentricity tide
2	1	0	$i\sqrt{6\pi/5}$	Obliquity tide
2	1	2	$i\sqrt{6\pi/5}$	Obliquity tide

with the rotating body A, the angular Doppler shift frequency $m\Omega_s$ coming from the azimuthal factor of the spherical harmonic $Y_l^m(\theta, \phi) \propto \exp(jm\phi)$. Therefore, the tidal frequencies $\hat{\omega}(n, m)$ represent the natural frequencies to be used when describing the action of the tidal potential in a reference frame rotating with the extended body A.

In Sect. 2.2.1 we made an approximation by considering only the large-scale deformation of body A and assuming a phase lag between the tidal potential and the hydrostatic (equilibrium) deformation of body A. In the reference frame of the fluid, the time-varying potential due to the orbiting companion produces a fluid motion across the tidal bulge and the dissipation of its kinetic energy is responsible for the tidal dissipation producing the lag angle α. In stars with an outer convection zone, Zahn (1977, 1989) modeled the tidal dissipation by considering the effects of turbulent convection on this tidal flow. This is called *the equilibrium tide* model and it does not represent the full solution of the hydrodynamic equations describing all the motions excited by the time-varying tidal potential inside body A.

To satisfy the equation of motion of the fluid, we must include both non-wave-like as well as wave-like motions (*dynamic tides*). In particular, body A may support different kinds of waves that may resonate with the time-varying tidal potential in the frame rotating with the fluid. Acoustic wave modes, also called *p-modes* in the case of stars, have periods on the order of a few minutes, while modes having buoyancy as their restoring force, called *g-modes*, have periods on the order of hours in main-sequence late-type stars. Therefore, they are generally not excited by the time-varying tidal potential whose characteristic periodicities correspond to the orbital and rotation period of the body A.[2] On the other hand, *inertial waves*, having the Coriolis force as their restoring force, can be excited by the tidal potential, provided that $|\hat{\omega}| < 2\Omega_s$.

---

[2] The situation is different in close binaries with an early-type component having an outer radiative envelope where g-modes can be excited close to the interface with the convective core and then radiatively damped close to the surface. Zahn (1975) recognized tidal g-mode dissipation as the dominant synchronization mechanism in binaries with early-type components, but we shall not consider those stars here because most of the currently known exoplanets orbit around late-type stars. In these stars, the excitation and dissipation of g-modes in the radiative core may play a relevant role in tidal evolution only for planets with a mass at least comparable to that of Jupiter and with orbital periods shorter than 1–2 days (see Barker 2020).

When they are excited, inertial waves make a very important contribution to the dynamical tides and can dominate the tidal dissipation both in the convection zones and in the radiative zones of late-type stars and giant planets. Note that, all the other parameters of the system being the same, the forcing of the inertial waves is proportional to the intensity of the Coriolis force, that is, to the spin angular velocity of the star Ω_s. Therefore, in a linearized theory, the amplitude of the waves is proportional to Ω_s and their dissipation rate to Ω_s^2 because their energy scales with the square of their amplitude (cf. Ogilvie 2013).

The dissipation of inertial waves is a very complicated process because most of it occurs in a chaotic fashion in wave attractors when the star or the planet has an inner discontinuity as in the case of the transition from a radiative interior to an outer convective envelope in late-type stars, thus a prediction of the efficiency of the tidal dissipation is not simple (see Ogilvie 2014 for a review). The change in the internal structure of a solar-like star from the pre-main sequence to the zero-age main-sequence and then along its main-sequence evolution modifies the dissipation of inertial waves in a remarkable way. Detailed numerical modeling of the dissipation along the entire evolution of a solar-like star is presently beyond our computing capabilities. Nevertheless, interesting indications have been obtained by Mathis (2015) by applying a simplified analytical model by Ogilvie (2013) to compute the inertial wave dissipation in stars with different core and envelope radii and masses.

A complete treatment of the dynamical tides is extremely complicated owing to the non-linear interactions of the different waves leading to their kinetic energy dissipation. A simpler approach is that of considering separately the effect of each Fourier component of the tidal potential of body B inside body A and quantify it by the perturbation it produces in the outer gravitational potential of body A (cf. Fig. 2.4). This approach is motivated by the fact that the tidal interaction between the two bodies A and B is mediated only by gravity, thus the perturbation of body A by the tidal potential of body B can be quantified by the resulting perturbation of the outer gravitational potential of body A. If the perturbation produced by the tidal potential component of body B is sufficiently small, the perturbation of the outer gravitational potential of body A will depend linearly on the former. Specifically, if the Fourier component of the tidal potential with frequency $\hat{\omega}$

$$\Re \left\{ \mathcal{A}(r/R)^l Y_l^m(\theta, \phi) \exp(-j\hat{\omega}t) \right\} \qquad (2.8)$$

is applied to the fluid body A with a mean radius R, the resulting deformation of A will generate an external gravitational potential perturbation whose component with the same spherical harmonic and time frequency will be

$$\Re \left\{ \mathcal{B}(R/r)^{l+1} Y_l^m(\theta, \phi) \exp(-j\hat{\omega}t) \right\}. \qquad (2.9)$$

The complex non-dimensional ratio

$$k_l^m(\hat{\omega}) \equiv \mathcal{B}/\mathcal{A} \qquad (2.10)$$

Fig. 2.4 Flowchart showing the perturbation of the outer gravitational potential of body A as produced by the tidal perturbing potential of body B. The ratio of the complex amplitude of the perturbation of the outer potential of body A to the complex amplitude of the perturbing potential produced by body B defines the Love number $k \equiv (\mathcal{B}/\mathcal{A})$ as explained in the text

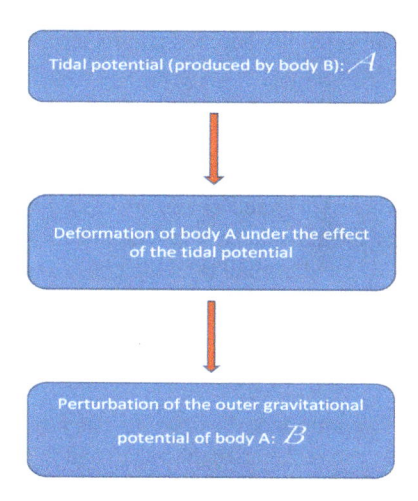

is a measure of the linear response of body A to the given perturbing tidal potential and is called *potential Love number* with order l and azimuthal degree m at the frequency $\hat{\omega}$.

The importance of the potential Love numbers comes from the possibility of expressing the tidal torque and the dissipation associated with each component of the tidal potential as simple functions of them (see, e.g., Ogilvie 2013 and Sect. 2.2 in Ogilvie 2014). Specifically, the component of the tidal torque in the fluid frame with frequency $\hat{\omega}$ is:

$$\mathcal{T} = \frac{(2l+1)R|\mathcal{A}|^2}{8\pi G} \Im\{k_l^m(\hat{\omega})\}, \tag{2.11}$$

where $\Im\{x\}$ is the imaginary part of the complex number x; while the corresponding contribution to the tidal dissipation is:

$$\mathcal{D} = \mathcal{T}\hat{\omega}. \tag{2.12}$$

We stress that this approach is valid in the limit of small perturbations of body A that allows a linear and independent treatment of the effects of the different components of the tidal potential in the reference frame of the fluid. Therefore, it represents a substantial improvement with respect to the simple model of Sect. 2.2.1, but it is still far from a complete treatment of tides. Nevertheless, it is the starting point for the definition of several useful parameters widely applied in the description of tides in close stellar binary and star-planet systems.

Since $\Re\{k_l^m(\hat{\omega})\}$ is a quantity of order unity, weakly dependent on m and $\hat{\omega}$, it can be approximated by its hydrostatic value k_l that does not depend on m. For a homogeneous body, it is

$$k_l \simeq k_l^{(\text{hom})} = \frac{3}{2(l-1)}. \tag{2.13}$$

The generally adopted parameterizations for $\Im\{k_l^m(\hat{\omega})\}$ are:

$$\Im\{k_l^m(\hat{\omega})\} = \frac{k_l}{Q(\hat{\omega})} = \frac{k_l^{(\text{hom})}}{Q'(\hat{\omega})}, \qquad (2.14)$$

where $Q(\hat{\omega})$ and $Q'(\hat{\omega})$ are the *tidal quality factor* and *modified tidal quality factor* for the considered tidal component (l, m) at the tidal frequency $\hat{\omega}$, respectively. For $l = 2$ and a small tidal lag phase angle α (see Sect. 2.2.1), Q is related to α as

$$\sin\alpha(\hat{\omega}) \simeq \frac{1}{Q(\hat{\omega})}. \qquad (2.15)$$

If we introduce the *time lag* τ as $\tau \equiv \alpha/|\hat{\omega}|$ and consider that $\sin\alpha \simeq \alpha$ since $\alpha \ll 1$, we can summarize the usually adopted parameterizations to compute the tidal torque and the tidal dissipation with Eqs. (2.11) and (2.12) as

$$\Im\{k_2^m(\hat{\omega})\} = \sigma \frac{k_2}{Q(\hat{\omega})} = \sigma \frac{3}{2Q'(\hat{\omega})} = k_2 \tau \hat{\omega}, \qquad (2.16)$$

where $\sigma \equiv \hat{\omega}/|\hat{\omega}| = \pm 1$ so that Q and Q' are always positive. Note that the introduced parameters are functions of the tidal frequency and the internal structure of body A. The dependence on the tidal frequency can be very complicated when the dissipation of inertial waves dominates with Q' wildly oscillating by several orders of magnitude in the interval $-2\Omega_s \leq \hat{\omega} \leq 2\Omega_s$ (cf. Ogilvie and Lin 2007; Ogilvie 2013).

2.2.2.1 Evolution of the Spin and Orbit Under the Effects of Tides

The tidal dissipation becomes zero when the spin of body A (and of body B, if it is considered as an extended body) becomes aligned with the orbital angular momentum, while the relative orbit becomes circular and synchronized with the spin of body A (and B). This corresponds to $\Omega_s = \Omega_0 = \Omega$, $e = 0$, and $i = 0$. This final state can be reached provided that the total angular momentum of the system L (spin + orbital) is larger than or equal to a minimum critical angular momentum L_c that corresponds to an equilibrium with $\Omega_s = \Omega_0 = \Omega_c$, where:

$$L_c = 4I\Omega_c \text{ with } \Omega_c = (GM_T)^{1/2}\left(\frac{M_R}{3I}\right)^{3/4}, \qquad (2.17)$$

where $M_T = M + \mathcal{M}$ is the total mass, $M_R = M\mathcal{M}/(M + \mathcal{M})$ the reduced mass, and $I = I_A + I_B$ the sum of the spin moments of inertia of the two bodies. If $L < L_c$, no equilibrium is possible and the two bodies will eventually collide. On the other hand, if $L > L_c$, there are two possible equilibria: the one with $\Omega_s = \Omega_0 < \Omega_c$ is stable because it corresponds to a minimum of the mechanical energy under the constraint of constant total angular momentum L; the other with $\Omega_s = \Omega_0 > \Omega_c$ is unstable (see Hut 1980).

The general direction of the tidal evolution can be predicted on the basis of the simplified model in Sect. 2.2.1. In the case of a circular and coplanar orbit, when the spin of body A is faster than the orbital motion, the orbiting body B lags behind the tidal bulge and angular momentum is transferred from the spin of body A to the orbit in an attempt to synchronize the system. On the other hand, when the spin of body A is slower than the orbital motion, angular momentum is transferred from the orbit to the spin of body A, again in an attempt to bring it into synchronism. As a consequence, the orbit shrinks and the orbital period decreases. If the total angular momentum is smaller than L_c, there is not enough angular momentum in the system to spin up body A to synchronism and the result will be a faster and faster decay of the orbit towards a final collision. If the orbit is eccentric, the tidal interaction is stronger at the periastron, given the dependence of the tidal torque on $(R/d)^6$ (cf. Eq. 2.5), thus tides will tend to synchronize the spin of body A with the orbital angular velocity at periastron or, more precisely, with a value that produces a zero net torque when averaged along the orbit (*pseudo-synchronization*).

It is possible to obtain closed-form expressions for the evolution of the orbital and spin elements only if some additional hypotheses are introduced, for example, if we assume a constant time lag τ all along the tidal evolution (e.g., Hut 1981; Eggleton et al. 1998; Leconte et al. 2010) or a constant modified tidal quality factor Q' (e.g., Mardling and Lin 2002). Assuming a constant time lag, the expression for the spin angular velocity when pseudosynchronization is reached is (Zahn 2008):

$$\frac{\Omega_s}{\Omega_0} = \frac{1 + (15/2)e^2 + (45/2)e^4 + (5/16)e^6}{(1-e^2)^{3/2}(1 + 3e^2 + 3e^4/8)}. \tag{2.18}$$

2.2.2.2 Comparison with Observations of Close Binary Systems and Hot Jupiters

Most of the planets known to date orbit around late-type main-sequence stars. The efficiency of tidal dissipation inside those stars can be inferred from the observations of close binary systems in open clusters of different ages, in particular looking for the maximum orbital period corresponding to a circular orbit or, better, considering the distribution of the orbital eccentricity in the binary population of a cluster of a given age. This allows us to avoid systematic errors due to the binaries that were born with a circular orbit and were not circularized as the result of tidal evolution (e.g., Meibom and Mathieu 2005).

Considering the data available from open cluster studies, Ogilvie and Lin (2007) find that the value of Q', averaged over the tidal evolution of the close binaries, is between 5×10^5 and 2×10^6, that is a value about 2–3 orders of magnitude smaller than that predicted by the dissipation of the equilibrium tides in the turbulent convection zones of late-type stars (Zahn 1977).[3] This suggests that the dissipation

---

[3] Note that the stronger the tidal dissipation, the smaller Q' as we can deduce from the larger value of the tidal lag angle α (cf. Eq. 2.15 and Mardling and Lin 2002; Zahn 2008).

of dynamical tides in close binaries that are synchronized is relevant. More precisely, Ogilvie and Lin (2007) propose that the dissipation of inertial waves excited in the radiative cores and in the convective envelopes of synchronized binaries can be the dominant effect. Note that for a synchronized or nearly synchronized binary with a nearly circular orbit, the dominant tide has $l = m = 2$ so its frequency in the fluid frame is $\hat{\omega}_{22} = 2(\Omega_s - \Omega_0) \ll 2\Omega_s$ because $\Omega_s \approx \Omega_0$, thus the condition for the excitation of inertial waves is satisfied. The above estimate of Q' refers to an average value along with the system evolution and the spin frequency of the system (consider that the dissipation rate of the inertial waves is proportional to Ω_s^2 as we saw in Sect. 2.2.2, therefore $Q' \propto \Omega_s^{-2}$, all the other parameters being the same).

In the systems formed by a late-type main-sequence star and a hot Jupiter, the total angular momentum is almost always insufficient to reach a final tidal equilibrium state (Levrard et al. 2009; see, however, Lanza et al. 2011). For very hot Jupiters, the remaining lifetime estimated by assuming $Q' = 10^6$ is 30–100 Myr, that is, much shorter than the main-sequence lifetime of their stars (Ogilvie and Lin 2007). Therefore, such a fast predicted orbital decay appears to be at variance with the observed existence of those systems. Nevertheless, most of the stars hosting hot Jupiters are far from synchronization with the planetary orbit, so the excitation of inertial waves is not possible in those stars, contrary to the case of close stellar binary systems. In hot Jupiter systems, only the dissipation of the equilibrium tide can produce the decay of the orbit and this leads to an average $Q' \sim 10^8$–10^9 ensuring a residual lifetime of hot Jupiters on the order of a few Gyrs, in agreement with their presumed ages (Ogilvie and Lin 2007).

2.2.2.3 Magnetic Fields and Rotation of Late-Type Stars

Late-type stars have outer convection zones that host magnetic fields. We can see them in detail in our Sun where they are responsible for active regions by modifying the transport of energy and momentum by convection and producing non-radiative heating in the outer layers of the atmosphere from the chromosphere to the corona. A nice introduction to the observations of magnetic fields and of the active region phenomena can be found in Chaps. 2–5 of Linsky (2019). For our discussion of tides, we are interested in the effects of magnetic activity on stellar rotation because tidal effects depend on the stellar spin.

The non-radiative heating of the outer atmosphere in the Sun and late-type stars produces a hot corona with temperatures up to $(1-3) \times 10^6$ K that cannot be confined by the gravitational field of the star, thus originating the solar and stellar winds.[4] The resulting mass loss carries angular momentum away from the star. Since the plasma

---

[4] The first model of the solar wind was proposed by Parker (1958) and we shall apply it to study the evaporation of the atmospheres of exoplanets in Sect. 2.4.1. For a comprehensive review of the solar and stellar winds, their evolution, and impact on exoplanets, see Vidotto (2021).

is frozen to the magnetic field[5] up to some radius r_A (called *Alfvén radius*), the spin angular momentum loss rate is:

$$\frac{dL_s}{dt} = \left(\frac{dM}{dt}\right) r_A^2 \Omega_s, \tag{2.19}$$

where dM/dt is the mass-loss rate in the wind. The Alfvén radius is larger than the surface radius R, $r_A \approx (10\text{--}100)\,R$, because the magnetic energy density in the corona is larger than the thermal and kinetic energy density of the plasma up to several stellar radii. This is a consequence of the very low density of the coronal plasma and of the relatively slow decrease with the distance of the large scale fields of the stars that can be approximated by dipole fields. Only in the outer corona, the magnetic energy density becomes so small that the kinetic energy of the wind flow is capable of carrying the field lines away with the wind and the plasma can move away from a star without extracting further angular momentum from the stellar spin. In other words, the rigid structure of the magnetic field up to the Alfvén radius r_A makes the sphere of radius r_A the effective source surface from which the angular momentum is extracted and carried away by the wind. In the absence of a magnetic field, the source of angular momentum would be at the surface of the star, that is, at $r = R$, producing a much smaller angular momentum loss rate.

The steady loss of angular momentum produced by the solar wind has significantly braked the Sun along its main-sequence lifetime. Stars of the same spectral type of the Sun and with an age of 1 Gyr rotate with a period of ≈ 12 days, that is about twice faster than the present Sun. With a present mass loss rate of $\sim 10^{-14}\,M_\odot\,\text{yr}^{-1}$, the total wind mass loss has been an insignificant $\approx 10^{-4}\,M_\odot$ along the past 3.5 Gyr, while $\sim 50\%$ of the angular momentum of the Sun at ~ 1 Gyr has been lost. A faster solar rotation in the past was associated with a higher level of coronal density and temperature and these produced a more intense solar wind because the intensity of the magnetic fields and their filling factors are an increasing function of the stellar spin Ω_s at least for values smaller than a certain saturation limit (e.g., Pizzolato et al. 2003). Therefore, most of the angular momentum loss occurs during the early stages of stellar evolution.

The observations of stellar rotation in associations and open clusters of different ages can be used to constrain the angular momentum loss rate and find parameterizations that can be inserted into models of tidal evolution of star-planet systems. During the pre-main sequence phase of their life, late-type stars have circumstellar discs with a mean lifetime of a few Myr. Before the dissipation of the disc, the rotation of the star is locked to the orbital motion of the inner edge of the disc by the strong magnetic fields present during the T Tauri phase, thus the rotation period stays constant in spite of the contraction of the star. After the disc dissipates, the contracting star is free to spin up as a consequence of the decrease of its moment of inertia because

---

[5] An introduction to the physics of magnetized plasmas as described by magnetohydrodynamics (MHD) can be found in Priest (1984) with the field frozen-in condition described in Chap. 2. For a general introduction to MHD see also Spruit (2013).

the stellar wind, although very strong, is not capable of counteracting this effect. Late-type stars reach the zero-age main-sequence (ZAMS) as fast rotators, if their discs were short-lived, while those with long-lived discs (up to ~10–12 Myr at most) reach the ZAMS as relatively slow rotators. Once on the main sequence, the moment of inertia becomes constant and the wind starts to brake the rotation of the star. But during the first phase of the main-sequence evolution, the surface rotation of the star can be modified by the coupling between the convection zone and the radiative core that provides an additional source of angular momentum to the convection zone that slows down the wind braking process. Only after an initial phase of 300–600 Myr, the exact duration of which depends on stellar parameters and initial conditions, the stellar rotation becomes a relatively simple function of the stellar age and mass.

The parameterization used by current models of angular momentum loss (AML) assume dL_s/dt to be a function of the stellar radius R, mass M, and angular velocity Ω_s. Starting from Kawaler (1988) and considering the saturation of coronal magnetic activity at some value of the stellar spin, Bouvier et al. (1997) considered a simple parameterization of the type:

$$\frac{dL_s}{dt} = \begin{cases} -K_W \Omega_s \Omega_{sat}^2 \left(\frac{R}{R_\odot}\right)^{1/2} \left(\frac{M}{M_\odot}\right)^{-1/2} & \text{for } \Omega_s \geq \Omega_{sat}, \\ -K_W \Omega_s^3 \left(\frac{R}{R_\odot}\right)^{1/2} \left(\frac{M}{M_\odot}\right)^{-1/2} & \text{for } \Omega < \Omega_{sat}, \end{cases} \quad (2.20)$$

where K_W is a calibration constant and Ω_{sat} is the angular velocity of saturation at which the dependence of the AML on the stellar spin changes. To reproduce the observations, Ω_{sat} must be a decreasing function of the stellar mass. More recent descriptions of the AML consider more sophisticated and detailed formulations, in some cases based on MHD models of stellar winds (e.g., Gallet and Bouvier 2013, 2015; Matt et al. 2015; Amard et al. 2016). This is a field of rapid development, thanks to the new measurements of stellar rotation periods in clusters and associations that are becoming available as a result of the search for transiting planets from the ground (e.g., Super-WASP, NGTS) and space (e.g., CoRoT, *Kepler*, K2, TESS) and the new stellar associations as revealed by Gaia astrometry.

Considering the simple case of a star settled on the main sequence and therefore with a constant moment of inertia I, and assuming that it is rotating rigidly with $\Omega_s < \Omega_{sat}$, the evolution of its spin in the absence of tidal interactions is given by:

$$\frac{d\Omega_s}{dt} = -\frac{K_W}{I} \left(\frac{R}{R_\odot}\right)^{1/2} \left(\frac{M}{M_\odot}\right)^{-1/2} \Omega_s^3, \quad (2.21)$$

that can be analytically integrated giving the solution $\Omega_s \propto t^{-1/2}$ for $t \gg t_0$, where t_0 is the initial time when this evolution equation becomes valid. This solution is the so-called *Skumanich law* for the evolution of stellar rotation.

To model the evolution of stellar rotation, Eq. (2.20) must be coupled with equations accounting for the variation of the moment of inertia along with the evolution and the internal mechanical coupling between different layers inside the star. In general, a simple parameterization is adopted to account for the coupling between the radiative interior and the convective envelope by expressing it through a characteristic timescale τ_c for the exchange of angular momentum between them (e.g., Spada et al. 2011). This so-called *two-zone model* (MacGregor and Brenner 2011) has the advantage of a simple parameterization of the complex and largely unknown processes that couple the dynamics of the radiative zone with the convection zone. It assumes that both domains rotate rigidly, albeit with different angular velocities. As a matter of fact, a significant differential rotation is seen in those models only during the first phase of the main-sequence evolution. After that phase, the internal differential rotation between the core and the envelope becomes negligible (see Gallet and Bouvier 2013, 2015).

A very interesting observational result, reproduced by the mentioned models of AML, is the dependence of Ω_s on the stellar age t and mass for stars with spectral types from mid-F to late K. This provides a method to estimate the age of a star from its rotation period (and mass) called *gyrochronology* (cf. Barnes 2003, 2007; Barnes and Kim 2010). In its simplest version, the age is proportional to Ω_s^{-2}, with the proportionality constant being a function of the stellar mass, as derived by inverting the Skumanich relationship. Gyrochronology cannot be applied to young stars (t smaller than about 500 Myr) because of the non-negligible effects of the initial conditions and a still poorly understood evolution of stellar rotation in the early phases of the main sequence. For older stars, it seems to be applicable at least up to the solar age (Barnes et al. 2016), although some authors have recently proposed that stars older than the Sun could be experiencing a reduced AML owing to a transition of their stellar dynamo towards a state of lower efficiency (van Saders et al. 2016; Metcalfe and Egeland 2019).

2.2.2.4 Modeling Tides and Stellar Rotation in Star-Planet Systems

In a system consisting of a late-type star and a close-by planet, the evolution of stellar rotation is ruled by the sum of the tidal torque and the magnetic wind torque that we have introduced in Sects. 2.2.2 and 2.2.2.3. In turn, the tidal interaction controls the evolution of the orbital elements of the planet and therefore the tidal torque. Because of the dependence of the stellar rotation on the torques acting on the star, this system has several feedbacks that can make its evolution non-monotone and remarkably complex. In this context, a fundamental role is played by the stellar hydromagnetic dynamo that produces the stellar field responsible for the wind carrying away stellar angular momentum.

Dynamos in late-type stars are based on the interaction between rotation and convection that produces differential rotation and the so-called α-effect, the basic ingredients of the so-called $\alpha\Omega$ dynamo. A general introduction can be found in textbooks describing solar dynamo such as Priest (1984) or Stix (2004), while a more detailed treatment can be found in, e.g., Moffatt (1978), Rüdiger and Hollerbach (2004), or Brandenburg and Subramanian (2005).

Differential rotation comes from the interaction of rotation and convection (Rüdiger and Hollerbach 2004) and produces the amplification of the magnetic field com-

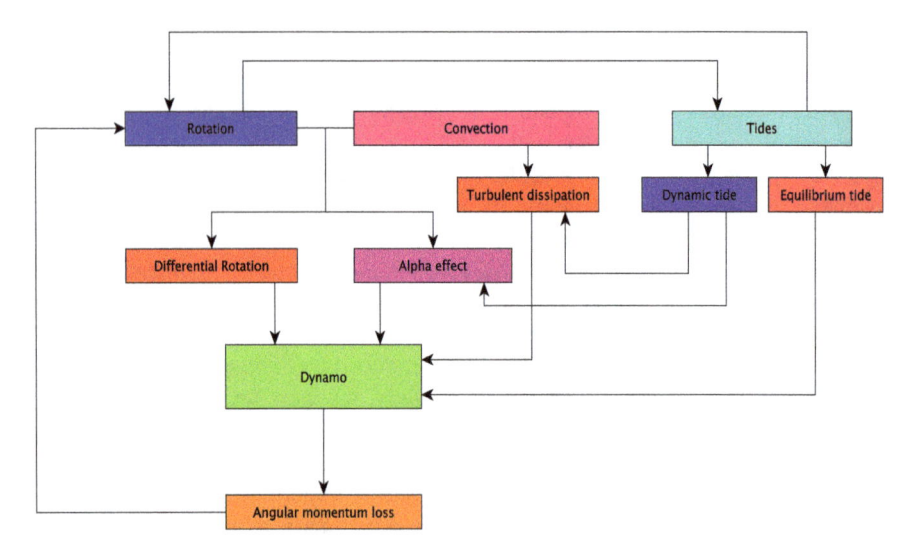

Fig. 2.5 Diagram showing the principal inter-connections among stellar rotation, tides, and hydromagnetic dynamo in late-type stars with close-by planets

ponent parallel to the equator (toroidal field component) starting from a much weaker component in the meridional plane (poloidal field component). The poloidal field is re-generated from the toroidal field thanks to the helicity of the convective motions produced by the Coriolis force that provides the α-effect, thus closing the dynamo cycle. In general, this kind of dynamo produces an oscillating field that is responsible for stellar activity cycles. An important role is played by turbulent convection that dissipates the magnetic fields, thus getting rid of the fields of previous cycles, and by the meridional circulation, which acts as a sort of conveyor belt changing the latitudinal distribution of the field components and making them interact with each other.

The sketch in Fig. 2.5 shows the complex interplay between tides, dynamo, and angular momentum loss by the stellar wind. Presently, we have only a limited understanding of all the inter-relationships because important aspects of the tidal dissipation, dynamo operation, wave-turbulence interaction, and stellar wind acceleration are unknown. Nevertheless, the first attempts to build models that include most of these effects, at least in parametrized and simplified ways, have appeared (see, e.g., Bolmont and Mathis 2016; Gallet et al. 2017, 2018; Benbakoura et al. 2019).

2.3 Tides and the Evolution of Exoplanets

The general introduction to tides and to the magnetic wind angular momentum loss provided in the previous Section will now be complemented with some applications to the study of star-planet systems. First of all, we have to include the effects of tides inside the planet. In the previous Section, we considered the secondary body B as a point mass. However, it is not difficult to include the effects of tides inside B by considering the same equations with the roles of the bodies exchanged, thus body A is now considered the source of the tidal potential and is regarded as a point mass M. Concerning the effects on the orbital elements, they will be the sum of the variations produced by tides inside body A and inside body B.

The equations for the evolution of the semi-major axis a and the orbital eccentricity e, in the simplified assumptions of constant modified tidal quality factors, orbital period much shorter than the stellar rotation period, planet spin synchronized with the orbit,[6] zero obliquity $i = 0$ and small eccentricity ($e < 0.1$), can be cast in the form (cf. Jackson et al. 2008):

$$\frac{1}{e}\frac{de}{dt} = -\left[\frac{63}{4}(GM_s^3)^{1/2}\frac{R_p^5}{Q_p'M_p} + \frac{171}{16}(G/M_s)^{1/2}\frac{R_s^5 M_p}{Q_s'}\right]a^{-13/2}, \tag{2.22}$$

$$\frac{1}{a}\frac{da}{dt} = -\left[\frac{63}{2}(GM_s^3)^{1/2}\frac{R_p^5}{Q_p'M_p}e^2 + \frac{9}{2}(G/M_s)^{1/2}\frac{R_s^5 M_p}{Q_s'}\right]a^{-13/2},$$

where G is the gravitation constant, M_s and R_s the mass and the radius of the star, M_p and R_p the mass and the radius of the planet, Q_s' and Q_p' the modified tidal quality factor of the star and the planet, respectively. These equations have been obtained by a truncation of the full equations to second-order in the eccentricity and are useful to discuss the contributions of the star and of the planet to the evolution of e and a, respectively. To this purpose, we explicitly write the ratios of the contributions to their variations produced inside the star and the planet:

$$\frac{\left(\dfrac{de}{dt}\right)_p}{\left(\dfrac{de}{dt}\right)_s} = \frac{28}{19}\left(\frac{M_s}{M_p}\right)^2\left(\frac{R_p}{R_s}\right)^5\frac{Q_s'}{Q_p'} \tag{2.23}$$

$$\frac{\left(\dfrac{da}{dt}\right)_p}{\left(\dfrac{da}{dt}\right)_s} = 7\left(\frac{M_s}{M_p}\right)^2\left(\frac{R_p}{R_s}\right)^5\frac{Q_s'}{Q_p'}e^2.$$

For a system consisting of a main-sequence late-type star and a hot Jupiter $M_s/M_p \sim 10^3$, $R_p/R_s \sim 0.1$, and $Q_s'/Q_p' \sim 1$–100 (see below for more details). Therefore, *the main contribution to the damping of the eccentricity comes from the dissipation*

---

[6] The timescale for the synchronization (or pseudosynchronization) of the planetary spin with the orbit is much smaller than the main-sequence lifetime of the star, usually not exceeding 0.1–1 Myr for Q_p' comparable to that of Jupiter (cf. Leconte et al. 2010). Therefore, we shall always assume the planet spin to be synchronized with the orbit.

of tides inside the planet, while for e<0.01–0.1 tidal dissipation inside the star dominates the evolution of the semi-major axis. The evolution of a is determined by the fact that the systems consisting of a star and a close-by planet generally do not have enough total angular momentum to reach tidal equilibrium, so planets will eventually be engulfed by their host stars (Levrard et al. 2009). This result was obtained without considering the angular momentum carried away by the magnetized stellar winds, which makes the situation even worse speeding up the tidal decay (e.g., Damiani and Lanza 2015; Gallet et al. 2018; Gallet and Delorme 2019).

An important question is the rate of decay of the orbits of close-by exoplanets because this is relevant to directly confirm the above prediction about the ultimate fate of those planets. From an observational point of view, transiting exoplanets offer the best opportunity to measure orbital decay. Specifically, we can measure the observed time of mid-transit O and compare it with the epoch C computed from an ephemeris with a constant period. The rate of decay of the orbital period $dP_{\rm orb}/dt$ is related to the $O - C$ difference by the formula:

$$O - C = \frac{1}{2} P_{\rm orb} \left(\frac{dP_{\rm orb}}{dt} \right) E^2, \qquad (2.24)$$

where E is the number of orbital periods elapsed from the initial epoch when $O = C$ and the orbital period is $P_{\rm orb}$. Therefore, the sensitivity to the orbital decay increases quadratically with time implying that a sufficiently long time baseline is required to measure it.

As an example of a system that displays a measurable $O - C$, I briefly discuss the case of WASP-12 (Hebb et al. 2011). It is a system consisting of an F-type star accompanied by a very hot Jupiter with a period $P_{\rm orb} = 1.09$ days that has been monitored for more than a decade. Its $O - C$ diagram shows a parabolic shape that implies $dP_{\rm orb}/dt = -29 \pm 3$ ms/year, if interpreted according to Eq. (2.24). This would imply a very fast orbital decay with a timescale $P_{\rm orb}/\dot{P}_{\rm orb} = 3.2$ Myr and a stellar modified tidal quality factor $Q'_{\rm s} = 2 \times 10^5$. However, an orbital decay is not the only possible explanation for the observed $O - C$. If the orbit of WASP-12 has a small eccentricity on the order of 0.002, the line of the apsides of the orbit can precess in the plane of the orbit itself (Patra et al. 2017; see also Damiani and Lanza 2011). Such a precession will produce oscillating and opposite $O - C$'s for the times of mid-transit and mid-occultation of the planet, in analogy with the case of close eclipsing binaries with an eccentric orbit (e.g., Wolf et al. 2013). The 1-σ upper limit for the orbital eccentricity coming from the available observations is 0.02 (Bonomo et al. 2017). Therefore, discriminating between the two possible explanations requires a longer time baseline. Patra et al. (2017) predict that an extension of the baseline by another ~ 10 year should be enough to distinguish a true orbital decay, for which the $O - C$ is the same for the transits and the occultations, from the precession of the apsidal line, for which the $O - C$'s are oscillating and have opposite signs for transits and occultations (cf. Fig. 2.6), because a precession period of ≈ 14 year is predicted by the present data. The recent investigation by Yee et al. (2020) suggests that neither the apsidal precession nor the possible presence of a third body are viable explanations,

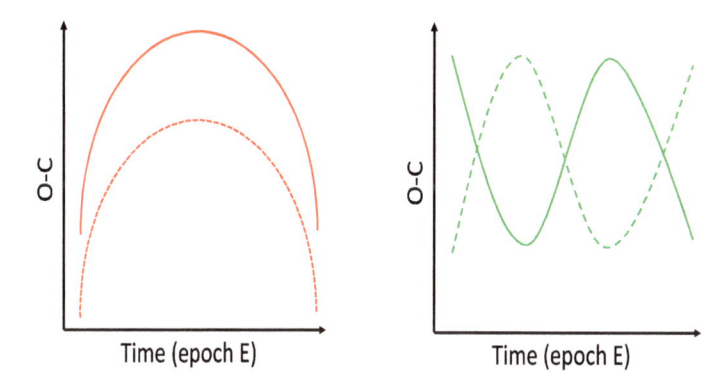

Fig. 2.6 Left panel: $O - C$ versus the time in the case of a planetary orbit experiencing tidal decay. The solid line indicates the $O - C$ time difference for the transits, while the dashed line indicates the $O - C$ for the occultations of the planet by the star (secondary eclipses). The two parabolae are actually superposed, but in this plot they have been vertically shifted for clarity. Right panel: same as the left panel, but in the case of precession of the apsidal line. Note the opposite oscillations of the $O - C$'s for the transits and the occultations

giving further support to a true orbital period decrease. Patra et al. (2020) and Turner et al. (2021a) support such a conclusion estimating $Q'_s = (1.6 \pm 0.2) \times 10^5$ and $(1.39 \pm 0.15) \times 10^5$, respectively. Nevertheless, the possibility that the long-term orbital period changes of WASP-12 are cyclic and come from an angular momentum exchange between the orbit and the rotation of the planet, has also been proposed (Lanza 2020).

2.3.1 Resonant Chains and Tides

The first example of a *resonant chain* was provided by the three inner Galilean moons of Jupiter, Io, Europa, and Ganymede. Their mean orbital motions ($n \equiv 2\pi/P_{orb}$, where P_{orb} is the orbital period) verify the relationship (known as *Laplace relationship*):

$$n_{Io} - 3n_{Europa} + 2n_{Ganymede} = 0, \qquad (2.25)$$

to within the observational error (10^{-9} deg/day). This resonance minimizes the tidal interactions between the three satellites because when two of them are aligned with Jupiter, the third is at least 60° away from conjunction (Murray and Dermott 1999). The Io-Europa resonance maintains the small eccentricity of the Io's orbit ($e = 0.041$) that is responsible for the tidal heating of the moon that was first observed by the Voyager I spacecraft in 1979 which detected an intense volcanic activity, three weeks after it was predicted by Peale et al. (1979). The tidally dissipated power P inside a body of modified tidal quality factor Q'_p on a moderately eccentric orbit ($e < 0.2$) is given by:

$$P = \frac{63}{4} \left[(GM_s)^{3/2} \left(\frac{M_s R_p^5 e^2}{Q'_p} \right) \right] a^{-15/2}, \tag{2.26}$$

where the symbols have been introduced above and the subscript s refers in this case to Jupiter or in general to the host star around which a planet is orbiting, while the subscript p refers to Io or in general to the planet (Miller et al. 2009).

An astonishing example of a resonant chain is provided by the planets orbiting the late M-type dwarf TRAPPIST-1. The star has a mass of only 0.09 $M_\odot$, a luminosity $\sim 5 \times 10^{-4} L_\odot$, and hosts a compact system of seven Earth-like transiting planets, the orbital periods of which are in the ratio 24 : 15 : 9 : 6 : 4 : 3 : 2 (Luger et al. 2017). These ratios minimize the tidal interactions between the planets, however their orbits maintain non-zero eccentricities that produce tidal heating inside them (Barr et al. 2018). In the case of TRAPPIST-1b, the tidal flux may reach values greater than about 10 W m^{-2} (about three times that of Io) during the periodic oscillations of its eccentricity (Luger et al. 2017). Note that the exact amount of dissipated heat depends on the modified tidal quality factor, that is on the order of 10^2–10^3 for telluric planets, and depends on their rheological properties. These are a complex function of the tidal frequency, the details of their internal structure and composition, and the presence of oceans on their surface (see Sect. 2.3.5). Therefore, an exact prediction is impossible because of our ignorance of the interior structure of those planets (cf. Tobie et al. 2019; Bolmont et al. 2020).

It is interesting to note that the resonant chain between the inner Galilean moons couples their tidal evolutions due to their spin-orbit angular momentum exchanges with Jupiter. The rotation period of Jupiter is shorter than the orbital period of Io, thus angular momentum is transferred from Jupiter spin to the orbit of Io to push it away from Jupiter. However, since Io is coupled to the other inner moons, the angular momentum is redistributed among all of them and tidal evolution occurs for the whole system. By measuring such a secular evolution, Lainey et al. (2009) determined the modified tidal quality factor of Jupiter as $k_2/Q = (1.102 \pm 0.203) \times 10^{-5}$ and of Io as $k_2/Q = 0.015 \pm 0.003$. For a recent review of the measurements of the planetary quality factors in the solar system, see Lainey (2016). When extrapolating Jupiter $Q' \sim 10^5$ to the case of hot Jupiters, it is important to keep in mind the difference in the rotation rate. For a similar planet rotating with a period of ~ 70 h, the inertial wave dissipation model considered in Sect. 2.2.2 predicts a Q' larger by a factor of ~ 50 because $Q' \propto \Omega_s^{-2}$.

2.3.2 Tides and the Formation of Hot-Jupiters

The existence of hot Jupiters is a challenge to our understanding of the formation of planetary systems. As discussed in Chap. 1, although a formation close to the star nearly their current orbital distances has been proposed (e.g., Bailey and Batygin 2018), the common view is that they form far from the star, beyond the snow line of the protoplanetary discs, where the formation of ice cores provides an easy way to rapid accretion of gas within the limited lifetimes of the protoplanetary discs. After their formation, those giant planets migrate towards the star reaching a separation on the order of 0.1 au. The migration can occur through angular momentum exchange with the disc with the planet moving on circular orbits of progressively smaller and smaller semi-major axis, or it can occur via tidal circularization of an initially highly eccentric orbit with the planet coming very close to the star at the periastron (cf. Dawson and Johnson 2018). The latter mechanism is usually called *high-eccentricity migration* and relies on the crucial effects of tides to shrink and circularize the orbit close to the star. The initial high-eccentricity orbit can be produced by planet-planet scattering, or Lidov-Kozai cycles, or secular chaos in a multiplanet system. In the case of planet-planet scattering, the process occurs during the early stages of the system evolution when there could be several massive planets on unstable orbits that can come very close to each other (cf. Marzari 2010). Lidov-Kozai cycles require the presence of a distant companion on a nearly circular and sufficiently inclined orbit ($i > 40°$) or on a low-inclination, but highly eccentric orbit (Naoz 2016). The periodic exchanges of angular momentum between the giant planet and the distant companion give rise to phases of high eccentricity in the orbit of the planet that let it come very close to the star where it can have a strong tidal encounter with the star itself. The result of this tidal encounter is a strong dissipation of the orbital kinetic energy inside the planet that circularizes and shrinks its orbit, thus making it a hot Jupiter. In the secular chaos model, there are long-term exchanges of orbital angular momentum among the planets leading to a very eccentric orbit for the inner planet that can finally be shrunk and circularized as in the previous mechanism forming a hot Jupiter (Wu and Lithwick 2011).

From a quantitative point of view, the orbital angular momentum is given by:

$$L_{\text{orb}} = \frac{M_s M_p}{M_s + M_p} \sqrt{G(M_s + M_p)a(1 - e^2)} \simeq M_p \sqrt{G M_s a(1 - e^2)} \text{ if } M_p \ll M_s, \tag{2.27}$$

while the mechanical energy of the orbit is given by:

$$E_{\text{orb}} = -\frac{G M_s M_p}{2a}, \tag{2.28}$$

where the meaning of the symbols is the same as in the previous sections. In the secular chaos or Lidov-Kozai cycles, the mechanical energy of the orbits is constant, while angular momentum is exchanged between the bodies orbiting the star. Therefore, only the eccentricity of the planet orbit varies with e approaching the unity when most of the orbital angular momentum of the planet is taken by the other bodies. When the planet comes very close to the star at the periastron, the tidal interaction becomes so strong that we can assume that the orbit is circularized at constant orbital angular momentum because we can assume that tides do not affect significantly the spin of the star. The strong *tidal dissipation inside the planet* reduces the orbital

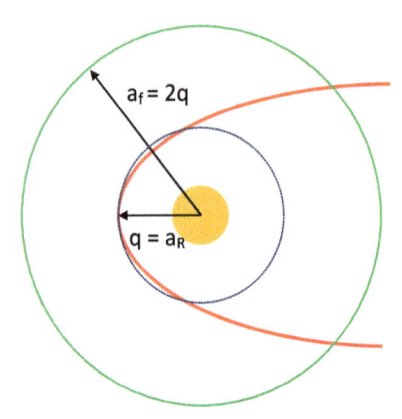

Fig. 2.7 Illustration of the formation of a hot Jupiter via high-eccentricity migration. The initial highly eccentric orbit of the planet is plotted in red and has a periastron distance $q = a_R$, where a_R is the separation corresponding to the planet coming in contact with its Roche lobe (dotted blue line). The final circular orbit of the planet, reached after the dissipation of the excess kinetic energy by the tides raised into the planet by the star, is plotted as a green line and has a radius a_f twice the periastron separation of the initial highly eccentric orbit

energy and therefore the orbital semi-major axis becomes smaller reaching a final value a_f corresponding to the circular orbit of the hot Jupiter. The conservation of the orbital angular momentum L_{orb} and of the mass of the planet implies (cf. Eq. 2.27):

$$a_f = a_i(1 - e^2), \quad (2.29)$$

where a_i is the initial semi-major axis of the highly eccentric orbit that leads to the tidal encounter between the planet and the star, and e is its eccentricity which is very close to the unity. Since $(1 - e^2) = (1 + e)(1 - e) \simeq 2(1 - e)$ and $a_i(1 - e) = q$, where q is the periastron distance on the highly eccentric orbit, we find:

$$a_f \simeq 2q, \quad (2.30)$$

in other words, the orbital separation of a hot Jupiter when it forms is twice the periastron distance of the highly eccentric orbit it had before coming close to the star (see Fig. 2.7). The minimum allowed value for q is given by the star-planet separation a_R corresponding to the planet in contact with its Roche lobe because a smaller separation would imply the transfer of mass from the planet to the star leading to a rapid engulfment of the planet.[7] Therefore, a prediction of the model of high-eccentricity migration is that the orbital semi-major axis is at least twice the Roche lobe separation a_R of the planet. A more detailed analysis by Ford and Rasio (2006) found that the distribution of the semi-major axis peaks at about three times the Roche lobe separation of the planet.

---

[7] The Roche lobe separation is given by: $a_R \simeq 2.16\, R_p (M_p/M_s)^{-1/3}$.

2.3.3 Constraints on Tidal Quality Factors from Exoplanet Demography

As we saw in Sect. 2.3, the attempts to directly measure the stellar modified quality factors Q'_s through the orbital decay of hot Jupiters are hampered by the possibility of alternative interpretations of the observed $O - C$ diagrams. A different approach is to make use of the orbital parameter distributions of hot Jupiters to derive constraints on Q'_s and Q'_p. This would provide the best results if the initial conditions and the ages of the systems were known because the present semi-major axes and eccentricities can be derived from their initial values by integrating the equations of tidal evolution (cf. Eq. 2.22). Unfortunately, we do not know the initial values of those parameters, so this approach can be pursued only by making some assumptions. Results on the stellar and planetary tidal dissipation efficiency were obtained in this way by, for example, Jackson et al. (2008); Matsumura et al. (2008); Husnoo et al. (2012); Hansen (2012). Here, I briefly report on two recent works that made use of extended samples of close-by planets.

Bonomo et al. (2017) re-determined in a homogeneous way the system parameters of a set of 231 transiting planets and used them to investigate the probable mechanisms of close planet formation and the tidal quality factors. The distribution of the orbital semi-major axes of these planets peaks at about three times a_R, in agreement with the prediction of the high-eccentricity migration model (see Sect. 2.3.2). Only a few systems have a semi-major axis smaller than $2a_R$ and these can be explained by the tidal decay of their orbits after circularization close to their stars.

A lower limit to the planet modified tidal quality factor Q'_p can be derived for the systems with an eccentric orbit, assuming that the eccentricity is established at the formation of the system and is not excited by a third body at any later time. A significant eccentricity requires that the timescale τ_e for the tidal decay of the eccentricity itself be longer than the age of the system τ_{age} that can be estimated by fitting isochrones to the position of the host star on the H-R diagram. Moreover, an eccentric orbit puts a constraint on the dissipation inside the star because it contributes to the decay of the eccentricity. In quantitative terms, for the systems with an eccentricity significantly different from zero, we can write (cf. Matsumura et al. 2008):

$$Q'_p > \frac{81}{2} n \left(\frac{M_s}{M_p}\right) \left(\frac{R_p}{a}\right)^5 F_p \tau_{age} \left[1 - \frac{81}{2} \frac{n}{Q'_s} \left(\frac{M_p}{M_s}\right) \left(\frac{R_s}{a}\right)^5 F_s \tau_{age}\right]^{-1},$$

$$Q'_s > \frac{81}{2} n \left(\frac{M_p}{M_s}\right) \left(\frac{R_s}{a}\right)^5 F_s \tau_{age},$$
(2.31)

where F_s and F_p are functions of the orbital eccentricity and the spin angular velocity of the star (and the planet) as given by Matsumura et al. (2008), while all the other parameters have been introduced before. On the other hand, the systems with an eccentricity not significantly different from zero, should have had time to circularize, if they were born with an eccentric orbit. This implies $\tau_e < \tau_{age}$ that sets an upper limit on Q'_p as:

$$Q'_p < \frac{81}{2} n \left(\frac{M_s}{M_p}\right) \left(\frac{R_p}{a}\right)^5 F_p \tau_{age}. \quad (2.32)$$

Using these relationships, Bonomo et al. (2017) derived both lower and upper limits on Q'_p. Most of the lower limits were derived for systems with eccentric orbits within 0.1 au from their hosts, giving $Q'_p > 10^5$–10^6. The very close system WAPS-18 has $Q'_p > 5 \times 10^7$, but its radial-velocity curve may indicate an apparently eccentric orbit because of the tides raised by the planet in the star that distort the spectral line profiles (cf. Maciejewski et al. 2020a, b). On the other hand, the upper limits for the planets with clearly circular orbits within 0.1 au show that $Q'_p < 10^6$–10^9 with the upper limit rapidly decreasing with increasing semi-major axis. These statistical results are derived assuming that the planets have similar tidal dissipation and do not take into account possible differences in their interior structure and rotation rate. For example, a planet without an inner core should not develop chaotic attractors producing rapid dissipation of inertial waves and should thus have a much higher Q'_p than a planet with a similar radius and a density discontinuity at its core. The existence of close-by planets with eccentric orbits poses a lower limit to the stellar Q'_s in the present assumptions, that Bonomo et al. (2017) found to be between 10^7 and 10^4 for the systems with $a < 0.1$ au with a rapid decrease with increasing semi-major axis.

A different approach was taken by Collier Cameron and Jardine (2018) who started by considering the distribution of the mass ratio M_p/M_s versus the relative semi-major axis a/a_R and found a lack of systems with relatively large M_p/M_s at $a/a_R < 2$–3. They interpreted this as the result of planetary engulfment due to the tidal decay of their orbits. By integrating the equation for the evolution of the semi-major axis from $a = 0$ to the present value a, they found the remaining lifetime of a planet τ_{rem} before it is engulfed (cf. Brown et al. 2011):

$$\tau_{rem} = \frac{2}{117} \frac{Q'_s}{n} \frac{M_s}{M_p} \left(\frac{a}{R_s}\right)^5. \quad (2.33)$$

For constant Q'_s, the value of τ_{rem} is smaller for the planets on the verge of the region without observed systems in the diagram, which motivated the proposed interpretation.

A more detailed analysis was performed by Collier Cameron and Jardine (2018) assuming an initially uniform distribution of $\log a$ for hot Jupiters and evolving it with a constant Q'_s. The observed distribution of a together with the measured parameters M_s, R_s, M_p, R_p, and the effective temperature of the star T_{eff}, can be used to put constraints on Q'_s for the population of stars hosting hot Jupiters. The most likely a posteriori distribution of Q'_s was derived using a Bayesian approach and considering the difference between systems where inertial waves can be excited inside the star ($|\hat{\omega}| \leq 2\Omega_s$) and those where they can not. For the former subset, Collier Cameron and Jardine (2018) found $\log Q'_s = 7.31 \pm 0.39$ from 19 systems, while for the latter $\log Q'_s = 8.26 \pm 0.14$ from 223 systems. The difference in the distributions of Q'_s is significant at the 2-σ level, providing support for the dynamic tide theory by Ogilvie and Lin (2007). No significant dependence of Q'_s on $T_{\rm eff}$ is found. A dependence of the results on the assumed initial distribution of the semi-major axis is present, but it does not have a major impact, except in the case of very small tidal dissipation rates because in that case, the evolution of the semi-major axis is limited. An interesting consequence of this model is that planets closer to their host stars are statistically younger than more distant planets.

A limitation of the above approaches is the assumption that Q'_s is constant along with the evolution of the systems. The role of the Q'_s evolution from the pre-main sequence to the end of the main sequence was investigated by, e.g., Mathis (2015), Bolmont and Mathis (2016), and Mathis (2018) who found that the strong tidal dissipation during the pre-main-sequence phase, coupled with a faster initial stellar rotation and a larger radius, push close-by planets away from their host star, thus putting them at a safe separation where tides can no longer produce their engulfment during the main-sequence lifetime of the system. On the other hand, models with a constant Q'_s starting from the pre-main-sequence phase leads to planet engulfment during the main-sequence lifetime of the star. The dependence of Q'_s on the internal structure and evolution of the stars has been recently discussed by Barker (2020) who confirmed the above results with a more sophisticated model of tidal dissipation.

2.3.4 Rotation of Planet-Hosting Stars

McQuillan et al. (2013) found a dearth of close-by planets around fast-rotating late-type stars using *Kepler* data. Their results refer to transiting planets, while the stellar rotation was measured from the modulation of the flux outside transits (see Fig. 2.8). Teitler and Königl (2014) interpreted this result as the consequence of the decay of the planet orbits under the effect of tides that leads eventually to engulfment of the close-by planets and a transfer of their angular momentum to the star that becomes a relatively fast rotator. Their model requires $Q'_s \leq 10^6$ to reproduce the observations, thus there is some tension with the larger values of Q'_s found by the statistical analyses of the exoplanet population mentioned in Sect. 2.3.3 that suggest $10^7 < Q'_s < 10^8$.

An alternative explanation has been introduced by Lanza and Shkolnik (2014) who propose that the innermost planets in the compact systems detected by *Kepler* experience secular chaotic interactions with distant, often undetected, companions that, in combination with tidal dissipation, can decrease their semi-major axes. They show that the probability for the innermost planet of coming closer and closer to the star increases with the age of the system. In turn, this leads to a star rotating more slowly, given the braking by the magnetic wind along with the main-sequence evolution of late-type stars. A simple quantitative model based on the above ingredients, i.e., chaotic evolution of the orbit of the innermost planet, star-planet tidal interaction, and the Skumanich law for the braking of stellar rotation, is capable of reproducing the observations.

The study of the rotation of stars hosting hot Jupiters suggests that those massive planets are affecting their hosts making them rotate faster than single stars of the same

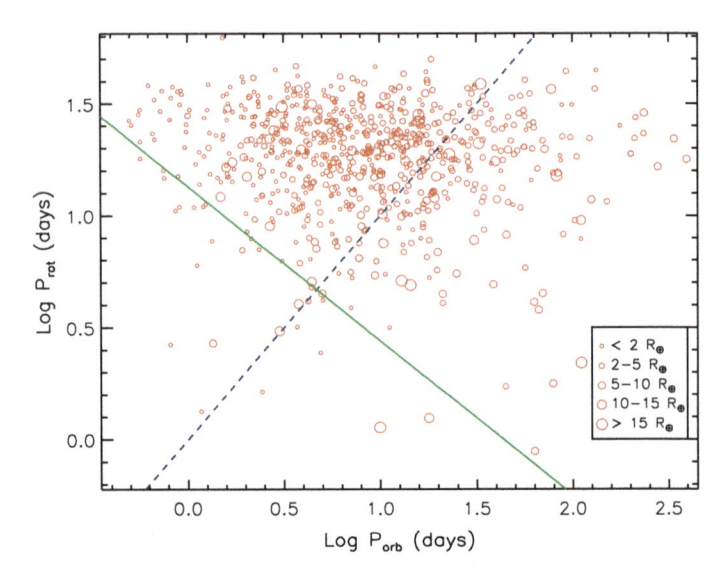

Fig. 2.8 Rotation period of the host star $P_{\rm rot}$ versus orbital period of its innermost transiting planet $P_{\rm orb}$ for the systems of McQuillan et al. (2013). The size of the red circle is proportional to the radius of the planet as indicated in the legend. The solid green line is the regression found by McQuillan et al. (2013) between the rotation period and the minimum orbital period of the innermost planet. The dashed blue line represents $P_{\rm rot} = P_{\rm orb}$, that is, the locus of the systems that have reached tidal synchronization

age (e.g., Pont 2009; Lanza 2010; Brown 2014; Maxted et al. 2015). Tidal spin-up has been proposed as a possible explanation, but it does not appear to work for all the systems (Lanza 2010; Maxted et al. 2015). Therefore, Lanza (2010) and Cohen et al. (2010) proposed that a close-by massive planet may interact magnetically with its host star reducing the efficiency of its magnetized wind and therefore the angular momentum loss rate along its main-sequence evolution. This will lead to a faster rotation of the star than predicted by gyrochronology according to its isochrone age.

The final word on these correlations will be said only after we get accurate ages for the planetary systems. The European space mission PLATO will provide them with a significant sample of transiting planets by applying asteroseismology to derive the ages of the host stars within 10% (Rauer et al. 2014). Moreover, it will provide accurate rotation periods for the host stars by measuring the rotational modulation of their optical flux.

2.3.5 Tides in Telluric Planets

Telluric planets have a solid interior and can have oceans and atmospheres on their surface. In the case of the Earth-Moon system, most of the tidal dissipation is due to the gravito-inertial waves that propagate in the oceans. Although their mass is only

0.023% of Earth's mass, oceans dominate tidal dissipation in our planet (Mathis 2018). For a distant observer outside the solar system who detects the Earth and the Moon by their transits across the Sun, oceans at Earth's surface would be undetected, thus any estimate of the tidal dissipation in our planet would be based on hypotheses on the rheology of its solid body (cf. Bolmont et al. 2020) and would be seriously wrong.

In addition to tides in the solid body of a telluric planet (Mathis 2018), a relevant role can be played by tides in its atmosphere excited by the stellar insolation. These *atmospheric tides* are generally not in phase with the gravitational tide acting on the solid body of the planet or on its oceans. To understand why, we can consider a naive model that assumes that the atmosphere consists of an ideal gas vertically stratified and without horizontal winds, so the pressure is the same at all the points at the same elevation above the surface of the planet. In this case, the region of the atmosphere that faces the star receives the maximum of insolation and is, therefore, hotter than the regions that are close to the terminator. Since the pressure p on a horizontal surface is the same at the point of maximum insolation and at the terminator, the product $\rho \times T$, where ρ is the gas density and T its temperature, is constant on that surface. Given that T is higher at the point of maximum insolation, the gas density will be at its minimum there, while it will be maximum on the terminator. In other words, the density profile of the atmosphere will show a bulge with the maximum density at the terminator, that is out of phase by ~90° with respect to the gravitational tidal bulge that points along the line joining the centers of the planet and of the star (see Fig. 2.9).

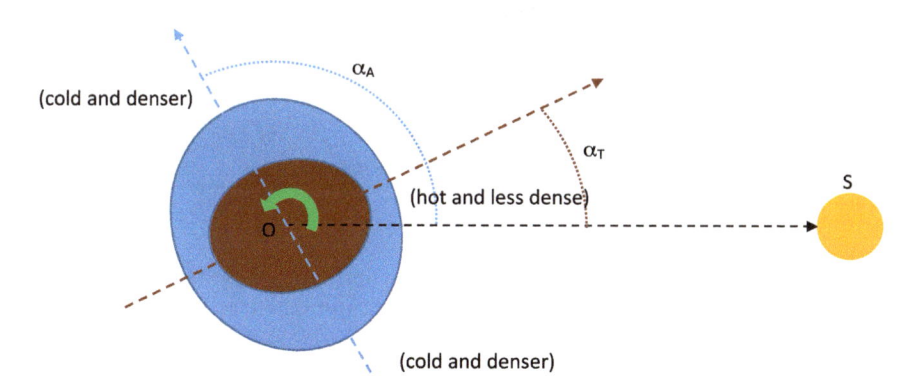

Fig. 2.9 Illustration of the atmospheric tide on a telluric planet. The gravity tide produces a deformation of the solid body of the planet (whose section is shown in brown) that is carried ahead of the line joining the center of the planet O with the star S by the rotation of the planet (indicated by the green arrow). The gravity tide lag angle is indicated by α_T. The insolation of the star on the planet's atmosphere (whose section is shown in blue) produces a less dense region at the substellar point and denser regions far away from that point. Therefore, the atmospheric tide lag angle α_A can be remarkably different from the gravity tide lag angle α_T. An equilibrium between the torques of the gravity tide and of the atmospheric tide can eventually be established, but with the planetary rotation out of synchronism with the orbit

The effects of the density bulge of the atmospheric tide can be as important as that of the gravitational bulge depending on the properties of its atmosphere and the rheology of the planetary rocky interior (Correia and Laskar 2010; Mathis 2018). In the solar system, the atmospheric tide of Venus is regarded as responsible for its retrograde rotation in spite of the fact that the mass of the atmosphere is only 0.01% of Venus' mass (e.g., Gold and Soter 1969; Correia and Laskar 2001).

In conclusion, tides in telluric planets are a complex phenomenon and we should be ready to accept that their rotation can be very different from our predictions based on simplified assumptions about their rheology and atmospheric properties. This can have an important role on the heating of the planet because a non-synchronous rotation produced by the atmospheric tides will lead to tidal dissipation in its interior due to the gravitational tides. For simplicity, the case of solid bodies having a permanent ellipsoidal deformation such as Mercury or the Moon is not considered here. The possibility of their capture into a spin-orbit resonance during their evolution is discussed in Murray and Dermott (1999).

2.4 Planet Evaporation and High-Energy Stellar Radiation

Stellar radiation provides a source of energy to the atmospheres of exoplanets that produces atmospheric dynamics, including evaporation, that can significantly affect the distribution of the mass and radius of the exoplanets, in particular of those smaller than 3–4 $R_\oplus$, during their lifetime. Therefore, it is important to discuss the impact of evaporation when studying the demography of exoplanets.

2.4.1 Evaporation of Planetary Atmospheres: A Simplified Model

A preliminary understanding of planetary evaporation can be based on a simplified model of the hydrodynamic wind escaping from an atmosphere and powered by the heating produced by stellar radiation.

We assume a one-dimensional model, a steady flow with uniform temperature (isothermal wind), and consisting of a single type of particles of mass m (e.g., protons). This simple model was introduced by Parker (1958) to explain the solar wind and has been applied to a variety of phenomena, for example, the evaporation of circumstellar discs or the winds produced by active galactic nuclei.

The basic equations of the model are:

$$4\pi r^2 n v = \text{constant} \quad \text{(mass continuity)}, \tag{2.34}$$

$$mnv\frac{dv}{dr} = -\frac{dp}{dr} - G\frac{M_p m n}{r^2} \quad \text{(momentum equation)}, \tag{2.35}$$

$$p = nk_B T \quad \text{(equation of state)}, \tag{2.36}$$

where r is the radial coordinate measured from the center of the planet, n the number density of particles of mass m, v their velocity, p the pressure, T the temperature, M_p the mass of the planet, k_B the Boltzmann constant, and G the gravitation constant. Note that only the pressure gradient and the planet gravity appear in the momentum equation, thus the wind is accelerated by a pressure gradient that opposes gravity. Since T is uniform, we can eliminate n between these equations and obtain:

$$\left(v - \frac{v_c^2}{v}\right) \frac{dv}{dr} = 2\frac{v_c^2}{r} - G\frac{M_p}{r^2}, \tag{2.37}$$

where

$$v_c \equiv \left(\frac{k_B T}{m}\right)^{1/2} \quad \text{(isothermal sound speed)}. \tag{2.38}$$

We define a critical radius r_c as

$$r_c \equiv GM_p/(2v_c^2), \tag{2.39}$$

thus Eq. (2.37) can be integrated to give:

$$\left(\frac{v}{v_c}\right)^2 - \ln\left(\frac{v}{v_c}\right)^2 = 4\ln\left(\frac{r}{r_c}\right) + 2G\frac{M_p}{rv_c^2} + C, \tag{2.40}$$

where C is a constant of integration.

Depending on the initial conditions and the value of C, there are different types of solutions of Eq. (2.40) that are illustrated in Fig. 2.10. We are interested in the solutions that connect the surface of the planet $r = R_p$ with the infinity because only those solutions represent a truly evaporative mass flow that is effectively lost by the body. Therefore, solutions of kind I and II are excluded. Similarly, we exclude solutions of kind III because they assume a high initial velocity of the flow at the surface of the planet. Solutions of type V, called *thermal breezes*, correspond to flows that escape from the planet, but that are initially accelerated and then decelerated, thus leading to a mass loss rate that goes to zero at infinity. The only acceptable solution is that labeled with IV that corresponds to the trajectory going through the point A of coordinates (r_c, v_c) (called *critical point*) in the r-v plane and that is obtained for a value of the integration constant $C = -3$. The flow starts with a very small velocity at the surface and is initially subsonic. When it crosses the critical point, it becomes supersonic and continues to accelerate under the effect of the steady pressure gradient. This solution, going through the critical point, corresponds to an *isothermal wind* producing evaporation and a mass loss of the planet (Priest 1984).

An important consequence of the transition from the subsonic to the supersonic regime is that any perturbation produced in the supersonic part of the flow cannot modify the mass loss. This happens because the mass loss is fixed by the boundary

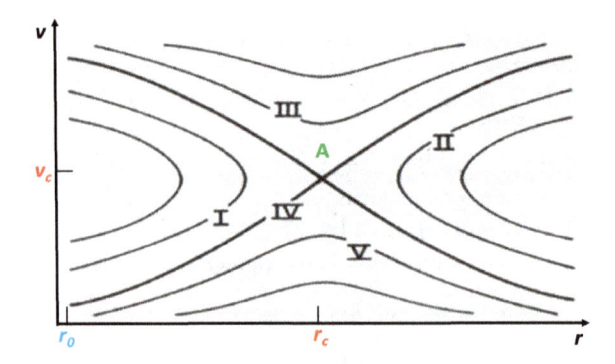

Fig. 2.10 Illustration of the different types of solutions of Eq. (2.40) in the r-v plane (after Priest 1984). The critical point is indicated by $A(r_c, v_c)$, where r_c is the critical radius and v_c the critical velocity corresponding to the isothermal sound speed given by Eq. (2.38). The radius r_0 from which the wind is accelerated is also indicated. See the text for a discussion of the different kinds of solutions labeled in the plot

conditions close to the surface of the planet and any perturbation propagates at the isothermal sound speed. Therefore, perturbations excited in the supersonic part of the flow are carried away by the flow itself without having the possibility to propagate down to the surface of the planet where the wind is launched.

It is not difficult to compute the velocity close to the base of the wind at $r = r_0$, where $v \ll v_c$ and $r_0 \ll r_c$. Note that r_0 is the level in the atmosphere where the wind is launched and that can be slightly larger than the planet radius R_p as we shall see below. We find the initial velocity:

$$v(r_0) = \exp(3/2) v_c \left(\frac{r_0}{r_c}\right)^2 \exp\left(-\frac{GM_p}{r_0 v_c^2}\right). \tag{2.41}$$

The mass loss rate in the evaporation is obtained by substituting the initial velocity in the equation of continuity:

$$\frac{dM_p}{dt} = 4\pi r_0^2 v(r_0) n(r_0) m \propto n(r_0) T^{5/2} \exp\left(-\frac{GM_p m}{r_0} \frac{1}{k_B T}\right), \tag{2.42}$$

where we made use of Eqs. (2.38) and (2.39) to express r_c and v_c in terms of the wind temperature.

Equation (2.42) allows us to compare the mass-loss rates that come from different levels in the atmosphere of a planet. In particular, we can compare the mass loss of a wind originating at the surface with that of a wind originating from the level where photoionization produced by the stellar ultraviolet radiation occurs, corresponding to an optical depth $\tau_{\text{EUV}} = 1$ in the extreme ultraviolet (EUV, see Fig. 2.11). The level corresponding to an optical depth of unity in the visible passband at $r = R_p$ in a strongly irradiated hot Jupiter reaches an equilibrium temperature of about ~ 2000 K

2 The Role of Interactions Between Stars ...

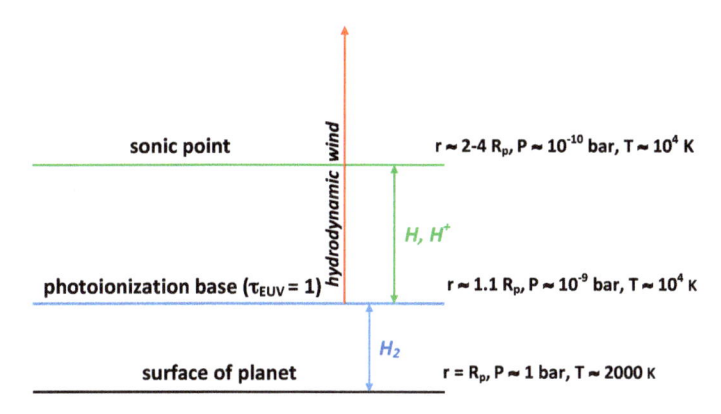

Fig. 2.11 Sketch of the different relevant levels in the atmosphere of an exoplanet (hot Jupiter) with indication of the typical radius r, pressure P, and temperature T (after Murray-Clay et al. 2009). The main chemical constituents of each layer are indicated. The red arrow indicates the wind accelerated from the photoionization level $\tau_{\rm EUV} = 1$ and crossing the sonic point

with a pressure of ∼1 bar, while the layer where photoionization occurs at $r \sim 1.1\,R_p$ has a temperature of about 10^4 K and a pressure of about 10^{-9} bar. Therefore, we can compare the mass-loss rates from the two levels as given by Eq. (2.42) assuming a ratio in the values of the base density $n(r_0)$ of 10^{-9} and temperatures of 2000 and 10^4 K, respectively. The result is that the mass-loss rate from the surface is ∼1.5 × 10^{-21} times that coming from the photoionization level owing to the dominant role played by the exponential factor in Eq. (2.42). In other words, the mass-loss rate from any thermal wind accelerated from the surface is completely negligible in comparison with the wind originating from the photoionization level.

The stellar radiation powering the evaporation consists of the EUV photons with a wavelength shorter than 91.2 nm which corresponds to the ionization threshold of hydrogen. Those photons ionize hydrogen and deposit their excess energy as heat in the plasma. Since the cross-section for bound-free absorption by hydrogen is proportional to λ^{-3}, where λ is the wavelength of the radiation, the ionization cross-section is maximum for photons close to the threshold and then decreases rapidly going towards the X-ray domain.

The heated plasma cools essentially by radiation because thermal conduction is not efficient at the low densities of planetary outer atmospheres. The radiation loss function of a plasma Λ is a function of its temperature and chemical composition and can be computed by assuming ionization and excitation equilibria for the ions (e.g., Raymond and Smith 1977; see Fig. 2.12). The energy radiated per unit volume in the low-density regime of the wind is $\Lambda(T)n^2$, where the dependence on the square of the density comes from the collisional excitation of the ions followed by their radiative de-excitation. This is similar to the case of stellar coronae that provides another example of a radiatively cooled low-density plasma. The radiative loss function $\Lambda(T)$ has a relative maximum at ∼(1–2) ×10^4 K associated with the strong Lyman-α radiation emitted by a plasma with a solar composition (see Fig. 2.12). Therefore,

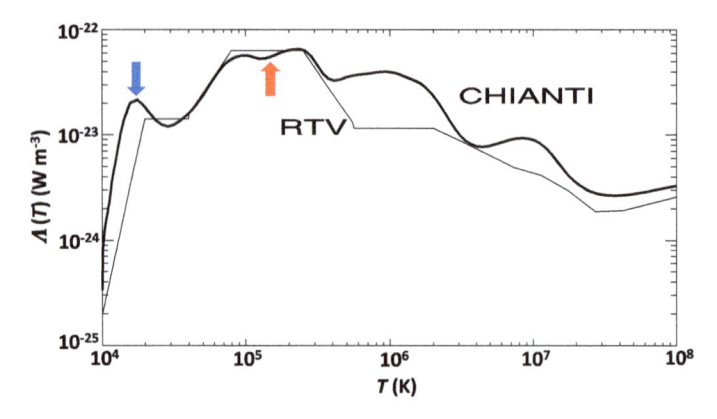

Fig. 2.12 The radiative loss function Λ of a high-temperature plasma with solar composition versus the temperature T after Reale and Landi (2012). The thin solid line labeled RTV is the parameterization adopted by Rosner et al. (1978), while the thick line is computed with version 7 of the CHIANTI spectral code by Landi et al. (2012). The vertical blue arrow marks the relative maximum around $\approx (1-2) \times 10^4$ K, associated with the emission of the Ly-α line, while the red vertical arrow marks the absolute maximum around $\approx (1-2) \times 10^5$ K due to the strong emissions in the lines of carbon, nitrogen, and oxygen

even a high flux of stellar EUV radiation is not capable of increasing the temperature of the wind beyond $\sim 10^4$ K because at higher temperatures the radiative loss rates will be smaller and not capable of balancing the heating by the EUV flux. In this way, the strong Ly-α emission acts as a sort of thermostat for the wind that becomes denser with the increase of the EUV flux in order to balance the greater heating by radiation.

Depending on the incoming stellar EUV flux, two regimes of evaporative mass loss are possible (cf. Murray-Clay et al. 2009). At relatively low EUV fluxes, most of the energy of the stellar radiation is used to lift the evaporating plasma out of the gravitational potential well of the planet. In this case, there is an approximate balance between the flux received by the planet and the work required to lift the evaporating mass outside the gravitational field of the planet (*energy-limited mass loss rate*), that is:

$$\dot{M}_p \sim \frac{\epsilon \pi F_{\text{EUV}} R_p^2}{G M_p / R_p} \sim 6 \times 10^6 \left(\frac{\epsilon}{0.3}\right) \left(\frac{R_p}{10^8 \text{m}}\right)^3 \left(\frac{0.7 \, M_J}{M_p}\right) \left(\frac{F_{\text{EUV}}}{0.45 \text{ W m}^{-2}}\right) \text{ kg s}^{-1}, \quad (2.43)$$

where $0 < \epsilon < 1$ is a factor specifying the fraction of the EUV flux F_{EUV} that is converted into work against gravity. In this regime, the mass loss rate is approximately proportional to the EUV stellar flux (see Murray-Clay et al. 2009 for details). On the other hand, when the EUV flux is very strong, we are in the *radiation-recombination limited mass loss rate* regime, where the wind is fully ionized and the radiative losses keep the temperature at $\approx 10^4$ K independently of the EUV flux. In this regime:

$$\dot{M}_\mathrm{p} \sim 4 \times 10^9 \left(\frac{F_\mathrm{EUV}}{500\,\mathrm{W\,m^{-2}}}\right)^{1/2} \mathrm{kg\,s^{-1}}. \qquad (2.44)$$

The dependence of the mass-loss rate on $F_\mathrm{EUV}^{1/2}$ comes from the balance between the radiative loss rate, that is proportional to $\Lambda(T)n^2$, and the incoming EUV flux at a fixed $T \sim 10^4$ K. Therefore, the base density and the mass-loss rate in the wind will be proportional to $n \propto F_\mathrm{EUV}^{1/2}$ in this regime.

The simplified model of planet evaporation presented above, based on the simple Parker isothermal wind, neglects several important effects. For example, the tidal potential due to the star modifies the gravitational field through which the evaporation flow moves, while the planet and stellar magnetic fields strongly affect the flow of an ionized plasma (e.g., Adams 2011; Trammell 2011). We neglected also the effect of the Coriolis force and the role played by the X-ray radiation.

Owen and Jackson (2012) discussed the different roles of the EUV and X-ray radiations due to their different penetration depths in the atmosphere. The EUV flux can penetrate down to the ionization level, while the X-rays can go much deeper because of their smaller cross-section. If the sonic point is below the ionization level, the EUV flux cannot power the wind because the mass loss is set below the sonic point. In this case, we have X-ray-driven evaporation with X-rays delivering energy below the sonic point to accelerate the wind. On the other hand, when the sonic point is beyond the ionization level, the wind is powered by the EUV flux, while the X-rays play a minor role as we considered in our model (cf. Fig. 2.11).

The maximum evaporation rates that can be reached in the case of hot Jupiters around active host stars are on the order of 10^9–10^{10} kg s^{-1}. Since the EUV flux emitted by the host stars decreases rapidly during the first phase of their main-sequence evolution (cf. Sect. 2.4.2), the impact on the hot Jupiters is generally regarded as very limited with no more than $\approx$5–10% of their mass being evaporated along their main-sequence lifetime. The situation is remarkably different for hot Neptunes and Earth-size planets because their mass is smaller than that of giant planets, therefore, we expect to see the most important effects, in particular on the planet radius, for those planets (see Sect. 2.4.4).

2.4.2 Stellar Extreme Ultraviolet and X-Ray Radiation

In this section, I shall briefly introduce the relevant properties of the outer stellar atmospheres of late-type stars that are necessary to understand planetary evaporation. For a general and complete introduction to the observations and models of the outer atmosphere of the Sun and of late-type stars, I suggest the textbooks by Priest (1984), Mariska (1992), Aschwanden (2005), and the recent work by Linsky (2019).

Main-sequence late-type stars host hydromagnetic dynamos that amplify and modulate magnetic fields in their interiors (cf. Sects. 2.2.2.3 and 2.2.2.4). Magnetic flux tubes emerge from the interior to the photosphere (e.g., Priest 1984) where they produce *active regions* with a rich variety of phenomena. Magnetic fields modify the transport of momentum and energy in plasmas and produce non-radiative heating of the outer atmospheres of late-type stars. The outer atmospheres are magnetically dominated because the ratio between the thermal energy density of the plasma and the magnetic energy density is remarkably smaller than the unity in the active regions. In particular, the geometry and the hydrodynamics of the plasma are dominated by the magnetic field, producing spatial inhomogeneities with strong departure from spherical symmetry, and time variability associated with the evolution of the fields. Transient phenomena, triggered by the fast dissipation of magnetic energy and its conversion into kinetic energy and particle acceleration, are also observed (flares and coronal mass ejections).

Considering the thermal profile of the outer atmosphere, we see that after reaching a minimum in the photosphere, the temperature starts to increase in the *chromosphere* (see Fig. 2.13). Chromospheric heating has a component coming from the dissipation of acoustic shock waves, excited by convection immediately below the photosphere, and another component produced by magnetic energy dissipation by magneto-sonic waves, Alfvén waves, and magnetic reconnection. The former component is independent of the magnetic fields and is considered responsible for the so-called *basal flux* of the chromosphere, while the latter magnetic component is responsible for the flux enhancement seen in the chromosphere of active regions. The temperature in the chromosphere rises rather rapidly with height up to about $(2-3) \times 10^4$ K. Above the chromosphere, the temperature continues to increase up to the *corona*, separated from the chromosphere by a thin *transition region*. Both the corona and the transition region are spatially inhomogeneous and variable in time because they are dominated by the magnetic field and there is no basal component in their emission attributable to non-magnetic processes.

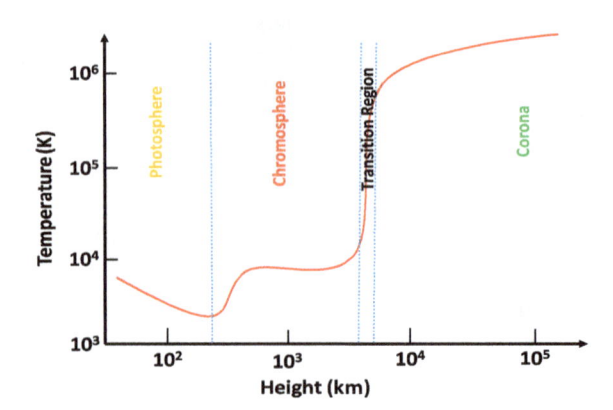

Fig. 2.13 Temperature versus height in a solar or stellar active region with the indication of the different atmospheric domains. This is a simple illustrative sketch and does not provide exact information on the run of the temperature that can change from one active region to the other or in the same active region as a function of the time. The height is purely indicative because the structure of the outer atmosphere is highly inhomogeneous

The corona is heated by magnetic fields through dissipation by reconnection and waves and reaches temperatures up to a few 10^7 K in the most active stars. In stars with a low activity level, comparable with the Sun, the corona reaches temperatures between $\sim(0.7$–$1.0) \times 10^6$ K in the so-called *coronal holes*, that have mostly open magnetic field lines, and $\sim 3 \times 10^6$ K in the cores of the brightest active regions. The flux emitted by the coronal plasma comes from collisional excitation, mainly by electron collisions, and the subsequent radiative decay of the excited ions or from bound-free and free-free emission of the plasma electrons. Since the plasma is optically thin and collisionally excited, the emitted flux depends on the square of the particle number density n and the radiative loss function $\Lambda(T)$ (see Fig. 2.12) and is emitted almost entirely in the X-ray owing to the high temperature T of the plasma: $F_X \propto n^2 \Lambda(T)$. In the *coronal loops* consisting of closed field lines, the density is about one order of magnitude higher than in the coronal holes, so they are about two orders of magnitude brighter. On the other hand, the density in coronal holes is kept low by the continuous outward flow of plasma forming the solar wind. The acceleration of the wind is essentially due to the thermal pressure as discussed in Sect. 2.4.1 where the Parker model was introduced.

In addition to the corona, emitting in the X-rays, the thin transition region emits a spectrum consisting of emission lines in the EUV domain (Linsky 2017). The small geometric thickness of the transition region is due to the profile of the radiative loss function $\Lambda(T)$ that reaches its absolute maximum around 10^5 K, that is, the temperature characteristic of the transition region (cf. Fig. 2.12). The associated intense radiative emission, mostly in the lines of ions of carbon, nitrogen, and oxygen, cannot be balanced by the magnetic heating in the transition region itself, so a steep temperature gradient develops across the transition region to allow a strong conductive flux of thermal energy from the corona in order to balance the radiative losses.[8] This explains the geometric thinness of the layers emitting the EUV radiation in late-type stars because thicker layers would not allow a sufficiently steep gradient.

As a consequence of the connection between the two domains, the coronal and transition region fluxes are strongly correlated with each other. The correlation is also strong with the component of the chromospheric emission in excess of the basal flux because it is also magnetically dominated. From a quantitative point of view, such correlations are expressed as power laws that link the X-ray flux with the fluxes in the different transition region and chromospheric lines (e.g., Oranje 1986; Piters et al. 1997; Pagano 2013).

Stellar X-ray emission can be detected from relatively distant stars because it is not strongly absorbed by the interstellar medium. On the other hand, the flux in the EUV below 91.2 nm is strongly absorbed by interstellar hydrogen, so we have very few direct measurements in the EUV band, only for stars that are very close to the Sun or that are intrinsically very active and for wavelengths shorter than ~ 35 nm. Therefore, the EUV flux must be reconstructed by indirect methods (Linsky et al.

---

[8] The thermal conductivity of the plasma scales as $T^{5/2}$, making thermal conduction efficient in the corona, especially along magnetic field lines, while the conductivity is much lower perpendicular to the field lines (e.g., Priest 1984).

2014; Linsky 2017). For example, one can use the above correlations between the X-ray flux and the EUV line fluxes. Another approach is that of Sanz-Forcada et al. (2011), who used a combination of X-ray observations and coronal loop models to reconstruct the EUV component of the stellar high-energy flux and study its evolution along the main-sequence lifetime.

The main parameter controlling the X-ray flux in late-type stars is rotation. Stars that rotate more rapidly have a larger X-ray flux up to reaching a saturation level. By introducing the Rossby number $Ro \equiv P_{\rm rot}/\tau_{\rm c}$, where $P_{\rm rot}$ is the rotation period and $\tau_{\rm c}$ the convective turnover time, empirically calibrated as a function of the $B - V$ color index, Pizzolato et al. (2003) found an approximately linear relationship between $L_{\rm X}/L_{\rm bol}$ and Ro^{-2} that saturates for $\log Ro < -0.8$.

Since the stellar rotation period increases rapidly owing to magnetic braking during the first phase of the main-sequence lifetime of a star, its X-ray and EUV emissions decay rapidly after the ZAMS and then more slowly during the remaining part of the main-sequence lifetime (e.g., Sanz-Forcada et al. 2011). In other words, most of the XUV radiation will reach the planets during the first 100–200 Myr of their evolution with a decrease in flux up to two orders of magnitude after that initial phase. This will have relevant consequences for the demography of exoplanets as we shall see in Sect. 2.4.4.

2.4.3 Signatures of Planetary Evaporation

Evidence of evaporation from a hot Jupiter was first provided by Vidal-Madjar et al. (2003) through observations of the stellar Ly-α line in HD 209458 with the STIS spectrograph on the Hubble Space Telescope. Although the core of the line is strongly absorbed by the Earth geocoronal hydrogen atoms, there was a clear difference in the line profile between in-transit and out-of-transit observations showing an excess absorption during transits that was attributed to hydrogen atoms in an extended atmosphere maintained by the planet evaporation. The number of transiting exoplanets with similar signatures of an extended atmosphere is still small because these observations require bright stars, such as HD 189733, WASP-12, or GJ 436 and Ly-α is not observable from the ground. For a recent review, see Bourrier and Lecavelier des Etangs (2018).

An intriguing observation that can be indirectly related to planetary evaporation was presented by Fossati et al. (2013) who found an anomalously low level of chromospheric emission in a sample of stars with transiting hot Jupiters. The chromospheric index $\log R'_{\rm HK}$, measuring emission in the cores of the Ca II H&K lines that are powered by chromospheric heating, is significantly lower in those stars in comparison with field stars of similar $B - V$ color. A detailed investigation of the WASP-12 system points to absorption by circumstellar plasma as the cause of the anomalously low level of flux in the cores of the Ca II H&K lines. This is not the only possible cause given that, in the case of WASP-18, the star was found to be intrinsically much less active than expected according to its rotation rate and age

(Pillitteri et al. 2014b; Fossati et al. 2018; cf. Sect. 2.5.4). Nevertheless, the presence of circumstellar plasma around WASP-12 can be associated with the high evaporation rate expected from its very close hot Jupiter, thus suggesting another indirect signature of planetary evaporation.

The possibility that the plasma evaporated from a close-by planet forms a torus around the star was suggested by some previous theoretical works showing that a closed magnetic field configuration with an azimuthal flux rope can be formed in the outer stellar corona, thanks to the continuous magnetic energy dissipation associated with the orbital motion of the planet (cf. Sect. 2.5.3 and Fig. 2.17). The flux rope can be filled by part of the evaporated plasma and helps to confine it around the star in its equatorial plane (Lanza 2009). Recent numerical models of the hydrodynamics of the evaporated plasma in WASP-12 confirm the possibility of forming a torus of plasma in the orbital plane of the planet around the star. This happens because the stellar wind is subsonic at the distance of the torus and kinetic and thermal pressures are not sufficient to blow away the plasma torus even without invoking the confining effect of the magnetic field (Debrecht et al. 2018).

An interesting consequence of having planet-evaporated absorbing material around stars with hot Jupiters is that their chromospheric activity indexes, notably $\log R'_{HK}$, would be affected if the plane of the torus contains the line of sight as happens in the case of transiting planets. Indeed, Hartman (2010) found a significant correlation between $\log R'_{HK}$ and the gravity of the planet,[9] for stars with $4200 < T_{eff} < 6200$ K, $M_p \geq 0.1\,M_{Jup}$, and semi-major axis $a \leq 0.1$ au. Specifically, stars hosting planets with a lower surface gravity have a statically lower chromospheric index, which can be explained because a lower planetary gravity favors the evaporation and therefore the formation of a more massive circumstellar torus leading to stronger absorption of the chromospheric line flux (see Lanza 2014 for details). Figueira et al. (2014) confirmed the results of Hartman (2010), while Fossati et al. (2015) found a bimodality in the correlation that was interpreted by their model on the basis of the observed bimodal distribution of the intrinsic level of chromospheric activity in late-type field stars (e.g., Vaughan and Preston 1980; Wright 2004). The additional absorption produced by circumstellar material replenished by the evaporation of close-by planets may represent a rather efficient way to select targets for planet search as well as for analyzing the composition of the evaporated planetary atmospheres (Haswell et al. 2020; Staab et al. 2020).

A different explanation for the correlation between planetary gravity and stellar chromospheric emission was suggested by Collier Cameron and Jardine (2018). They noted that surface gravity correlates strongly with planet mass, while the remaining lifetime of a hot Jupiter scales with the inverse planet mass in Eq. (2.33). Their analysis also suggests that any planet, whose remaining life expectancy is significantly shorter than the main-sequence lifetime of its star, is more likely to be seen when it is young if the planetary birth rate is more or less uniform. A younger planet implies

---

[9] The surface gravity of a transiting planet can be derived directly from radial-velocity and transit observations, independently of stellar parameters (Sozzetti et al. 2007).

a younger and more active star, thus qualitatively accounting for the observed correlation. In summary, more massive planets with a larger gravity are expected to be found around younger and more active stars.

2.4.4 Consequences of Stellar Irradiation and Planet Evaporation for Planet Demography

The *Kepler* space telescope allows measuring the radius of transiting planets down to about 1.0–1.2 $R_\oplus$. An accurate determination of the planet radius requires an accurate value of the stellar radius that was not generally the case for the values listed in the *Kepler* Input Catalogue. However, more recent work by the California *Kepler* Survey project led to a much more homogeneous and accurate determination of stellar parameters that unveiled a clear dichotomy in the distribution of the planet radii for $R_\mathrm{p} < 4.5$ $R_\oplus$ in a sample of 2025 planets (see Fulton et al. 2017, and Chap. III, Sects. 3.2 and 4.2). Using stellar parameters derived with asteroseismic techniques, Van Eylen et al. (2018) confirmed the existence of the dichotomy, although for a much smaller sample of 117 stars. Clearly, PLATO will provide a major advance in this field thanks to the systematic application of asteroseismic techniques that will allow to obtain measurements of the stellar radii with an accuracy of 3% in the case of Sun-like stars and the discovery of many more transiting small planets (Rauer et al. 2014).

A simple model to account for the observed dichotomy in the radius distribution of small planets was proposed by Owen and Wu (2017). They interpret the reduced fraction of planets with a radius around 2 $R_\oplus$ as the effect of the evaporation of the planet outer envelope under the action of the stellar EUV irradiation leaving a bare core with a smaller radius peaking around 1.3–1.5 $R_\oplus$. Specifically, this model assumes that a planet initially consists of a core of uniform composition and an envelope of hydrogen and helium with outer radii R_core and R_p, respectively. The mass of the core is M_core, while that of the envelope is $M_\mathrm{env} \equiv X M_\mathrm{core}$, where $X < 1$. The self-gravity of the envelope is neglected and the envelope is considered to be adiabatic in the deep interior, while it is roughly isothermal (radiative) near the surface. The planet luminosity arises from the gravitational contraction of the envelope.

The model by Owen & Wu allows to compute X as a function of the radii R_core, R_p, and the chemical composition of the core. Assuming a given flux of stellar EUV photons and an energy-limited evaporation rate, the model allows to compute the mass loss timescale of the planets that peaks when the planet radius is doubled by the presence of the H/He envelope, corresponding to $X \approx 0.01$ (see Fig. 2.14). When the envelope is less massive ($X < 0.01$), the planet radius remains approximately constant, thus its surface gravity is smaller, which reduces the evaporation timescale. Similarly, when the envelope is more massive ($X > 0.01$), the planet swells up faster than the accumulation of envelope mass, thus intercepting a larger EUV flux from the star and reducing its evaporation timescale (Owen and Lai 2018). Since the stellar EUV flux decreases fast with increasing stellar age, planets that have $X \approx 0.01$

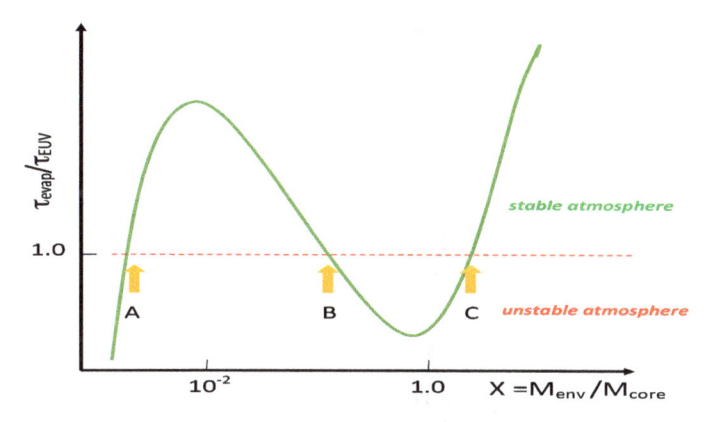

Fig. 2.14 Ratio of the evaporation time scale τ_{evap} to the typical duration τ_{EUV} of the phase of EUV strong emission in late-type stars versus X, the ratio of the mass of the atmospheric envelope M_{env} to the core mass M_{core} of an exoplanet. The mass of the core is fixed. If $\tau_{evap}/\tau_{EUV} < 1$, the strong EUV flux has time to completely evaporate the envelope, while the envelope is retained when $\tau_{evap}/\tau_{EUV} > 1$. The red horizontal dashed line marks the separation between these photo-evaporation stable and unstable envelopes. The two stable configurations A and B are accessible to low-mass planets and correspond to $X \ll 10^{-2}$, that is, to a core almost entirely stripped of its envelope, and to $X > 10^{-2}$, respectively, both having a radius smaller than the maximum radius (about twice the radius of the core, reached at $X \sim 10^{-2}$). The stable configuration C is characteristic of more massive planets having most of their mass in the envelope such as giant planets

experience only a limited amount of evaporation of their envelopes and keep a radius larger than $2\,R_\oplus$ at the end of the evaporation phase. On the other hand, planets with $X \ll 0.01$ or $X \gg 0.01$ will rapidly lose their envelopes and expose their bare cores displaying a significantly smaller radius than the planets which could retain much of their envelopes. The duration τ_{EUV} of the initial phase of intense evaporation is on the order of only 100–150 Myr because of the fast decay of the level of EUV and X-ray emission in main-sequence stars. Therefore, for planets orbiting field stars with an average age of some Gyr, the photoevaporation phase was completed long ago.

The radius dichotomy disappears for orbital periods longer than about 20 days (Fulton et al. 2017; Owen and Wu 2017), thus supporting an interpretation based on photoevaporation because the stellar EUV flux becomes too small to produce a significant mass loss when we move sufficiently away from the star. The model by Owen & Wu allows also to put constraints on the composition of the core. A homogeneous core consisting of pure water ice is excluded by the observations, while a core consisting of a mixture of iron and silicates is preferred, although the mass fraction of the two main constituents cannot be precisely constrained.

A precise modeling of the planet evaporation and of the consequent radius evolution should go beyond a simplified approach based on the energy-limited evaporation, (see, e.g., Kubyshkina et al. 2018), and consider also the contribution of the internal heat of the planet (e.g., Kubyshkina et al. 2020).

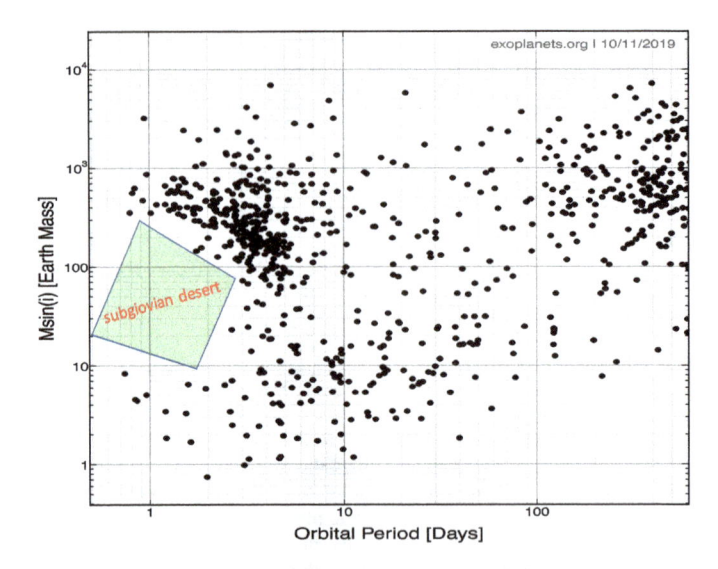

Fig. 2.15 Projected planet mass versus orbital period showing the lack of planets in the domain called sub-giovian desert (labeled). Credits: exoplanets.org database; Han et al. (2014)

In a diagram plotting the planet mass versus the orbital period, there is a remarkable paucity of objects with mass <0.5–0.7 $M_{\rm Jup}$ in a wedge-shaped domain running from periods shorter than ~3 days for the more massive planets to ~20 days for the less massive (cf. Fig. 2.15). This feature of the mass-period diagram has been called *sub-Jovian desert*. Owen and Lai (2018) presented an evolutionary model to explain this feature in the exoplanet distribution, based on the high-eccentricity migration and the tidal disruption barrier in the case of massive planets ($M_{\rm p} \geq 0.2 - 0.3\, M_{\rm Jup}$), and on the photoevaporation in the case of low-mass planets ($M_{\rm p} \leq 0.1 - 0.2\, M_{\rm Jup}$). The shape of the sub-Jovian desert is interpreted as the effect of photoevaporation on small-mass planets being much more efficient close to their star, while the orbit of a more massive giant planet is circularized closer to the star because the Roche lobe semi-major axis $a_{\rm R}$ decreases with planet mass and high-eccentricity migration produces a peak of the planetary separation at ~$3a_{\rm R}$ (see Sect. 2.3.2). This model suggests that the evolution of planets in the mass-period diagram can be fast for small planets because most of the photoevaporation occurs during the first 100–150 Myr of the stellar evolution on the main-sequence, while the high-eccentricity migration of giant planets can take much longer, up to Gyrs in the case of a dynamic chaotic evolution involving distant companions.

Finally, I mention the discrepancy between the radii of hot Jupiters as measured by means of their transits and those expected on the basis of their structure models (the so-called *radius anomaly* of hot Jupiter). There is a correlation between the radius excess and the stellar flux received by the planet as quantified by its equilibrium temperature with hotter planets usually showing a greater radius anomaly. In principle, a source of heating in the deep planetary interior, that has a nearly adiabatic stratification, can explain a larger radius, provided that its power is a few percent of the stellar insolation. Several models have been proposed to account for this extra heating, but no clearly definite mechanism has been proposed to explain how the heat deposited by the stellar radiation in the planet atmosphere could be transferred into the deep interior to inflate the adiabatically stratified layers of the planet (see Laughlin 2018; Sainsbury-Martinez et al. 2019, for details).

2.5 Star-Planet Magnetic Interactions

Magnetic fields of extrasolar planets have not yet been measured directly as it has been possible in our solar system. Nevertheless, they are expected to play a very important role because they can reduce the evaporation rate (e.g., Adams 2011), protect planetary atmospheres from the impacts of stellar coronal mass ejections, affect the habitability of planets (e.g., Airapetian et al. 2019), and provide information on planetary interiors where the fields are generated (e.g., Driscoll 2018). In this section, I shall discuss the possibility of inferring information on planetary fields by observing star-planet interaction effects in their host stars.

2.5.1 Stellar Winds and Magnetic Perturbations by Close-By Planets

In the solar system, planets are in a region of space where the velocity of the solar wind v_w is greater than the local *Alfvén velocity* $v_\mathrm{A} = B/\sqrt{\mu \rho}$, where B is the magnetic field intensity, μ the magnetic permeability, and ρ the density of the plasma wind. The Alfvén velocity is the velocity of propagation of the perturbations along the magnetic field lines[10] (e.g., Priest 1984; Spruit 2013). The *super-Alfvénic regime*, characteristic of the solar wind at the distance of the planets ($v_\mathrm{w} > v_\mathrm{A}$), implies that any perturbation produced by the planets in the interplanetary field cannot propagate down to the Sun, but it is blown away with the wind. In the case of our Earth, a bow shock is generally present at the boundary between the solar wind and the magnetosphere of our planet and the surface of the shock can be regarded as a surface of discontinuity separating the wind field from the magnetospheric field with little reconnection occurring between them.

In the case of close-by planets observed around other stars, the situation is probably different. A planet orbiting within 0.1–0.2 au from a solar-type star is likely located in a region where the stellar wind is still accelerating, characterized by a *sub-Alfvénic regime*, that is with $v_\mathrm{w} < v_\mathrm{A}$ (Preusse et al. 2005). Therefore, perturbations excited

---

[10] The Lorentz force in an incompressible plasma provides tension along the magnetic field lines, similar to the tension along an elastic string, that allows the propagation of transverse waves along the field lines (*Alfvén waves*). Their phase velocity is the Alfvén velocity.

by the orbital motion of a close-by planet in the interplanetary field can travel down to the star and produce observable effects in its atmosphere, at least in principle (cf. Preusse et al. 2006; Kopp et al. 2011; Saur et al. 2013; Saur 2018). Specifically, Alfvén waves will propagate through a moving medium, that is, the stellar wind. The solutions of the wave equation indicate that they will travel along *characteristics* given by $\mathbf{c}_A^{\pm} = \mathbf{v}_w \pm \mathbf{v}_A$, where $\mathbf{v}_w$ is the velocity of the wind, $\mathbf{v}_A = \mathbf{B}/\sqrt{\mu\rho}$ the vector Alfvén velocity, and the $\pm$ sign indicates the two characteristics that originate at the planet. One of the two characteristics points away from the star, while the other generally points towards it, thus allowing the Alfvén waves traveling along it to reach the star (Saur et al. 2013).

An example of a system where Alfvén waves propagate from an orbiting body down to the atmosphere of the central body is provided by Jupiter with its inner Galilean moons, Io, Europa, and Ganymede. They orbit inside the magnetosphere of Jupiter and excite Alfvén waves that travel down to the planet along characteristics dissipating their energy in its atmosphere and producing hot spots in the UV that move in phase with the orbital motion of the moons. Spectacular images of these phenomena have been obtained by the Hubble Space Telescope and show the hot spots produced by the planet-moon magnetic interaction moving inside the auroral oval of Jupiter well evident in the UV band. A model of the electromagnetic interaction between Jupiter and its moons, in particular Io, was first proposed by Neubauer (1980) (the so-called *Alfvén-wing model*). Note that the excitation of Alfvén waves does not require an intrinsic magnetic field in Io (actually the satellite does not have one), but only a conductive layer in its atmosphere or interior to couple with the magnetic field lines of Jupiter's field and excite their perturbations.

Inspired by these observations of the Jupiter system, Shkolnik et al. (2003, 2005) observed a sample of stars hosting hot Jupiters and found some evidence of chromospheric hot spots rotating in phase with the orbital motion of the planet rather than with the (different) rotation period of the host star. Successive observations provided further evidence of the phenomenon in the same stars (Shkolnik et al. 2008; Gurdemir et al. 2012; Cauley et al. 2018), but revealed that it is clearly non-stationary because it appears in some observing seasons, while it is absent in others. Moreover, the enhancement in the chromospheric flux associated with the hot spots is of only $\approx$0.5–1% requiring telescopes at least of the 2–4-m class and a dedicated intensive campaign to properly sample the rotational and orbital phases even in the case of bright targets. Therefore, progress in the acquisition of new data has been slow.

Theoretical models of the star-planet interaction, based on the excitation and dissipation of Alfvén waves by hot Jupiters in the framework of the Alfvén-wing model, have recently been proposed by Saur et al. (2013) and Saur (2018). They predict dissipated powers up to 10^{19} W in the case of the hot Jupiter systems that appear to be at least one order of magnitude smaller than the power emitted in the hot spot of HD 179949 as observed by Shkolnik et al. (2005).

Numerical models of the Alfvén wings in the case of magnetized or unmagnetized planets have been computed by, e.g., Strugarek et al. (2014), Strugarek et al. (2015), and Strugarek (2016). They show the difference between the cases in which the planet has no intrinsic magnetic field (unipolar inductor) and when it has one (dipolar inductor) with the strength of the interaction being generally greater in the latter case because of the larger effective cross-section provided by the planetary magnetosphere. In the case of the dipolar inductor, the relative orientation of the planet and wind magnetic fields is relevant making the available power change by at least one order of magnitude between the aligned and anti-aligned configurations. The interaction not only dissipates energy in the atmosphere of the star but also transfer angular momentum between the stellar spin and the orbit, thus potentially affecting the evolution of stellar rotation and star-planet separation with the most relevant effects concentrated in the pre-main-sequence and early main-sequence phases of the evolution of the star when its magnetic field is stronger (Lovelace et al. 2008; Strugarek et al. 2017). A review of the results of these numerical models has been provided by Strugarek (2018). They confirm the results of the analytical models by Saur et al. (2013) regarding the power made available to account for chromospheric hot spots in the Alfvén-wing scenario, thus suggesting that a different process is necessary to interpret the observations by Shkolnik et al. (2005).

2.5.2 Stellar Coronal Magnetic Fields

Motivated by the limitation of the Alfvén-wing models to account for the power irradiated in chromospheric hot spots, we investigate a different class of models considering a non-wave-like interaction between the stellar and the planetary magnetic fields.[11] In the coronae of late-type stars, the magnetic field is dominating the energetics and the dynamics of the plasma in the closed loops and in general up to a distance from the star where $\beta \equiv 2\mu p/B^2 < 1$, with p the thermal pressure of the plasma. Above a certain distance from the stellar center, the β parameter becomes greater than the unity and the plasma flow takes control of the geometry of the field lines that are frozen in it owing to the extremely high electric conductivity of the plasma at coronal temperature (Priest 1984; Wiegelmann et al. 2017). In the case of the Sun, this is estimated to happen at a radius of about $\sim 2.5\ R_\odot$ and, above that level, the coronal field lines become open because the plasma pressure cannot be counterbalanced by the Lorentz force. However, a complete control of the plasma flow over the magnetic field is reached only at the Alfvén radius, which is estimated to be at least on the order of 10–15 $R_\odot$ in the case of the Sun (e.g., Weber and Davis 1967).

In this simplified model, the solar magnetic field can be considered to have closed field lines up to about $\sim 2.5\ R_\odot$ and radial field lines above this level, called the *source surface* in coronal field models. The simplest model is obtained by considering the minimum-energy configuration of the field corresponding to the boundary conditions fixed at the photosphere and assuming the continuity of the radial field across the source surface (because $\nabla \cdot \mathbf{B} = 0$) and a purely radial field above it. Mathematically, the minimum-energy configuration corresponds to a potential field that satisfies the Laplace equation $\nabla^2 \mathbf{B} = 0$ with the boundary conditions fixed at the photosphere

---

[11] For the case of planets without an intrinsic field see, e.g., Laine and Lin (2012).

and at the source surface where the field becomes purely radial. This simple model, called *potential field source-surface* (PFSS) model, has been widely applied to compute the global magnetic field configuration of the Sun starting from photospheric magnetograms (Altschuler and Newkirk 1969; Wiegelmann et al. 2017). In more active stars, the radius of the source surface can be remarkably larger than in the case of the Sun because the magnetic field is stronger (e.g., See et al. 2017).

The main limitation of the PFSS model is the assumption of a potential coronal field, that is, the minimum energy field for the given boundary conditions. The true coronal field is not potential, otherwise there would be no energy to power phenomena such as flares and coronal mass ejection. Although the global field geometry is reproduced rather well by the PFSS model, it is useful to consider a class of more general non-potential fields for the coronae of stars hosting exoplanets, at least for the regions closer to the star where the magnetic field dominates the energetics and the dynamics of the plasma.

We adopt the simplifying assumptions of neglecting the thermal pressure of the plasma p, its kinetic pressure $\rho v^2/2$, and the gravity, so the only remaining force is the Lorentz force, the expression of which is $\mathbf{J} \times \mathbf{B}$ per unit volume, where $\mathbf{J} = (\nabla \times \mathbf{B})/\mu$ is the current density as given by the Ampere law because the displacement current is negligible in MHD given that the plasma velocity v is much smaller than the speed of light. An equilibrium configuration of the field then requires the vanishing of the Lorentz force because it cannot be balanced by any other force in the regime characteristic of a stellar corona where the β parameter is much smaller than the unity (see Priest 1984; Wiegelmann et al. 2017 for more details):

$$(\nabla \times \mathbf{B}) \times \mathbf{B} = 0. \tag{2.45}$$

The magnetic field satisfying equation (2.45) is said to be *force-free*. It satisfies also the equivalent equation:

$$\nabla \times \mathbf{B} = \alpha \mathbf{B}, \tag{2.46}$$

where α is a scalar function of the position. In other words, the general solution of Eq. (2.45) implies that the current density be parallel to the field in the domain where the force-free condition applies.[12] By applying the solenoidal condition ($\nabla \cdot \mathbf{B} = 0$) to both sides of Eq. (2.46), we get:

$$\mathbf{B} \cdot \nabla \alpha = 0 \tag{2.47}$$

implying that α is constant along each field line. Therefore, two general kinds of solutions to Eq. (2.45) are possible: those where α is the same along all the field lines, that is, it is a constant (*linear force-free fields*) and those where α is different along different field lines (*non-linear force-free fields*).

---

[12] Chandrasekhar (1961) showed that the only field that is force-free everywhere is a vanishing field ($\mathbf{B} = 0$). Therefore, the force-free condition (2.45) cannot be satisfied everywhere and the field must be stressed at some boundary.

Linear force-free fields satisfy the vector Helmoltz equation (e.g., Morse and Feshbach 1953) that is obtained by taking the curl of Eq. (2.45), applying the solenoidal condition, and Eq. (2.46):
$$\nabla^2 \mathbf{B} + \alpha^2 \mathbf{B} = 0. \tag{2.48}$$

A general analytic solution of Eq. (2.48) in spherical geometry has been given by Chandrasekhar and Kendall (1957), thanks to its linearity. On the other hand, no general solution is known for the non-linear force-free equation (2.46), despite its disarming simplicity (cf. Priest 1984).

Further progress in the study of force-free fields is possible by considering the role of *magnetic helicity*. It is a quantity that is conserved in closed MHD systems as a consequence of the gauge invariance in the definition of the magnetic vector potential, that is, the vector field $\mathbf{A}$ such that $\mathbf{B} \equiv \nabla \times \mathbf{A}$. If the magnetic field lines are confined inside a closed surface Σ enclosing a finite volume V_Σ, the quantity:

$$H(\mathbf{B}) = \int_{V_\Sigma} \mathbf{A} \cdot \mathbf{B} \, dV \tag{2.49}$$

called *magnetic helicity* of the field $\mathbf{B}$, is conserved in ideal MHD, that is when the conductivity of the plasma is infinite and the frozen-in condition is perfectly satisfied (cf. Priest 1984). Moreover, the minimum-energy configuration for the given helicity is a linear force-free field with that helicity (Priest 1984). Nevertheless, a real stellar corona is not a closed domain and the conductivity of the plasma, although very high, is finite. Berger and Field (1984) and Berger (1985) defined a *relative helicity* to overcome the first problem. It is defined in an open volume such as a stellar corona and it is based on the comparison between the given field and the potential field $\mathbf{B}_p \equiv \nabla \times \mathbf{A}_p$ having the same normal component at the surface of the star:

$$H_R(\mathbf{B}) = \int_V (\mathbf{A} + \mathbf{A}_p) \cdot (\mathbf{B} - \mathbf{B}_p) \, dV, \tag{2.50}$$

where V is the infinite volume outside the stellar surface. The effects of the finite conductivity and the magnetic energy dissipation in the current sheets of a stellar corona were found to be negligible, thus we can assume that the relative magnetic helicity is conserved during the evolution of the coronal field with fixed boundary conditions at the photosphere (cf. Berger 1984; Heyvaerts and Priest 1984).

The minimum-energy property of the linear force-free fields was extended by Heyvaerts and Priest (1984) to stellar coronal fields by means of the definition of relative helicity, in the sense that the minimum energy configuration in a confined domain with a given relative helicity is a linear force-free field. We can assume that, for fixed photospheric boundary conditions, the coronal field evolves to reach that minimum-energy configuration by making a transition from an initially non-linear force-free configuration to the linear one with the given relative helicity. Note that the absolute minimum energy configuration for the assigned boundary conditions is a potential field, but it cannot be reached unless the coronal field can get rid of its helicity because the relative helicity of the potential field is zero by definition. Such a transition can occur in major solar flares where the coronal field opens and a coronal mass ejection carries away most of the helicity of the initial field, thus allowing a transition to the absolute minimum-energy potential configuration (see, e.g., Lanza 2018 for more details).

2.5.3 Interactions Between the Stellar Coronal Fields and a Close-By Planet

Considering the simplified force-free model for the stellar coronal field introduced in Sect. 2.5.2 and assuming that the planet has its own magnetic field, we can have two kinds of stationary configurations. If the stellar coronal field is potential, that is, the force-free parameter $\alpha = 0$ in Eqs. (2.46) and (2.48), it is possible to form a magnetic loop interconnecting the star and the planet because the planetary field is potential, given that its atmosphere is largely neutral and dissipates electric currents, so that $\mathbf{J} = 0$. Therefore, given that α must be constant along a given field line, $\alpha = 0$ on the field lines interconnecting the star and the planet. Conversely, if the stellar field is force-free and non-potential, the two flux systems of the star and of the planet reconnect at the boundary where they come into contact, but no steady interconnecting loop can be formed because the condition $\alpha \neq 0$ in the stellar corona cannot match the potential field condition $\alpha = 0$ close to the planet (cf. Lanza 2009, 2012, 2013). This is illustrated in Fig. 2.16.

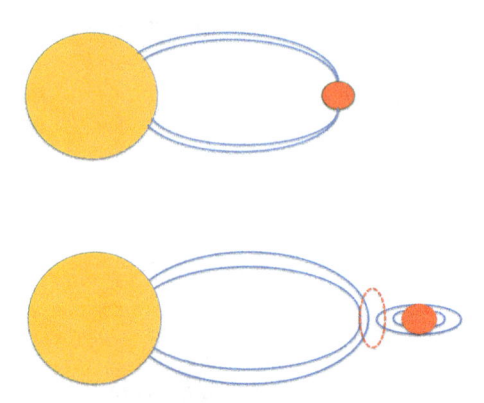

Fig. 2.16 Magnetic field configurations, in the case of an interconnecting magnetic flux system with $\alpha = 0$ (upper panel), or in the case of two topologically separated star-planet flux systems (lower panel). Illustrative field lines are plotted in blue, while the star is in orange-yellow and the planet in red. The red dashed ellipse encircles the region of magnetic reconnection between the stellar coronal field and the planetary field

In the case of an interconnecting loop, we can estimate the maximum power that is conveyed into the loop by the orbital motion of the planet by considering the flux of the Poynting vector across the base of the loop itself. We consider the simplified model by Adams (2011) and assume that there is no ionosphere, thus the field at the base of the loop is the same as that at the surface of the planet. The power made available by the orbital motion of the planet is:

$$P = \frac{2\pi}{\mu} f_{\rm AP} R_{\rm p}^2 B_{\rm p}^2 v_0, \qquad (2.51)$$

where μ is the magnetic permeability of the plasma, $f_{\rm AP}$ the fraction of the hemisphere of the planet covered by the interconnecting field lines, $R_{\rm p}$ the radius of the planet, $B_{\rm p}$ the intensity of the planetary field at its poles, and v_0 the relative velocity between the footpoints of the interconnecting loop (see Adams 2011; Lanza 2013). The storage of magnetic energy into the loop must be accompanied by a simultaneous dissipation that keeps it in the minimum energy potential state, otherwise no stationary interconnecting configuration is possible, at least in this simplified model.

On the other hand, in the case that the two flux systems of the planet and of the star are topologically separated, magnetic energy can be released by reconnection only at the boundary between the planetary field and the stellar field due to the orbital motion of the planet. In this case, we have (cf. Lanza 2009, 2012):

$$P = \gamma \frac{\pi}{\mu} \left[B_{\rm s}\left(a, \frac{\pi}{2}\right)\right]^2 R_{\rm m}^2 v_{\rm rel}, \qquad (2.52)$$

where $0 < \gamma < 1$ is a factor depending on the relative orientation of the field lines in the reconnection region having its maximum value for oppositely directed lines, $B_{\rm s}(a, \pi/2)$ the stellar coronal field at the star-planet separation a on the stellar equatorial plane, $R_{\rm m}$ the radius of the planet magnetosphere, and $v_{\rm rel}$ the relative velocity between the two flux systems at the reconnection interface. The radius of the magnetosphere $R_{\rm m}$ is derived from the force balance at the boundary between the stellar field and the planetary field (see Lanza 2009, 2012).

The observations of chromospheric hot spots moving in phase with the orbital motion of hot Jupiters can be calibrated to derive the power dissipated by the spots themselves. Assuming that the process of energy dissipation is similar to that of solar flares, Cauley et al. (2019) derived the powers released by the magnetic star-planet interaction in the case of the hot Jupiter systems HD 179949, HD 189733, τ Bootis, and υ Andromedae. They found dissipated powers between 0.4×10^{20} and 1.5×10^{20} W.

These stars have measurements of their surface fields, obtained by means of spectropolarimetry (Moutou et al. 2018), that range between 2.5 and 27 G. Since the orbital elements are well known and the radius of the hot Jupiters can be estimated, it is possible to compare the outcomes of different star-planet interaction models in terms of available power. The Alfvén-wing model requires a planetary field at least 4,000 times greater than Jupiter's field to deliver enough power. Similarly, unrealistically large planetary fields are required by the reconnection model in the case of two separated flux systems as in Eq. (2.52). Among the proposed models, only the interconnecting loop with steady dissipation appears capable of providing the required powers with realistic magnetic field strengths.

By applying Eq. (2.51), Cauley et al. (2019) found magnetic field strengths at the poles of these hot Jupiters between $\sim$20 and $\sim$120 G. These are about $10 - 100$ times larger than the values predicted by classic dynamo models, but they are in agreement with recently proposed scaling laws relating magnetic field strength with the internal heat flux in giant planets that have been suggested considering the radius inflation observed in hot Jupiters. Therefore, the observations of chromospheric hot spots, in addition to provide the first (indirect) measurements of the magnetic fields in hot Jupiters, have the potential to shed light on the mysterious phenomenon of radius inflation in those close-by planets (Cauley et al. 2019). Note that a direct measure of the magnetic field of hot Jupiters may come from radio observations (e.g., Zarka 2018; Trigilio et al. 2017; Vidotto and Donati 2017; Trigilio et al. 2018). In particular, observations in the low frequency bands (150–300 MHz) accessible to LOFAR promise to provide relevant results (Turner et al. 2021b; de Gasperin et al. 2021). Related observations include the detection of electron cyclotron maser radio emission from late-type stars, possibly powered by magnetic star-planet interactions of the kinds introduced above (Vedantham et al. 2020; Pérez-Torres et al. 2021).

An alternative approach to estimate the magnetic fields of hot Jupiters has been proposed by Vidotto et al. (2010b) and is based on the asymmetric transit profile observed in WASP-12 in the UV with the Hubble Space Telescope showing an early ingress with respect to the optical transit (Fossati et al. 2010). The early ingress has been attributed to the extra absorption of a bow shock formed in front of the planet, owing to its fast orbital motion through the stellar wind, and the stand-off distance, measured from the time delay between the UV and optical transit profiles, has been used to estimate a planetary magnetic field strength of $\approx$25 G in the hot Jupiter (Vidotto et al. 2011b).[13]

A consequence of the energy released by magnetic reconnection is an increase in the evaporation rate of the planet because such an energy can be conveyed by thermal conduction and accelerated particles into the planet's atmosphere where it is added to the energy deposited by the stellar EUV flux. The contribution from magnetic star-planet interaction may dominate over the stellar EUV flux for planets closer than 0.03–0.05 au, depending on the level of stellar activity of the host, the geometry of its coronal field, and the stratification of the planetary atmosphere (Buzasi 2013; Lanza 2013). Kawahara et al. (2013) suggested that this magnetically powered evaporation can be modulated by the uneven longitudinal distribution of the hosting star active regions and provided a possible example of this phenomenon. Furthermore, star-planet magnetic reconnection is important for the chemistry of the atmospheres of

---

[13] Another indirect method to measure the magnetic fields of hot Jupiters from the longitudinal temperature distributions in their atmospheres, derived from their phase curves, has been proposed by Rogers (2017).

close-by planets because it accelerates particles that can produce chemical reactions leading to the formation of several molecules (see, e.g., Rimmer et al. 2014).

The steady reconnection produced by the orbital motion of the planet inside the outer stellar corona produces a continuous energy dissipation. Therefore, it is conceivable that the coronal field can adjust itself in order to minimize this continuous dissipation reaching the minimum energy configuration compatible with its given magnetic helicity within a finite volume, that is a linear force-free field. Lanza (2009) and Lanza (2012) have explored such minimum energy configurations. They have several interesting properties, including the possibility of azimuthally twisted magnetic field lines that can account for the observed phase lag between planets and chromospheric hot spots, not explained by axisymmetric potential field models (cf. also Lanza 2008), and the possibility of storing plasma evaporated from the planet into azimuthal magnetic flux ropes encircling the star, thus explaining the observations of circumstellar absorption in systems such as WASP-12 (see Fig. 2.17). Finally, a global transition of the stellar field from a non-linear to a linear force-free configuration triggered by a close-by planet can release energy to power large stellar flares (cf. Lanza 2009, 2012, 2018). Flare activity phased with the hot Jupiter orbit has indeed been proposed in HD 189733 (Pillitteri et al. 2011, 2014a, 2015; Gao 2021) and HD17156 (Maggio et al. 2015).

Fig. 2.17 Meridional section of a linear force-free field in the corona of a star with a close-by planet indicated by the orange dot. The field lines are plotted in green, while in red is the matter evaporated from the planet and accumulated in the minimum of the gravitational potential along the field lines that come close to the star. The coordinates x in the equatorial plane of the star and z along its rotation axis are given in units of the stellar radius R_*. (Copyright by ESO, figure from Lanza, A. F., 2014, *Astronomy & Astrophysics*, vol. 472, L6)

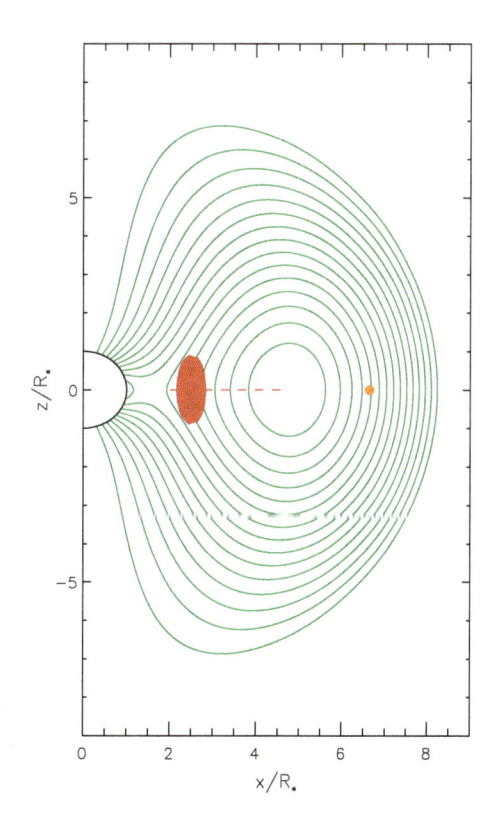

Numerical models overcome the simplifications of the analytical models considered above and can include the effects of plasma pressure and gravity. Cohen et al. (2009, 2011a, b) considered models of interactions based on the measured magnetic fields in the photosphere of HD 189733, confirming that magnetic field configurations interconnecting the star with the planet can exist and release a large amount of power when they are stressed by the relative motion of their footpoints. Models of stellar magnetized winds and their interactions with close-by planets have been proposed by, for example, Vidotto et al. (2009, 2010a, 2012, 2014); Vidotto and Bourrier (2017); Vidotto (2018); Garraffo et al. (2016, 2017); Cohen et al. (2014); Strugarek et al. (2017). They show the effects of the different MHD regimes (super-Alfvénic and sub-Alfvénic) that can be experienced by a planet orbiting close to its star because of the changing intensity and geometry of the stellar photospheric fields that can be mapped by spectropolarimetry. The impact of the different regimes on planetary evaporation can be relevant because in the super-Alfvénic case the planetary magnetosphere is open in the direction opposite to the bow shock facing the star, thus enhancing the escape of atmospheric gas (see Sect. 2.4). Conversely, when the regime is sub-Alfvénic, the planetary magnetosphere is almost closed and the atmospheric escape is reduced.

A numerical study of the different kinds of star-planet interactions, including tidal effects, planetary evaporation, and magnetic fields, was presented by Matsakos et al. (2015). They found four different regimes according to the ratios of the different length scales, that is, the equilibrium distance between the stellar and planetary winds R_w, the radius of planetary magnetosphere R_m, and the Roche-lobe radius of the planet R_t. When the stellar wind dominates and the planet evaporation is moderate, a bow-shock develops at the boundary of the planetary magnetosphere facing the star, while an evaporation tail forms along the orbit of the planet. On the other hand, if the planetary evaporation is strong, accretion of the evaporated material onto the star and a dense plasma tail along the orbit of the planet develop. The material falling on the star does not follow a radial trajectory, but there is a shift in phase between the planet and the impact hot spot on the surface of the star, even larger than 90°. This mechanism has been considered as an alternative to the purely star-planet magnetic interaction previously proposed to account for flares phased to the orbital motion of the hot Jupiter in HD 189733 (Pillitteri et al. 2015).

2.5.4 Statistical Analyses of Star-Planet Interactions and Some Intriguing Cases

In principle, if close-by planets are capable of increasing the level of activity of their host stars either directly by star-planet magnetic interaction or indirectly by preventing the braking of stellar rotation (cf. Sect. 2.3.4), a statistical study should provide evidence of such an enhancement. However, the enhancement in the chromospheric emission is tiny with chromospheric hot spots providing less than 1% excess flux (Shkolnik and Llama 2018) and the effects on the coronal and transition region emissions are difficult to be identified unambiguously because of several observational biases and systematic effects that tend to produce spurious correlations (e.g., Poppenhaeger et al. 2010; Poppenhaeger and Schmitt 2011; Miller et al. 2015; France et al. 2018).

The conclusion would be that there is no clear general statistical evidence for star-planet interactions, although Miller et al. (2015) found some evidence of a correlation between the X-ray luminosity and the parameters measuring the strength of magnetic and/or tidal interaction in a subsample of systems with close-by massive planets. Nevertheless, there are some intriguing cases that do not fit into this general scheme and suggest that the reality can be more complicated than the simple view generally adopted when we look for this kind of statistical correlations.

CoRoT-2 and HD 189733 are two stars accompanied by hot Jupiters that have rotation periods of 4.52 and 11.8 days, respectively, and levels of activity in agreement with their rotation rates. They have visual companions that are M-type dwarfs the activity levels of which can be used to assign them ages of several Gyrs. This is in remarkable tension with the younger ages attributed to CoRoT-2 and HD 189733 on the basis of their rotation and level of activity (Poppenhaeger and Wolk 2014). CoRoT-2 and HD 183733 therefore could be more active than they would be if they had no close-by massive planets.

To make things even more complicated there is the case of WASP-18, an F6V star with a very massive ($10.4 \pm 0.4\,M_J$) hot Jupiter on an extremely close orbit with $a = 0.0205$ au. The age of the star is estimated between 0.5 and 2.0 Gyr on the basis of isochrone fitting and lithium absorption in the spectrum, thus the star should display easily detectable coronal and chromospheric emissions. Contrary to this prediction, WASP-18 is unusually inactive with a chromospheric activity index $\log R'_{HK} = -5.15$, indistinguishable from the basal level, and undetected coronal EUV and X-ray emissions (Pillitteri et al. 2014b; Fossati et al. 2018). The expected level of coronal emission, based on the observations of stars of similar spectral type and rotation period ($v \sin i = 10.9 \pm 0.7$ km/s for WASP-18), is two orders of magnitude greater than the observed upper limit. The low level of chromospheric emission could be attributed to a circumstellar torus of plasma absorbing in the chromospheric lines as in the case of WASP-12, but the torus cannot absorb EUV and X-ray radiations. Therefore, the inescapable conclusion is that WASP-18 is intrinsically much less magnetically active than expected for a star with its parameters. This suggests that the very close-by and massive planet is somehow responsible for shutting off the stellar dynamo, although a physical model for this process has not been proposed yet (see discussion in Pillitteri et al. 2014b and Fossati et al. 2018).

In conclusion, the observations provide examples where a close-by massive planet is probably enhancing stellar activity, such as in CoRoT-2 and HD 183733, and cases where a close-by massive planet is shutting off the stellar dynamo as in WASP-18. Clearly, a much wider and well-characterized sample of hot Jupiter systems is required before we can draw definitive conclusions. For the moment, the tentative conclusion is that close-by massive planets can act differently in different systems, thus complicating the search for any statistical evidence of star-planet interaction affecting stellar magnetic activity.

Acknowledgements The author is grateful to the Scientific Organizing Committee of the Third Advanced School on Exoplanetary Science on "Demographic of Exoplanetary Systems" for their kind invitation to talk about star-planet interactions and their warm hospitality. Special thanks are due to Dr. Katia Biazzo for a critical reading of the manuscript and several valuable suggestions that helped to improve this contribution. Support from INAF through the *Progetto Premiale Frontiera* (PI Dr. Isabella Pagano) is gratefully acknowledged. The author is grateful to his colleagues participating in the research project GAPS (Global Architecture of Planetary Systems) for discussions on star-planet interactions on several occasions. This research has made use of the Exoplanet Orbit Database and the Exoplanet Data Explorer at exoplanets.org.

References

Adams, F.C.: Astrophys. J. **730**, 27 (2011)
Airapetian, V.S., Barnes, R., Cohen, O., et al.: (2019). https://ui.adsabs.harvard.edu/abs/2020IJAsB..19..136A/abstract
Altschuler, M.D., Newkirk, G.: Solar Phys. **9**, 131 (1969)
Amard, L., Palacios, A., Charbonnel, C., et al.: Astron. Astrophys. **587**, A105 (2016)
Aschwanden, M.J.: Physics of the Solar Corona. Springer Internation AG (2005)
Bailey, E., Batygin, K.: Astrophys. J. Lett. **866**, L2 (2018)
Barker, A.J.: Monthly Not. R. Astron. Soc. **498**, 2270 (2020)
Barnes, S.A.: Astrophys. J. **586**, 464 (2003)
Barnes, S.A.: Astrophys. J. **669**, 1167 (2007)
Barnes, S.A., Kim, Y.-C.: Astrophys. J. **721**, 675 (2010)
Barnes, S.A., Weingrill, J., Fritzewski, D., et al.: Astrophys. J. **823**, 16 (2016)
Barr, A.C., Dobos, V., Kiss, L.L.: Astron. Astrophys. **613**, A37 (2018)
Benbakoura, M., Réville, V., Brun, A.S., et al.: Astron. Astrophys. **621**, A124 (2019)
Berger, M.A.: Geophys. Astrophys. Fluid Dyn. **30**, 79 (1984)
Berger, M.A.: Astrophys. J. **59**, 433 (1985)
Berger, M.A., Field, G.B.: J. Fluid Mech. **147**, 133 (1984)
Bolmont, E., Mathis, S.: Celestial Mech. Dyn. Astron. **126**, 275 (2016)
Bolmont, E., Breton, S.N., Tobie, G., et al.: Astron. Astrophys. **644**, A165 (2020)
Bonomo, A.S., Desidera, S., Benatti, S., et al.: Astron. Astrophys. **602**, A107 (2017)
Bourrier, V., Lecavelier des Etangs, A.: Characterizing evaporating atmospheres of exoplanets. In: Deeg, H., Belmonte, J.A. (eds.) Handbook of Exoplanets. Springer International Publishing AG, id.148 (2018)
Bouvier, J., Forestini, M., Allain, S.: Astron. Astrophys. **326**, 1023 (1997)
Brandenburg, A., Subramanian, K.: Phys. Rep. **417**, 1 (2005)
Brown, D.J.A., Collier Cameron, A., Hall, C., Hebb, L., Smalley, B.: Monthly Not. R. Astron. Soc. **415**, 605 (2011)
Brown, D.J.A.: Monthly Not. R. Astron. Soc. **442**, 1844 (2014)
Buzasi, D.: Astron. J. Lett. **765**, L25 (2013)
Cauley, P.W., Shkolnik, E.L., Llama, J., et al.: Astron. J. **156**, 262 (2018)
Cauley, P.W., Shkolnik, E.L., Llama, J., Lanza, A.F.: Nat. Astron. **3**, 1128 (2019)
Chandrasekhar, S.: Hydrodynamic and Hydromagnetic Stability. Clarendon Press (1961)
Chandrasekhar, S., Kendall, P.C.: Astrophys. J. Lett. **126**, 457 (1957)
Cohen, O., Drake, J.J., Kashyap, V.L., et al.: Astrophys. J. Lett. **704**, L85 (2009)
Cohen, O., Drake, J.J., Kashyap, V.L., et al.: Astrophys. J. Lett. **723**, L64 (2010)
Cohen, O., Kashyap, V.L., Drake, J.J., et al.: Astrophys. J. **733**, 67 (2011)
Cohen, O., Kashyap, V.L., Drake, J.J., et al.: Astrophys. J. **738**, 166 (2011)
Cohen, O., Drake, J.J., Glocer, A., et al.: Astrophys. J. **790**, 57 (2014)

Collier Cameron, A., Jardine, M.: Monthly Not. R. Astron. Soc. **476**, 2542 (2018)
Correia, A.C.M., Laskar, J.: Tidal evolution of exoplanets. In: Exoplanets, p. 239. University of Arizona Press, Seager (2010)
Correia, A.C.M., Laskar, J.: Nature **411**, 767 (2001)
Damiani, C., Lanza, A.F.: Astron. Astrophys. **535**, A116 (2011)
Damiani, C., Lanza, A.F.: Astron. Astrophys. **574**, A39 (2015)
Dawson, R.I., Johnson, J.A.: Ann. Rev. Astron. Astrophys. **56**, 175 (2018)
de Gasperin, F., Williams, W.L., Best, P., et al.: Astron. Astrophys. **648**, A104 (2021)
Debrecht, A., Carroll-Nellenback, J., Frank, A., et al.: Monthly Not. R. Astron. Soc. **478**, 2592 (2018)
Driscoll, Peter E.: Planetary interiors, magnetic fields, and habitability. In: Deeg, H., Belmonte, J.A. (eds.) Handbook of Exoplanets. Springer International Publishing AG, id.76 (2018)
Eggleton, P.P., Kiseleva, L.G., Hut, P.: Astrophys. J. **499**, 853 (1998)
Figueira, P., Oshagh, M., Adibekyan, VZh., Santos, N.C.: Astron. Astrophys. **572**, A51 (2014)
Ford, E.B., Rasio, F.A.: Astron. J. Lett. **638**, L45 (2006)
Fossati, L., Haswell, C.A., Froning, C.S., et al.: Astron. J. Lett. **714**, L222 (2010)
Fossati, L., Ayres, T.R., Haswell, C.A., et al.: Astron. J. Lett. **766**, L20 (2013)
Fossati, L., Ingrassia, S., Lanza, A.F.: Astron. J. Lett. **812**, L35 (2015)
Fossati, L., Koskinen, T., France, K., et al.: Astron. J. **155**, 113 (2018)
France, K., Arulanantham, N., Fossati, L., et al.: Astron. J. **239**, 16 (2018)
Fulton, B.J., Petigura, E.A., Howard, A.W., et al.: Astron. J. **154**, 109 (2017)
Gallet, F., Bouvier, J.: Astron. Astrophys. **556**, A36 (2013)
Gallet, F., Bouvier, J.: Astron. Astrophys. **577**, A98 (2015)
Gallet, F., Delorme, P.: Astron. Astrophys. **626**, A120 (2019)
Gallet, F., Bolmont, E., Mathis, S., et al.: Astron. Astrophys. **604**, A112 (2017)
Gallet, F., Bolmont, E., Bouvier, J., et al.: Astron. Astrophys. **619**, A80 (2018)
Gao, Y.: Astrophys. J. **161**, 259 (2021)
Garraffo, C., Drake, J.J., Cohen, O.: Astrophys. J. Lett. **833**, L4 (2016)
Garraffo, C., Drake, J.J., Cohen, O., et al.: Astrophys. J. Lett. **843**, L33 (2017)
Gold, T., Soter, S.: Icarus **11**, 356 (1969)
Gurdemir, L., Redfield, S., Cuntz, M.: PASA **29**, 141 (2012)
Han, E., Wang, S.X., Wright, J.T., et al.: Proc. Astron. Soc. Pac. **126**, 943 (2014)
Hansen, B.M.S.: Astrophys. J. **757**, 6 (2012)
Hartman, J.D.: Astrophys. J. Lett. **717**, L138 (2010)
Haswell, C.A., Staab, D., Barnes, J.R., et al.: Nat. Astron. **4**, 408 (2020)
Hebb, L., Collier-Cameron, A., Loeillet, B., et al.: Astrophys. J. **693**, 1920 (2009)
Heyvaerts, J., Priest, E.R.: Astron. Astrophys. **137**, 63 (1984)
Husnoo, N., Pont, F., Mazeh, T., et al.: Monthly Not. R. Astron. Soc. **422**, 3151 (2012)
Hut, P.: Astron. Astrophys. **92**, 167 (1980)
Hut, P.: Astron. Astrophys. **99**, 126 (1981)
Jackson, B., Greenberg, R., Barnes, R.: Astrophys. J. **678**, 1396 (2008)
Kawahara, H., Hirano, T., Kurosaki, K., et al.: Astrophys. J. Lett. **776**, L6 (2013)
Kawaler, S.D.: Astrophys. J. **333**, 236 (1988)
Kopp, A., Schilp, S., Preusse, S.: Astrophys. J. **729**, 116 (2011)
Kubyshkina, D., Fossati, L., Erkaev, N.V., et al.: Astrophys. J. Lett. **866**, L18 (2018)
Kubyshkina, D., Vidotto, A.A., Fossati, L., et al.: Monthly Not. R. Astron. Soc. **499**, 77 (2020)
Laine, R.O., Lin, D.N.C.: Astrophys. J. **745**, 2 (2012)
Lainey, V.: Celestial mechanics and dynamical Astronomy **126**, 145 (2016)
Lainey, V., Arlot, J.-E., Karatekin, O., et al.: Nature **459**, 957 (2009)
Landi, E., Del Zanna, G., Young, P.R., et al.: 744, 99 (2012)
Lanza, A.F.: Astron. Astrophys. **487**, 1163 (2008)
Lanza, A.F.: Astron. Astrophys. **505**, 339 (2009)
Lanza, A.F.: Astron. Astrophys. **512**, A77 (2010)

Lanza, A.F.: Astron. Astrophys. **544**, A23 (2012)
Lanza, A.F.: Astron. Astrophys. **557**, A31 (2013)
Lanza, A.F.: Astron. Astrophys. **572**, L6 (2014)
Lanza, A.F.: Astron. Astrophys. **610**, A81 (2018)
Lanza, A.F.: Monthly Not. R. Astron. Soc. **497**, 3911 (2020)
Lanza, A.F., Shkolnik, E.L.: Monthly Not. R. Astron. Soc. **443**, 1451 (2014)
Lanza, A.F., Damiani, C., Gandolfi, D.: Astron. Astrophys. **529**, A50 (2011)
Laughlin, G.: Mass-radius relations of giant planets: the radius anomaly and interior models. In: Deeg, H., Belmonte, J.A. (eds.) Handbook of Exoplanets. Springer International Publishing AG, id.1 (2018)
Leconte, J., Chabrier, G., Baraffe, I., Levrard, B.: Astron. Astrophys. **516**, A64 (2010)
Leto, P., Trigilio, C., Buemi, C.S., et al.: Monthly Not. R. Astron. Soc. **469**, 1949 (2017)
Levrard, B., Winisdoerffer, C., Chabrier, G.: Astrophys. J. Lett. **692**, L9 (2009)
Linsky, J.: Lecture Notes in Physics. Springer International AG, 955 (2019)
Linsky, J.L.: Ann. Rev. Astron. Astrophys. **55**, 159 (2017)
Linsky, J.L., Fontenla, J., France, K.: Astrophys. J. **780**, 61 (2014)
Lovelace, R.V.E., Romanova, M.M., Barnard, A.W.: Monthly Not. R. Astron. Soc. **389**, 1233 (2008)
Luger, R., Sestovic, M., Kruse, E., et al.: Nat. Astron. **1**, 0129 (2017)
MacGregor, K.B., Brenner, M.: Astrophys. J. **376**, 204 (1991)
Maciejewski, G., Knutson, H.A., Howard, A.W., et al.: Acta Astron. **70**, 1 (2020)
Maciejewski, G., Niedzielski, A., Villaver, E., et al.: Astrophys. J. **889**, 54 (2020)
Maggio, A., Pillitteri, I., Scandariato, G., et al.: Astrophys. J. Lett. **811**, L2 (2015)
Mardling, R.A., Lin, D.N.C.: Astrophys. J. **573**, 829 (2002)
Mariska, J.T.: The Solar Transition Region. Cambridge University Press (1992)
Marzari, F.: Planet-planet gravitational scattering. In: Barnes, R. (ed.) Formation and Evolution of Exoplanets, Wiley, p. 223 (2010)
Mathis, S.: Tidal star-planet interactions: a stellar and planetary perspective. In: Deeg, H., Belmonte, J.A. (eds.) Handbook of Exoplanets. Springer International Publishing AG, id.24 (2018)
Mathis, S.: Astron. Astrophys. **580**, L3 (2015)
Matsakos, T., Uribe, A., Königl, A.: Astron. Astrophys. **578**, A6 (2015)
Matsumura, S., Takeda, G., Rasio, F.A.: Astrophys. J. Lett. **686**, L29 (2008)
Matt, S.P., Brun, A.S., Baraffe, I., et al.: Astrophys. J. Lett. **799**, L23 (2015)
Maxted, P.F.L., Serenelli, A.M., Southworth, J.: Astron. Astrophys. **577**, A90 (2015)
McQuillan, A., Mazeh, T., Aigrain, S.: Astrophys. J. Lett. **775**, L11 (2013)
Meibom, S., Mathieu, R.D.: Astrophys. J. **620**, 970 (2005)
Metcalfe, T.S., Egeland, R.: Astrophys. J. **871**, 39 (2019)
Metcalfe, T.S., Egeland, R., van Saders, J.: Astrophys. J. Lett. **826**, L2 (2016)
Miller, N., Fortney, J.J., Jackson, B.: Astrophys. J. **702**, 1413 (2009)
Miller, B.P., Gallo, E., Wright, J.T., et al.: Astrophys. J. **799**, 163 (2015)
Moffatt, H.K.: Magnetic Field Generation in Electrically Conducting Fluids. Cambridge University Press (1978)
Morse, P.M., Feshbach, H.: Methods of Theoretical Physics. McGraw-Hill Book Company (1953)
Moutou, C., Fares, R., Donati, J.-F.: Magnetic fields in planet-hosting stars. In: Deeg, H., Belmonte, J.A. (eds.) Handbook of Exoplanets. Springer International Publishing AG, id.21 (2018)
Murray, C.D., Dermott, S.F.: Solar System Dynamics. Cambridge University Press (1999)
Murray-Clay, R.A., Chiang, E.I., Murray, N.: Astrophys. J. **693**, 23 (2009)
Naoz, S.: Ann. Rev. Astron. Astrophys. **54**, 441 (2016)
Neubauer, F.M.: J. Geophys. Res. **85**, 1171 (1980)
Ogilvie, G.I.: Monthly Not. R. Astron. Soc. **429**, 613 (2013)
Ogilvie, G.I.: Ann. Rev. Astron. Astrophys. **52**, 171 (2014)
Ogilvie, G.I., Lin, D.N.C.: The. Astrophys. J. **661**, 1180 (2007)
Oranje, B.J.: Astron. Astrophys. **154**, 185 (1986)
Owen, J.E., Jackson, A.P.: Monthly Not. R. Astron. Soc. **425**, 2931 (2012)

Owen, J.E., Lai, D.: Monthly Not. R. Astron. Soc. **479**, 5012 (2018)
Owen, J.E., Wu, Y.: Astrophys. J. **847**, 29 (2017)
Pagano, I: Stellar activity. In Oswalt, T.D., Barstow, M.A. (eds.) Planets, Stars and Stellar Systems, Vol. 4, Springer Science+Business Media Dordrecht, p. 485 (2013)
Parker, E.N.: Astrophys. J. **128**, 664 (1958)
Patra, K.C., Winn, J.N., Holman, M.J., et al.: Astron. J. **154**, 4 (2017)
Patra, K.C., Winn, J.N., Holman, M.J., et al.: Astron. J. **159**, 150 (2020)
Peale, S.J., Cassen, P., Reynolds, R.T.: Science **203**, 892 (1979)
Pérez-Torres, M., Gómez, J.F., Ortiz, J.L., et al.: Astron. Astrophys. **645**, A77 (2021)
Pillitteri, I., Günther, H.M., Wolk, S.J., et al.: Astrophys. J. Lett. **741**, L18 (2011)
Pillitteri, I., Wolk, S.J., Lopez-Santiago, J., et al.: Astrophys. J. **785**, 145 (2014)
Pillitteri, I., Wolk, S.J., Sciortino, S., Antoci, V.: Astron. Astrophys. **567**, A128 (2014)
Pillitteri, I., Maggio, A., Micela, G., et al.: Astrophys. J. **805**, 52 (2015)
Piters, A.J.M., Schrijver, C.J., Schmitt, J.H.M.M., et al.: Astron. Astrophys. **325**, 1115 (1997)
Pizzolato, N., Maggio, A., Micela, G., et al.: Astron. Astrophys. **397**, 147 (2003)
Pont, F.: Monthly Not. R. Astron. Soc. **396**, 1789 (2009)
Poppenhaeger, K., Schmitt, J.H.M.M.: Astrophys. J. **735**, 59 (2011)
Poppenhaeger, K., Wolk, S.J.: Astron. Astrophys. **565**, L1 (2014)
Poppenhaeger, K., Robrade, J., Schmitt, J.H.M.M.: Astron. Astrophys. **515**, A98 (2010)
Preusse, S., Kopp, A., Büchner, J., Motschmann, U.: Astron. Astrophys. **434**, 31191 (2005)
Preusse, S., Kopp, A., Büchner, J., Motschmann, U.: Astron. Astrophys. **460**, 317 (2006)
Priest, E.R.: Solar Magneto-Hydrodynamics. D. Reidel Publishing Company (1984)
Rauer, H., Catala, C., Aerts, C., et al.: Exp. Astron. **38**, 249 (2014)
Raymond, J.C., Smith, B.W.: Astrophys. J. **35**, 419 (1977)
Reale, F., Landi, E.: Astron. Astrophys. **543**, A90 (2012)
Rimmer, P.B., Helling, Ch., Bilger, C.: Int. J. Astrobiol. **13**, 173 (2014)
Rogers, T.M.: Nat. Astron. **1**, 0131 (2017)
Rosner, R., Tucker, W.H., Vaiana, G.S.: Astrophys. J. **220**, 643 (1978)
Rüdiger, G., Hollerbach, R.: The Magnetic Universe: Geophysical and Astrophysical Dynamo Theory. Wiley-VCH, Weinheim (2004)
Sainsbury-Martinez, F., Wang, P., Fromang, S., et al.: Astron. Astrophys. **632**, A114 (2019)
Sanz-Forcada, J., Micela, G., Ribas, I., et al.: Astron. Astrophys. **532**, A6 (2011)
Saur, J.: Electromagnetic coupling in star-planet systems. In: Deeg, H., Belmonte, J.A. (eds.) Handbook of Exoplanets. Springer International Publishing AG, id.27 (2018)
Saur, J., Grambusch, T., Duling, S., et al.: Astron. Astrophys. **552**, A119 (2013)
See, V., Jardine, M., Vidotto, A.A., et al.: Monthly Not. R. Astron. Soc. **466**, 1542 (2017)
Shkolnik, E., Bohlender, D.A., Walker, G.A.H., Collier Cameron, A.: Astrophys. J. **676**, 628 (2008)
Shkolnik, E.L., Llama, J.: Signatures of star-planet interactions. In: Deeg, H., Belmonte, J.A. (eds.) Handbook of Exoplanets. Springer International Publishing AG, id.20 (2018)
Shkolnik, E., Walker, G.A.H., Bohlender, D.A.: Astrophys. J. **597**, 1092 (2003)
Shkolnik, E., Walker, G.A.H., Bohlender, D.A., et al.: Astrophys. J. **622**, 1075 (2005)
Sozzetti, A., Torres, G., Charbonneau, D., et al.: Astrophys. J. **664**, 1190 (2007)
Spada, F., Lanzafame, A.C., Lanza, A.F., et al.: Monthly Not. R. Astron. Soc. **416**, 447 (2011)
Spruit, H.C.: (2013) arXiv:1301.5572
Staab, D., Haswell, C.A., Barnes, J.R., et al.: Nat. Astron. **4**, 399 (2020)
Stix, M.: The Sun: An Introduction. Springer, Berlin (2004)
Strugarek, A.: Models of star-planet magnetic interaction. In: Deeg, H., Belmonte, J.A. (eds.) Handbook of Exoplanets. Springer International Publishing AG, id.25 (2018)
Strugarek, A.: Astrophys. J. **833**, 140 (2016)
Strugarek, A., Brun, A.S., Matt, S.P., Réville, V.: Astrophys. J. **795**, 86 (2015)
Strugarek, A., Brun, A.S., Matt, S.P., Réville, V.: Astrophys. J. **815**, 111 (2015)
Strugarek, A., Bolmont, E., Mathis, S., et al.: Astrophys. J. Lett. **847**, L16 (2017)
Teitler, S., Königl, A.: Astrophys. J. **786**, 139 (2014)

Tobie, G., Grasset, O., Dumoulin, C., et al.: Astron. Astrophys. **630**, A70 (2019)
Trammell, G.B., Arras, P., Li, Z.-Y.: Astrophys. J. **728**, 152 (2011)
Trigilio, C., Umana, G., Cavallaro, F., et al.: Monthly Not. R. Astron. Soc. **481**, 217 (2018)
Turner, J.D., Ridden-Harper, A., Jayawardhana, R.: Astron. J. **161**, 72 (2021a)
Turner, J.D., Zarka, P., Grießmeier, J.-M., et al.: Astron. Astrophys. **645**, A59 (2021b)
Van Eylen, V., Agentoft, C., Lundkvist, M.S., et al.: Monthly Not. R. Astron. Soc. **479**, 4786 (2018)
van Saders, J.L., Ceillier, T., Metcalfe, T.S., et al.: Nature **529**, 181 (2016)
Vaughan, A.H., Preston, G.W.: Proc. Astron. Soc. Pac. **92**, 385 (1980)
Vedantham, H.K., Callingham, J.R., Shimwell, T.W., et al.: Nat. Astron. **4**, 577 (2020)
Vidal-Madjar, A., Lecavelier des Etangs, A., Désert, J.-M., et al.: Nature **422**, 143 (2003)
Vidotto, Aline A.: Stellar coronal and wind models: impact on exoplanets. In: Deeg, H., Belmonte, J.A. (eds.) Handbook of Exoplanets. Springer International Publishing AG, id.26 (2018)
Vidotto, A.A.: Living reviews in solar physics **18**, 3 (2021)
Vidotto, A.A., Bourrier, V.: Monthly Not. R. Astron. Soc. **470**, 4026 (2017)
Vidotto, A.A., Donati, J.-F.: Astron. Astrophys. **602**, A39 (2017)
Vidotto, A.A., Opher, M., Jatenco-Pereira, V., Gombosi, T.I.: Astrophys. J. Lett. **703**, 1734 (2009)
Vidotto, A.A., Opher, M., Jatenco-Pereira, V., Gombosi, T.I.: Astrophys. J. Lett. **720**, 1262 (2010)
Vidotto, A.A., Jardine, M., Helling, Ch.: Astrophys. J. Lett. **722**, L168 (2010)
Vidotto, A.A., Jardine, M., Helling, Ch.: Monthly Not. R. Astron. Soc. **411**, L46 (2011)
Vidotto, A.A., Jardine, M., Helling, Ch.: Monthly Not. R. Astron. Soc. **414**, 1573 (2011)
Vidotto, A.A., Fares, R., Jardine, M., et al.: Monthly Not. R. Astron. Soc. **423**, 3285 (2012)
Vidotto, A.A., Jardine, M., Morin, J., et al.: Monthly Not. R. Astron. Soc. **438**, 1162 (2014)
Weber, E.J., Davis, L.: Astrophys. J. **148**, 217 (1967)
Wiegelmann, T., Petrie, G.J.D., Riley, P.: Space Sci. Rev. **210**, 249 (2017)
Wolf, M., Zasche, P., Kučáková, H., et al.: Astron. Astrophys. **549**, A108 (2013)
Wright, J.T.: Astron. J. **128**, 1273 (2004)
Wu, Y., Lithwick, Y.: Astrophys. J. **735**, 109 (2011)
Yee, S.W., Winn, J.N., Knutson, H.A., et al.: Astrophys. J. Lett. **888**, L5 (2020)
Zahn, J.-P.: Astron. Astrophys. **41**, 329 (1975)
Zahn, J.-P.: Astron. Astrophys. **57**, 383 (1977)
Zahn, J.-P.: Astron. Astrophys. **220**, 112 (1989)
Zahn, J.-P.: EAS Publ. Ser. **29**, 67 (2008)
Zarka, P.: Star-planet interactions in the radio domain: prospect for their detection. In: Deeg, H., Belmonte, J.A. (eds.) Handbook of Exoplanets. Springer International Publishing AG, id.22 (2018)

Part III
Close-In Exoplanets

Chapter 3
The Demographics of Close-In Planets

K. Biazzo, V. Bozza, L. Mancini, and A. Sozzetti

Abstract The large sample of presently-known exoplanetary systems orbiting within a few au from their parent stars has enabled detailed studies of their demographics, which provide essential constraints for improved understanding of the processes at work in planet formation and evolution. In this Chapter, we first summarize the strengths and weaknesses of the two detection methods that have unveiled the population of short- and intermediate-separation exoplanets, i.e. the radial velocity (RV) and photometric transit techniques. Secondly, we review the wealth of information on the statistical properties of close-in exoplanets gathered from Doppler and transit surveys, focusing on occurrence rates as a function of two fundamental physical properties, mass and radius, and orbital separation. Next, we discuss statistical trends and correlations in the orbital and structural properties of two classes of planets that are not found in our Solar System, super Earths and sub-Neptunes. Then, we present an overview of the global and planet-to-planet patterns in the demographics of multiple-plane systems. We conclude by discussing the (primarily dynamical)

K. Biazzo
INAF—Rome Astronomical Observatory, Via di Frascati, 33, 00078 Monte Porzio Catone, RM, Italy
e-mail: katia.biazzo@inaf.it

V. Bozza
Department of Physics "E.R. Caianiello", University of Salerno, Via Giovanni Paolo II 132, 84084 Fisciano, SA, Italy
e-mail: valboz@sa.infn.it

L. Mancini
Department of Physics, University of Rome "Tor Vergata", Via della Ricerca Scientifica 1, 00133 Rome, Italy
e-mail: lmancini@roma2.infn.it

A. Sozzetti (✉)
INAF—Turin Astrophysical Observatory, Via Osservatorio 20, 10025 Pino Torinese, TO, Italy
e-mail: alessandro.sozzetti@inaf.it

© Springer Nature Switzerland AG 2022
K. Biazzo et al. (eds.), *Demographics of Exoplanetary Systems*, Astrophysics and Space Science Library 466, https://doi.org/10.1007/978-3-030-88124-5_3

information content of the orbital architectures of close-in planets (eccentricities, obliquities), and by highlighting the peculiar properties of the class of ultra-short period exoplanets. Throughout this Chapter, we underline key aspects of the fundamental link between planetary and host stars' properties and stress relevant elements of the mutual feedback between observations and theory (This Chapter is based on the lecture series given in May 2019 by A. Howard.).

3.1 Introduction

Out of the > 4300 known exoplanets to-date, for about 3800 (87%) the measured orbital period is $P \lesssim 1$ year.[1] Of course, this reflects an observational bias intrinsic to the two most successful planet detection techniques, i.e. photometric transits and radial velocities, which are primarily sensitive to close-in exoplanets. Furthermore, the very slow orbital motion of planets in wide orbits makes the reconstruction of the orbital parameters a very time-taking endeavor.

Whatever the reason, we can clearly see that the available statistics for close-in exoplanets is nearly an order of magnitude richer than that of exoplanets on distant orbits. With thousands of planets available, we can thus not only distinguish basic exoplanetary populations by radius, mass, density, incident stellar flux, but we are also starting to be sensitive to fine structures in the distributions (e.g., the famous 'radius valley', see Sect. 3.4.2) that can help us understand the formation scenarios of different classes of exoplanets. Furthermore, the numerous detections of multiple systems provide precious examples of the possible final outcomes of the formation of planetary systems. The existence of planets on high eccentric orbits with high obliquities with respect to the host star's spin is indicative of non-trivial dynamical interactions among the planets in the same disk. Every single system is thus important in its own regard as an individual case study, besides adding one more brick to all distributions of the various orbital and physical parameters.

3.2 Radial Velocity and Transit Measurement Techniques

Here we briefly review the basic concepts of the two main detection methods responsible for uncovering the present-day sample of close-in exoplanets, which has enable the demographics studies discussed in the following sections of this chapter. We will refer the reader to more specialized sources when necessary.

---

[1] Source: http://exoplanet.eu.

3.2.1 Radial Velocities

3.2.1.1 Doppler Measurements

The first successful observational campaign that found a planet around a solar-type star was based on high-precision RV monitoring of 51 Peg (Mayor and Queloz 1995). The principle of the Doppler technique had been well-known for more than a century and had been used to detect and characterize spectroscopic binaries (Pickering 1890). The orbital motion of two bodies around the common center of mass determines a periodic modulation of the radial velocity v_r, which can be detected to very high accuracy in terms of the Doppler shift z in the absorption lines in the spectrum:

$$z \equiv \frac{\lambda - \lambda_0}{\lambda_0} = \gamma(1 + v_r/c) - 1, \qquad (3.1)$$

where λ_0 is the rest wavelength and γ is the relativistic Lorentz factor. For stellar binaries with sufficiently bright components, we can follow the absorption lines of both stars as they move periodically in opposite directions. In the case of a planetary system, we can only measure the RV variations of the host star, with a semi-amplitude $K_\star$ that is related to that of the planet by the law of barycenter: $K_\star = (M_p/M_\star) K_p$. If the radial velocity of a close-in terrestrial planet around a solar-type ($M_p/M_\star \simeq 10^{-5}$) has $K_p \sim 10^5$ m/s, the reflex velocity of the star would be of the order of $1\,\mathrm{m\,s^{-1}}$, which poses a severe challenge to current spectroscopic facilities.

In order to resolve such tiny displacements in the spectra, state-of-the-art instruments are typically echelle spectrographs (Bouchy et al. 2001) working at very high spectral resolution (i.e. with resolving power $R \sim 100{,}000$). It is interesting to note that a displacement of $30\,\mathrm{cm\,s^{-1}}$ in the spectrum may correspond to just a few atoms on the detector surface! Therefore, such displacements can only be detected by means of cross-correlation techniques on broad regions of the spectrum (Fellgett 1955; Griffin 1967; Queloz 1995). Besides being very precise, the measurement apparatus must be very stable over long timescales, in order to enable firm detection of longer-period planets. For example, the motion of the Sun induced by Jupiter has a semi-amplitude of $12\,\mathrm{m\,s^{-1}}$ over a period of 12 years! For more information on the details of RV measurements, we refer the reader to specific reviews (e.g., Hatzes 2016; Perryman 2018; Wright 2018).

3.2.1.2 Form of the Signal

If we neglect the mutual interactions between the planets of the system, the measured radial velocity of the host star will be the sum of the individual Keplerian contributions. For n planets, we have

$$v_r(t) = \sum_{j=1}^{n} K_j \left[\cos(\omega_j + f_j(t)) + e_j \cos \omega_j \right] + \gamma + \dot{\gamma}(t - t_0), \quad (3.2)$$

where t_0 is a reference epoch, e_j is the eccentricity of planet j, ω_j is the argument of the periastron, f_j is the true anomaly (angular position of the planet taken from the periastron), K_j is the semi-amplitude of the modulation, as detailed below. Equation (3.2) also contains a constant offset γ that corresponds to the stellar barycentric motion with respect to the Solar System, and a linear trend $\dot{\gamma}$, which takes into account possible systematic drifts in the measurement apparatus or long-term RV signals induced by long-period companions that cannot be modeled in detail within a finite duration campaign.

The semi-amplitude of the modulation is a function of the masses, period, eccentricity and inclination of the system. For a planet with mass M_p, in convenient units, it can be cast in the form

$$K_\star = \frac{28.4 \text{m s}^{-1}}{\sqrt{1 - e^2}} \frac{M_p \sin i}{M_{\text{Jup}}} \left(\frac{M_p + M_\star}{M_\odot} \right)^{-2/3} \left(\frac{P}{1 \text{ year}} \right)^{-1/3}, \quad (3.3)$$

where $M_\odot$ is the solar mass and $M_{\text{Jup}} \simeq 10^{-3} M_\odot$ the Jupiter mass.

This formula shows how the semi-amplitude of the RV signal decreases for longer period planets, following Keplerian laws. If the mass of the planet is negligible compared to the mass of the star, the third factor only depends on the mass of the host star. This implies that a conversion of the RV measure into a planetary mass requires accurate knowledge of the mass of the host star. This is typically obtained by means of techniques that match properties of the observed spectra to stellar model libraries (e.g., Torres et al. 2012). Of course, this procedure leaves some systematic uncertainty that is reflected in the ultimate error budget of the planetary mass.

Finally, we clearly see the well-known degeneracy between the mass of the planet and the inclination of the system. With RVs we can only derive a lower limit for the mass of the planet by fixing $\sin i = 1$ (edge-on system). However, without independent knowledge of the inclination, the mass of the planet can be anything from this lower limit to infinity ($\sin i = 0$ corresponds to a face-on system).

For hypothetical extrasolar analogues of the planets in our system, we get $K = 12 \text{ m s}^{-1}$ for Jupiter and $K = 0.09 \text{ m s}^{-1}$ for the Earth. Close-in planets are mildly favored by Kepler's third law. In fact, for a Jupiter-mass planet orbiting a solar-mass star with a period of 3 days we would have $K_\star = 140 \text{ m s}^{-1}$, while an Earth-mass planet orbiting in one day would have $K = 0.6 \text{ m s}^{-1}$. A more typical Super-Earth with $M_p = 10 M_\oplus$ and a period of 10 days would give $K = 3 \text{ m s}^{-1}$. We conclude that it is relatively easy to detect hot giants, while we need better than 1 m s^{-1} accuracy to reach the regime of detectability of Earths and Super-Earths or to detect planets in wide orbits.

The shape of the radial velocity modulation, as shown in Eq. (3.3), is sinusoidal in the true anomaly. This angle is a linear function of time only for zero eccentricity. In general, in order to disclose the time-dependence of our signal, we must recall the

3 The Demographics of Close-In Planets

relation between the true anomaly and the eccentric anomaly E_j

$$\tan \frac{f_j(t)}{2} = \sqrt{\frac{1+e_j}{1-e_j}} \tan \frac{E_j(t)}{2}. \quad (3.4)$$

The eccentric anomaly as a function of time can be obtained by numerically solving the Kepler equation

$$E_j(t) - e_j \sin E_j(t) = \frac{2\pi(t - t_{p,j})}{P_j} \equiv M_j(t), \quad (3.5)$$

where $t_{p,j}$ is the epoch of periastron passage and the last quantity is called the mean anomaly. Planets in eccentric orbits will move faster when they are at the periastron and slower at the apoastron. Therefore, the sine wave will be compressed at the periastron epoch and stretched at the apoastron. The position of the periastron is given by ω_j. In Fig. 3.1 we show how the shape of the Keplerian RV signal changes for different values of e and ω.

3.2.1.3 Signal Analysis

A generic, unevenly sampled RV time series will contain periodic modulations as described above superposed to instrumental and astrophysical noise. The latter mainly comes from the stellar activity of the target star, which generates oscillations

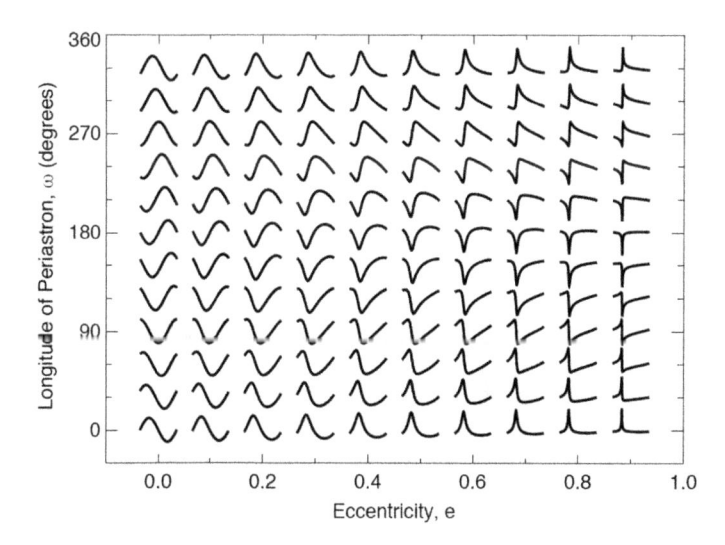

Fig. 3.1 Shapes of the radial velocity signals for various eccentricities and periastron arguments. *Note that higher eccentricities lead to very rapid variations*

in the RVs due to a variety of time-variable stellar surface structures (Queloz et al. 2001; Fischer et al. 2016). In order to extract the planetary signal, one may use the (Generalized) Lomb-Scargle periodogram (Lomb 1976; Scargle 1982; Zechmeister and Kürster 2009), which evaluates the χ^2 improvement coming from the fit of a sine wave compared to a constant function using a figure of merit called periodogram power computed as a function of a grid of trial periods. In such diagrams, the periodic signals arise as sharp peaks in power over the background. Furthermore, the periodogram may exhibit peaks produced by the cadence of the observations (the so-called window function), dictated, for example, by Earth's rotation and revolution period, and lunar cycles. Periodic RV signals induced by stellar activity will also be revealed in a periodogram. For in-depth discussions of how to unveil planetary signals in the presence of stellar activity see, for example, Haywood et al. (2014); Hatzes (2016).

When a planetary signal is identified, it can be subtracted from the data and the analysis can be repeated to search for progressively lower signals from other planets until the noise level is reached (Fischer et al. 2008). A variety of sophisticated methods for Keplerian orbit fitting are utilized for the task (for a review, see e.g. Dumusque et al. 2017)

The efficiency of a radial velocity campaign can be evaluated by injecting simulated signals with known characteristics in the data and then running the analysis pipeline to recover the signal Howard and Fulton (2016). Figure 3.2 illustrates the efficiency for a given Doppler survey calculated in this way as a function of the planetary mass and semi-major axis. Indeed the noise level constrains the efficiency from below, while longer period planets cannot be detected with a finite duration campaign.

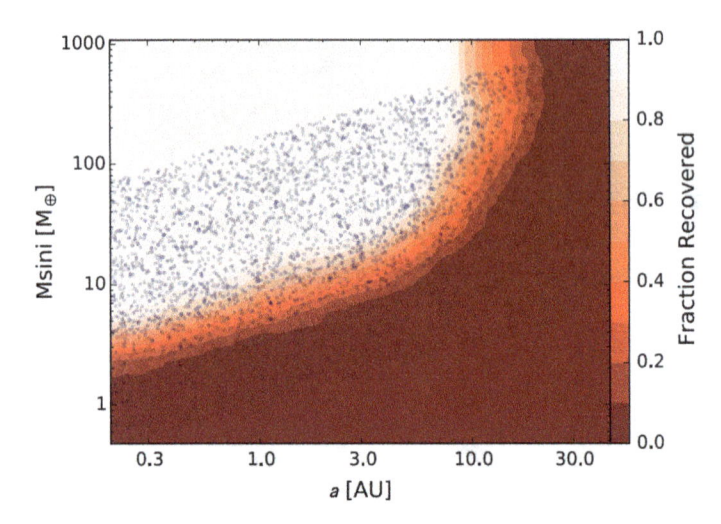

Fig. 3.2 Completeness of the RV sample of planets in a Doppler survey estimated by injection/recovery of simulated signals in the data

3.2.1.4 RV Surveys Highlights

Besides being the first method successfully employed for the detection of a planet around a normal star, the RV technique has also been able to provide the detection of the first multiple planetary systems (Butler et al. 1999), the detection of long-period planets (Howard et al. 2010a), and, enabled by m s^{-1}-level precision, the discovery of the first low-mass planets, Neptunes and Super Earths (Lovis 2006; Mayor et al. 2009). Although next-generation spectrographs (e.g., ESPRESSO, Pepe et al. 2021; EXPRES, Jurgenson et al. 2016; for a review see Fischer et al. 2016) are approaching a nominal precision of 10 cm s^{-1}, the main limitation comes from the modeling of stellar activity, which can easily produce RV signals with the dominant amplitude when compared to those of low-mass planets (Dumusque et al. 2017).

3.2.2 Transits

About 70% of the currently known exoplanets have been discovered by the transit method. Most of these discoveries come from the *Kepler* mission, to which we owe the possibility to speak in so great detail about the statistics of close-in exoplanets, but many ground-based surveys have contributed to the discovery of interesting systems.

3.2.2.1 Transit Geometry

As depicted in Fig. 3.3, planets on nearly edge-on orbits may transit in front of their parent star, blocking the light coming from an area corresponding to their projected disk. In a first approximation, the flux detected by the observer drops by a fraction

$$\delta \simeq \frac{R_p^2}{R_\star^2}, \tag{3.6}$$

which means that, if we know the radius of the star by its position on the HR diagram, we can immediately infer the radius of the planet (Seager and Mallén-Ornelas 2003). More refined modeling of the transit light curve may take into account the limb darkening profile of the star (Mandel and Agol 2002), the thermal flux from the night side of the planet (Charbonneau 2003), oblateness (Carter and Winn 2010) and so on (Winn 2010; Perryman 2018).

In general, the planet will also pass behind the star. In this secondary eclipse, the light of the day-side of the planet is blocked, giving rise to a much shallower event that, when observed, preciously increases our knowledge on the planet, as it allows to further constrain the orbital eccentricity, but most importantly opens the door to measurements of the properties of its atmosphere (Deming et al. 2005).

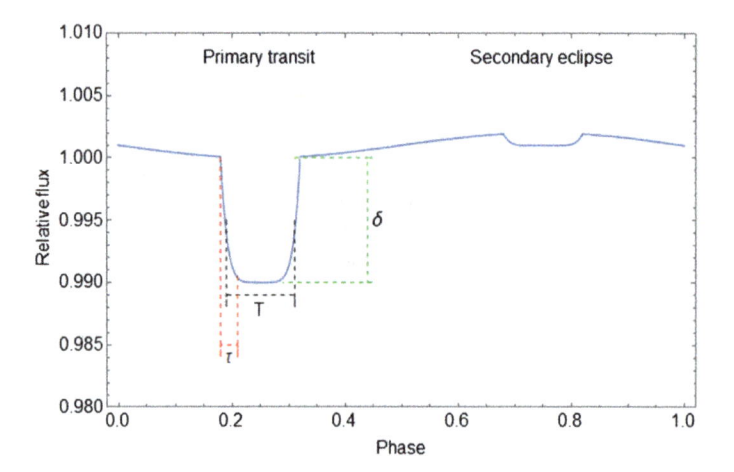

Fig. 3.3 Light curve of a system with a transiting planet. The primary transit (planet in front of the host star) is characterized by a duration T, a descent time τ and a deficit flux δ. The secondary eclipse occurs when the planet dayside light is blocked by the star

The direct observables in a transit are the orbital period P, the depth δ, the duration T and the time τ taken by the planet to cross the limb of the star (see also Chap. 4, Sect. 4.4). Simple formulae relate these to physical parameters such as the ratio of the radii $R_p/R_\star$, the semi-major axis in units of the star radius $a/R_\star$, the impact parameter b, and the stellar density $\rho_\star$ (Seager and Mallén-Ornelas 2003). Finally, after the use of stellar models to derive the host star parameters, we may get R_p, a and the orbital inclination i (for details, see e.g. Seager and Mallén-Ornelas 2003; Collier Cameron 2016).

If the host star is bright enough, spectroscopic follow-up with high-precision RVs and subsequent detection of the RV semi-amplitude can provide, combined with the i value from the transit, a determination of the true planet mass. For these precious planets, we can derive their bulk density and finally gain insights concerning their internal composition (see Sect. 3.4.3).

Transiting planets obviously orbit on nearly edge-on configurations. Assuming an isotropic distribution for the orbital planes, the probability that a generic planet on a circular orbit transits in front of its star is roughly (Borucki and Summers 1984)

$$P_{\text{tra}} \approx \frac{R_\star}{a} = 0.005 \left(\frac{R_\star}{R_\odot}\right)\left(\frac{a}{1\text{AU}}\right)^{-1}, \tag{3.7}$$

which means that we need to survey thousands of stars before finding one transiting planet. The probability decreases relatively fast with the orbital distance a, biasing the technique toward the detection of very close-in planets.

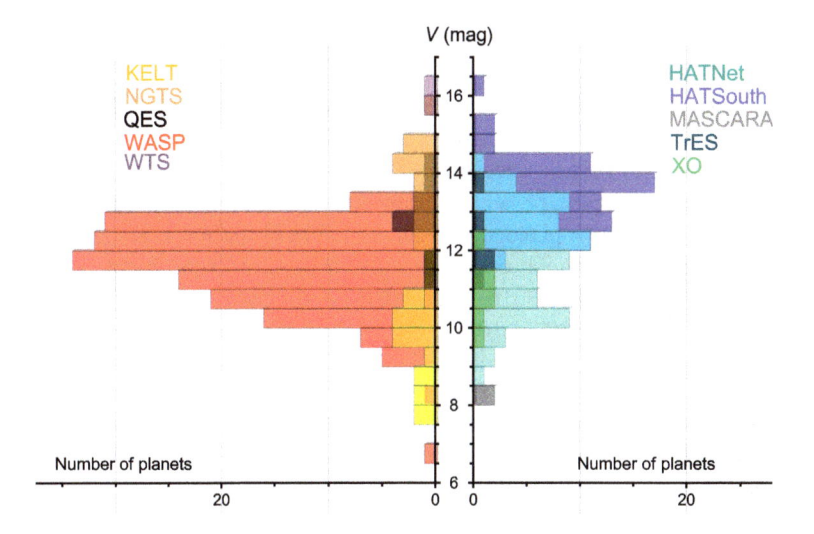

Fig. 3.4 Histograms of the number of planets detected by the most successful ground-based all-sky transit surveys versus host-star V mag. Data taken from the Transiting ExoPlanet Catalogue (TEPCat; www.astro.keele.ac.uk/jkt/tepcat) as of May 2021

3.2.2.2 Transit Campaigns and Analysis

Surveys for the discovery of transiting exoplanets typically monitor large fields in the sky with percent-level, high-cadence photometry. Ground-based surveys of F-G-K-type stars have often used arrays of wide-field cameras on the same mount. Here we mention HATNet (Bakos et al. 2004), WASP/SuperWASP (Pollacco et al. 2001; Wheatley 2015), OGLE (Udalski et al. 2002), TrES (Alonso et al. 2004), XO (McCullough et al. 2005), KELT (Pepper et al. 2007), MASCARA (Snellen et al. 2012), QES (Alsubai et al. 2013), HATSouth (Bakos et al. 2013), WTS (Kovács et al. 2013), NGTS (Wheatley et al. 2013). Note that depending on the details of the technical implementations, choices of exposure time and observing strategy, and camera sensitivity, all these surveys have differing sky coverages, limiting magnitudes for faint targets and a minimum magnitude for bright stars leading to saturation, translating into a broad visual magnitude distribution of the detected companions (see Fig. 3.4; for a review, see Collier Cameron 2016).

Other ground-based transit surveys (MEarth, Charbonneau et al. 2009; APACHE, Sozzetti et al. 2013; TRAPPIST, Gillon et al. 2012; ExTrA, Bonfils et al. 2015; SPECULOOS, Delrez et al. 2018; EDEN, Gibbs et al. 2020) have instead focused on late-type M and ultra-cool dwarfs, adopting a one (or few) target(s) per field approach and 40-cm to 1-m class telescope arrays. Such experiments exploit the possibility to reach detection of short-period transiting sub-Neptunes, super Earths, and even temperate Earth-sized, rocky planets from the ground with modest-size telescopes, which is enabled by the small radii of low-mass stars, leading to deep transits (Δmag $\gtrsim 0.005$ mag), and by the low temperatures of the primaries, leading to much closer-in habitable zones that those of solar-type stars, and therefore increased likelihood of observing transits of temperate planets.

Space-based transit photometry allows obtaining uninterrupted, high-cadence light curves that can reach better than 10^{-4} precision, which is the level required for detection of transits of Earth-sized planets across Sun-like stars. Figure 3.5 depicts the time progression of past, present and future transiting-planet space missions over a quarter-century. The first planetary transit discovery in space was made by CoRoT (Auvergne et al. 2009), which operated in the 2007–2013 time frame, but it was indeed with *Kepler* (Borucki 2016) that the field experienced a dramatic revolution. *Kepler* observed the same 10° × 10° field on the Galactic plane between the Cygnus and Lyra constellations for 4 years (2009–2013), leaving us with more than 4000 transiting planet candidates. Most of them have either been statistically validated or confirmed as true planets via dynamical mass measurements. Even in its refurbished version (renamed K2) after the failure of the reaction wheels, the *Kepler* spacecraft continued (in the period 2014–2018) to deliver transiting candidates distributed all along the ecliptic plane (Vandenburg et al. 2016).

More recently the TESS mission, launched in 2018, is performing an all-sky survey of bright stars to look for transiting super-Earths and sub-Neptunes, particularly around late-type dwarfs (Ricker et al. 2015). TESS is designed to be a primary provider of bright transiting planets for spectroscopic follow-up measurements with JWST (Greene et al. 2016), which, due to launch in December 2021, is expected to provide an in-depth characterization of their atmospheres. The sample of bright transiting planets will also be the key input list for ESA's Ariel mission (launch expected in 2029), which is expected to survey $\sim$ 1000 of them for the purpose of atmospheric characterization (Tinetti et al. 2018). CHEOPS (launched in 2019) is starting to observe known transiting planets to derive much more accurate radii (Broeg et al. 2013). Finally, PLATO (launch expected in 2026) is designed to monitor approximately 50% of the visible sky and discover Earth-like transiting planets in the habitable zone of Sun-like stars (Rauer et al. 2014).

Independent of the cadence and precision of the light curve, the search for planetary transits is prone to a large number of false positives, i.e. transit-like events caused by phenomena other than planetary transits. Astrophysical false positives are mostly produced by a variety of configurations of stellar systems (for details on the subject see e.g., Collier Cameron 2016).

Transit detection is performed by fitting a periodic box-like function (the Box Least Squares technique; Kovács et al. 2002) to the photometric time-series phase-folded over a large grid of trial frequencies. The frequency spectrum produced by the box-fitting algorithm is then inspected for identification of a statistically significant peak, corresponding to the period of the transit-like feature. Many sophisticated algorithms are then able to fit for the detailed shape of the transit light curve and retrieve the planet and stellar parameters. For more details, we direct the interested readers to the review by Collier Cameron (2016), and also the works by, e.g., Mandel and Agol (2002), Fulton et al. (2011), Eastman et al. (2013) and Southworth (2013).

3 The Demographics of Close-In Planets

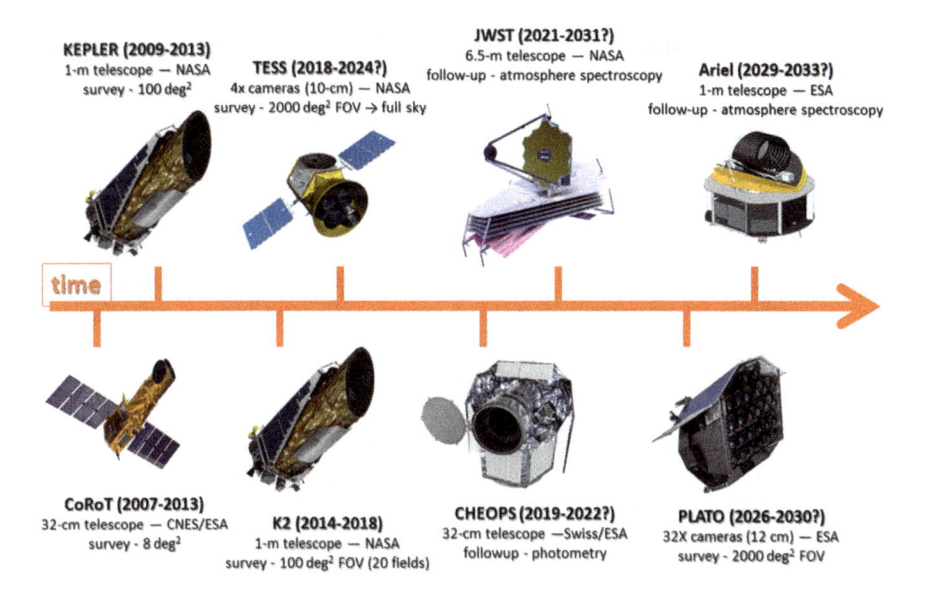

Fig. 3.5 Timeline of NASA and ESA space missions devoted to the detection and characterization of transiting exoplanets

3.2.2.3 Transit Surveys Highlights

The first discovered transiting exoplanet, HD 209458b (Henry et al. 1999; Charbonneau et al. 2000), was a typical hot Jupiter, and provided the final proof that the Doppler-detected companions since 1995 were, in fact, exoplanets. Ground-based transit photometry allowed to discover GJ 436b, the first Neptune-sized planet (Butler et al. 2004; Gillon et al. 2007), and GJ 1214b, the first transiting Super Earth orbiting a low-mass star (Charbonneau et al. 2009). However, in order to find the first bona-fide rocky planets (CoRoT-7b, Kepler-10b), space missions were needed (Léger et al. 2009; Batalha et al. 2011). *Kepler* has been able to find hundreds of multi transiting-planet systems (see Sect. 3.5), containing up to six planets within Mercury's orbit, like Kepler-11 (Lissauer et al. 2011a). *Kepler* also unveiled a class of transiting giant planets orbiting short-period eclipsing binaries, i.e. circumbinary planets, Kepler-16b being the first of its kind (Doyle et al. 2011. For a review see Welsh and Orosz 2018). Besides, ground-based surveys have continued to provide spectacular results, such as the TRAPPIST-1 multi-planet system (Gillon et al. 2017), composed of 7 transiting Earth-sized, Earth-mass planets orbiting an ultra-cool dwarf. Probably three of these planets lie in the habitable zone and may harbor water on their surface. Finally, transiting exoplanets have allowed us to enter the era of atmospheric characterization of close-in planets orbiting main-sequence stars (see, e.g., the review by Sing 2018).

3.3 Mass, Size, and Period Distributions

Modern Doppler and transit surveys have shown the prevalence of planets at short and intermediate separations, which we will operationally define here as having a semi-major axis $a < 5$ au. The RV method has driven the field during its first decade of development, while ground-based and space-borne transit surveys (CoRoT, *Kepler*) have provided a major boost in the second decade. The initial focus of such investigations was on the discovery of new exoplanets and on improving the surveys' sensitivities in order to expand the region of accessible parameter space. When the sample of detected planets in individual surveys became large enough, the first statistical analyses were performed.

The basic aim of such studies was to determine *a*) the intrinsic frequency of stars with planets or *b*) the mean number of planets per star. These are two ways of determining occurrence rates of exoplanets having properties (such as mass, radius, and orbital distance) within a specified range. The key calculation to perform is the following:

$$\text{Occurrence Rate} = \frac{\# \text{planets}}{\# \text{stars}}, \quad (3.8)$$

where # planets is the number of detected planets in the survey that have the stipulated properties and # stars is the number of stars in the survey for which such planets could have been detected. This is however not a trivial task, as an accurate measurement of occurrence rate requires a large sample of stars that have been searched for planets and the proper understanding of observational biases/selection effects that favor the discovery of certain types of planets. For instance, in the representation of Fig. 3.6 one can appreciate how the most sensitive transit survey to-date, the *Kepler* mission, can discover planets with a very wide range of planetary radii up to a separation of about 1 au, beyond which it is essentially blind. Conversely, the Doppler domain can

Fig. 3.6 Relevant domains of planet detection techniques in the mass (o radius) versus orbital period (or semi-major axis) parameter Space. Figure showed by A. Howard in May 2019 during his lecture

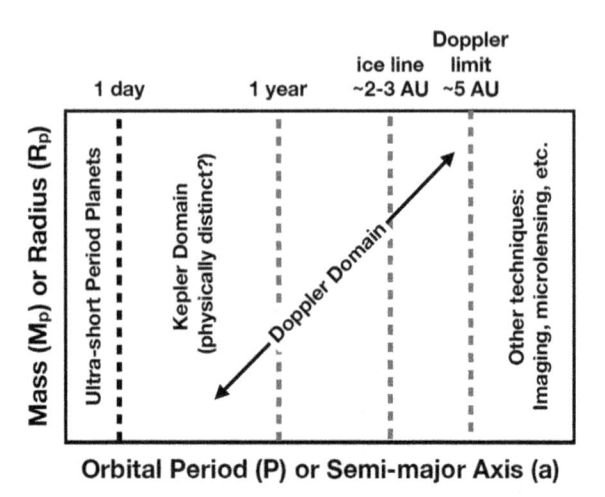

Fig. 3.7 The observed histogram (black line) of planetary masses compared with the equivalent histogram after correcting for the detection bias (gray line). Figure adapted from Mayor et al. (2011)

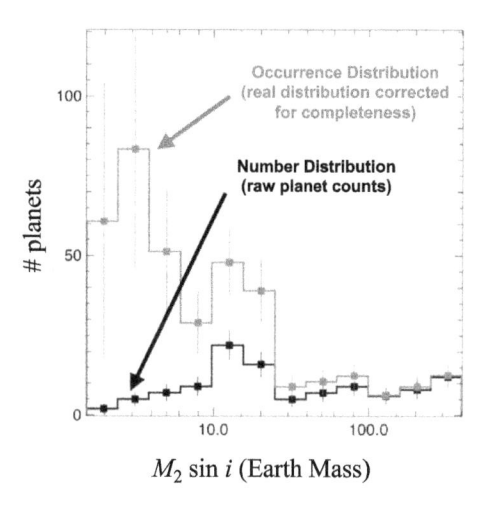

$M_2 \sin i$ (Earth Mass)

extend to significantly larger semi-major axes, but with variable sensitivity in terms of accessible mass range.

In order to effectively compute occurrence rates, the strategy to adopt includes modeling all effects, which cause planets to be missed during the detection process, and correct for those as much as possible, so that remaining errors are small compared to counting statistics. This implies achieving the best-possible understanding of the underlying observational biases/selection effects that affect a survey's completeness. In this way one moves from a measurement of raw planet counts (number distribution) to a representation of the true underlying distribution (see Fig. 3.7). When biases are not sufficiently well-understood, one is expected to honestly, clearly state caveats and limitations of the performed analyses.

In the remainder of this Section we will focus on understanding how occurrence is computed and then discuss major trends in occurrence at short and intermediate separations for Doppler-detected and transiting planets, using mass, size, and orbital period/semi-major axis as proxies of the accessible parameter space.

3.3.1 Doppler Surveys

The results from three major Doppler surveys have been used to address the issue of planet occurrence around F-G-K-type stars in the solar neighborhood ($d \lesssim 30\,\text{pc}$). The analysis of the Keck RV survey, which was performed by Cumming et al. (2008), included 475 F-G-K dwarfs monitored for about 8 years and found to host a total of 48 planets (including confirmed companions and candidates). Cumming et al. (2008) determined the completeness-corrected occurrence rate of giant planetary companions with minimum mass in the range $0.3\,M_{\text{Jup}} < m_p \sin i < 10\,M_{\text{Jup}}$ and orbital period $P < 2000\,\text{days}$. Limits on companion detectability were determined based

Fig. 3.8 Giant-planet occurrence based on the Keck survey consisting of 475 FGK-dwarf stars. Figure adapted from Cumming et al. (2008), © The Astronomical Society of the Pacific. Reproduced by permission of IOP Publishing. All rights reserved

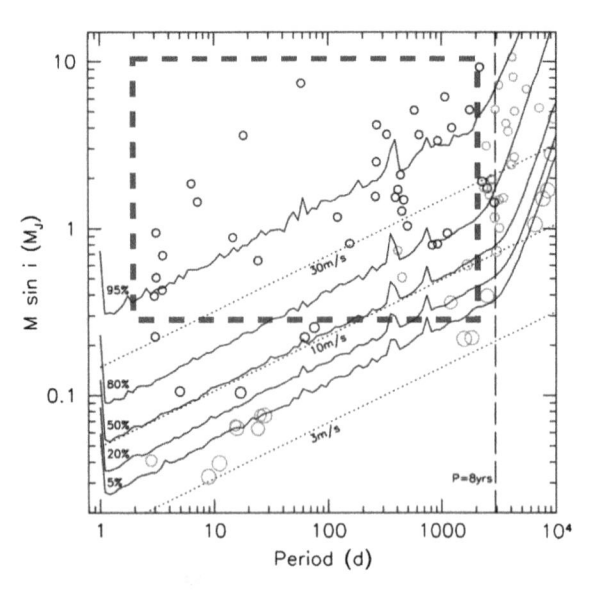

on a Lomb-Scargle (Lomb 1976; Scargle 1982) periodogram analysis to identify significant periodicities in the RV time-series. Signals with given RV amplitude K and P producing a false-alarm probability (FAP; Horne and Baliunas 1986) below 10^{-3} were considered as detections and the fraction of stars, for which such an instance occurred, recorded. Any detected RV amplitude K was transformed into a corresponding minimum companion mass by inverting Eq. 3.2. Cumming et al. (2008) found that companions with $K > 30$ m s^{-1} would have been detected around most of the stars in their sample. Corrections for incompleteness became important to characterize the loss in sensitivity below that threshold. This was done in practice by computing a missed planet correction factor $F(P, m_p \sin i)$ for each detected planet (both confirmed and candidates). If in a given region of the mass-period plane, planets can be ruled out for a fraction C of stars, the best estimate of the number of planets is $1/C$ times the number of detections, and $1/C-1$ planets remain undetected. The corresponding completeness-corrected map produced by Cumming et al. (2008) is shown in Fig. 3.8.

Cumming et al. (2008) then employed a maximum likelihood technique to fit a simple power-law function to the exoplanet minimum mass-period distribution, finding:

$$\frac{dN}{d\log m_p \sin i \, d\log P} = C(m_p \sin i)^\alpha P^\beta \propto (m_p \sin i)^{-0.31 \pm 0.20} P^{0.26 \pm 0.10}. \quad (3.9)$$

The normalization constant was chosen such that $\sim 10\%$ of solar-type stars host at least one planet with mass and period in the above range. This functional form to represent the occurrence rate of planets eventually became a benchmark for many

3 The Demographics of Close-In Planets

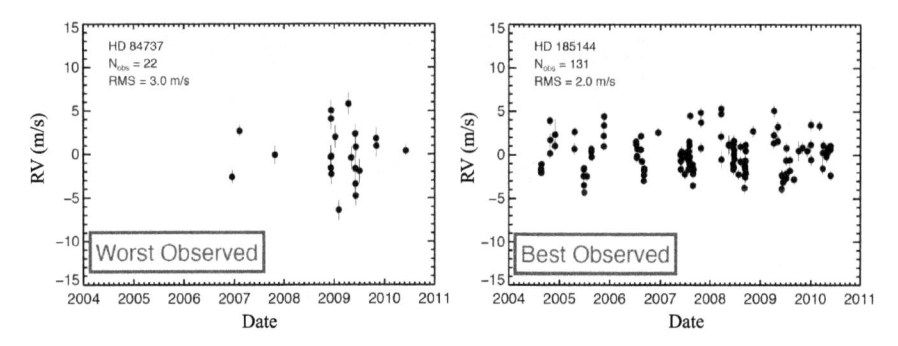

Fig. 3.9 Examples of RV time-series for targets in the Keck Eta-Earth survey with sparse and dense sampling of the observations (left and right panel, respectively). Figure showed by A. Howard in May 2019 during his lecture

subsequent studies.[2] The mass and period power-law coefficients indicate that small-mass planets are more common, and longer-period planets are more common. Cumming et al. (2008) also extrapolated the occurrence rate of giant planets at wider separations, estimating that $\sim 20\%$ of solar-type stars host a giant planet with mass $> 0.3\,M_{Jup}$ within 20 au.

The period distribution studied by Cumming et al. (2008) showed additional structure, not directly captured by the power-law fits. They found evidence of a pile-up of hot Jupiters (gas giants operationally defined as having $P < 10$ days) at $P \simeq 3$ days, followed by a 'period valley' of lower probability in the approximate period range $10 - 100$ days, and then by a sharp rise in occurrence for separations $a \gtrsim 1$ au. These findings are entirely in line with those first discussed by Udry et al. (2003). In the Keck RV survey, the fraction of F, G, and K dwarfs in the solar neighborhood hosting hot Jupiters was found to be $1.5\% \pm 0.6\%$, a number in line with those later obtained by Mayor et al. (2011), using CORALIE and HARPS data ($0.89\% \pm 0.36\%$), and by Wright et al. (2012), based on data from the Keck and Lick observatories ($1.2\% \pm 0.38\%$).

The Keck Eta-Earth survey (Howard et al. 2010b) and the CORALIE+HARPS survey (Mayor et al. 2011) provided the first opportunity to explore the domain of low-mass (3–30 $M_\oplus$) planets with periods $\lesssim 50$ days. The two programs surveyed 166 G-K-M dwarfs and 376 G-K dwarfs, respectively. Most of the stars in both samples were selected based on low levels of chromospheric activity, allowing to achieve a typical *rms* dispersion in the Keck Eta-Earth and CORALIE+HARPS RV datasets of $2 - 4\,\mathrm{m\,s^{-1}}$ and $1 - 3\,\mathrm{m\,s^{-1}}$, respectively. Figure 3.9 shows detection limits for two individual objects with low and high cadence RV observations in the Keck Eta-Earth survey.

The two surveys achieved comparable sensitivity to close-in planets with $P < 50$ days (an example is shown in Fig. 3.10), and when accounting for the high degree of incompleteness (10%–30%) for minimum masses in the super-Earth regime ($m_p \sin i < 10\,M_\oplus$), the findings from both experiments provide a convergent picture:

---

[2] More details on the mathematical formalism of the technique are provided in Sect. 2 of Chap. 4.

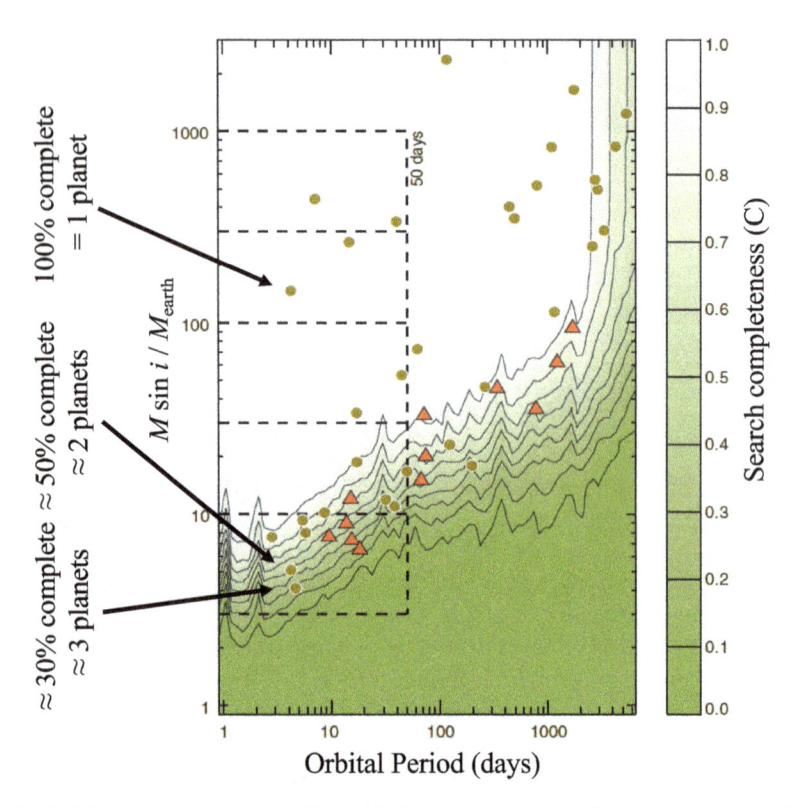

Fig. 3.10 Minimum mass versus orbital period for the detected (yellow circles) and candidate planets (FAPs ∼ 1 − 5%; red triangles) from the Eta-Earth survey. Contours refer to the *search completeness*, that is the fraction of stars with enough RV measurements to be sensitive to planets in circular orbits of a given orbital period and minimum mass. Dashed lines refer to five different mass domains out to 50 day orbits. The arrows identify high, intermediate, and low completeness. Figure adapted from Howard et al. (2010b) and reproduced with permission from A. Howard

the planet mass function rises steeply towards lower-mass planets, such that the occurrence of planets with $3\, M_\oplus \lesssim m_p \sin i \lesssim 30\, M_\oplus$ is found to be around 20%–25%.[3] These planets are at least an order of magnitude more common than giant planets in the same period range around solar-type stars. The results from both surveys were also broadly consistent in indicating a lack of hot Neptunes ($10\, M_\oplus \lesssim m_p \sin i \lesssim 30\, M_\oplus$) with $P < 10$ days, a population otherwise conspicuous at longer periods.

The findings of the Keck Eta-Earth and CORALIE+HARPS surveys were at the time very informative for planet formation theory. Early attempts at producing models of synthetic planet populations (e.g., Ida and Lin 2005, 2008; Mordasini et al. 2009a, b) predicted a "planet desert" in the super-Earth mass domain for $P < 50$ days or so, precisely where the two programs determined the highest val-

---

[3] Similar findings were also presented by Wittenmyer et al. (2011), who reported that 17.4% of stars host a planet with $m_p \sin i < 10\, M_\oplus$ and $P < 50$ days.

Fig. 3.11 A population-synthesis model superimposed on the plot shown in Fig. 3.10. Figure adapted from Mordasini et al. (2009b) and Howard et al. (2010b) and showed by A. Howard in May 2019 during his lecture

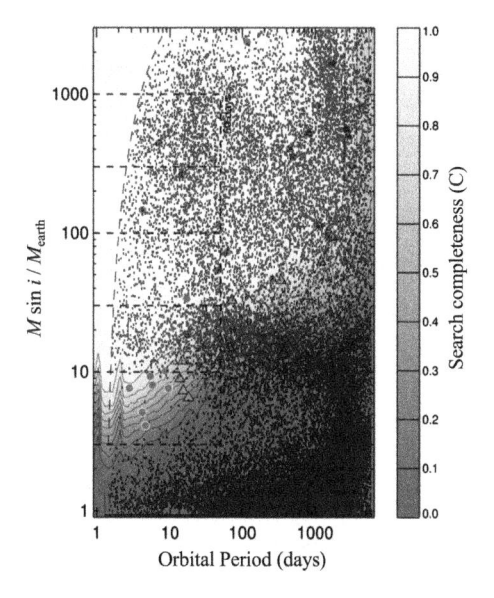

ues of completeness-corrected planet frequency (see Fig. 3.11). The observational evidence put forth motivated significant improvements in population synthesis models of planet formation (e.g., Ida and Lin 2010, 2013; Mordasini et al. 2012a,b,c; Paardekooper et al. 2011, 2013).

At intermediate separations, the Eta-Earth and CORALIE+HARPS surveys confirmed the trend of increasing frequency of giant planets with increasing orbital period, up to ~ 5 year (~ 3 au). This roughly corresponds to the location of the snow line in protoplanetary disks around solar-type stars (e.g., Mulders et al. 2015b; Morbidelli et al. 2016). Beyond 3 au or so, the agreement of Doppler surveys appears to degrade sharply with orbital separation. Recent studies indicate the presence of a decline in giant planet occurrence with increasing separation (Fernandes et al. 2019), others (Wittenmyer et al. 2020) find no evidence of a turnover in giant planet frequency at the snowline (see top and bottom panels of Fig. 3.12), and in general observational uncertainties (e.g., sample sizes and their exact definition in terms of mass interval, survey duration, sampling) do not allow yet achieving consensus on the exact regime of orbital separations at which the occurrence rate of gas giants might start declining as well as its possible dependence on planetary mass (Fulton et al. 2021). The connection between occurrence rates at close-in, intermediate and wide separations is further discussed in Chap. 4 (Sect. 5).

Finally, a large number of investigations has used the results from the above-mentioned RV surveys and other Doppler programs to explore the dependence of planet occurrence on stellar properties. In particular, giant planet frequency increases sharply with increasing stellar metallicity for F-G-K dwarfs (Santos et al. 2004b; Fischer and Valenti 2005; Sozzetti et al. 2009; Mortier et al. 2012), while whether small-planet occurrence correlates positively with metallicity in the same stellar samples is still a matter of debate (Sousa et al. 2008; Courcol et al. 2016; Sousa et al.

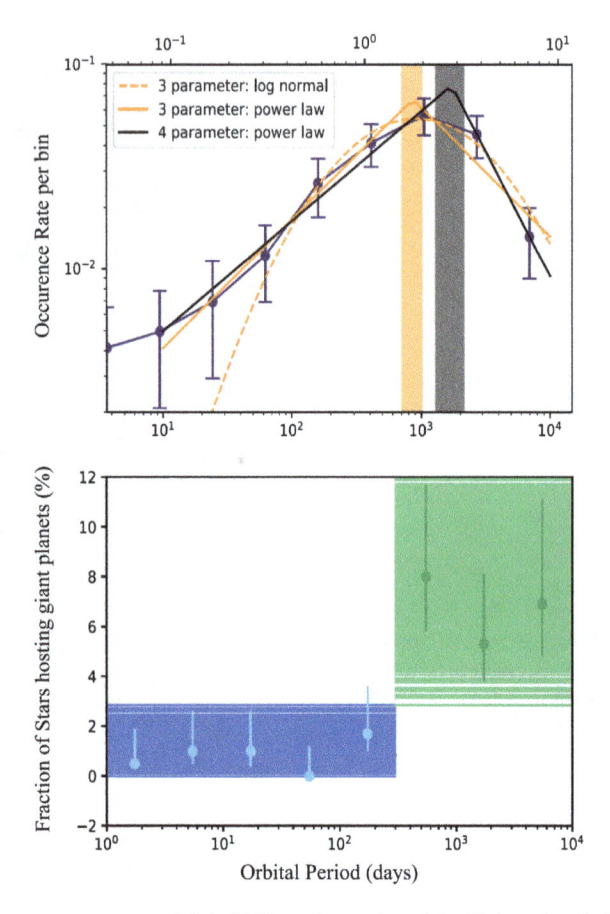

Fig. 3.12 *Top*: occurrence rate of 0.1–20 M_{Jup} planets (purple) with best-fit relations beyond 10 days: asymmetric broken power law (solid black line), symmetric broken power law (solid yellow line), and log-normal (dotted yellow line). The location of the *break* is shown as a shaded region (gray for the asymmetric broken power law and yellow for the other two fits). Figure adapted from Fig. 3 of Fernandes et al. (2019) and reproduced with permission from R. Fernandes. *Bottom*: frequency of giant planets as a function of orbital period. Figure adapted from Fig. 1 of Wittenmyer et al. (2020)

2019; Bashi et al. 2020). The occurrence rate for giant planets within 3 au increases with host star mass up to $\sim 2\,M_\odot$ (Johnson et al. 2010; Reffert et al. 2015; Ghezzi et al. 2018), with low-mass M dwarfs hosting as few as 3–10 times less Jovian mass companions than solar-type stars (e.g., Butler et al. 2004; Endl et al. 2006; Bonfils et al. 2013; Tuomi et al. 2014). On the contrary, the occurrence rate of close-in super Earths around M dwarfs could be higher than that for F-G-K stars by a factor of 2–3 (Bonfils et al. 2013) and it presents weak or no correlation with mass and metallicity, respectively (Maldonado et al. 2020). The theoretical interpretation of the above trends in terms of planet formation and evolution processes and models is provided in Chap. 1.

3 The Demographics of Close-In Planets

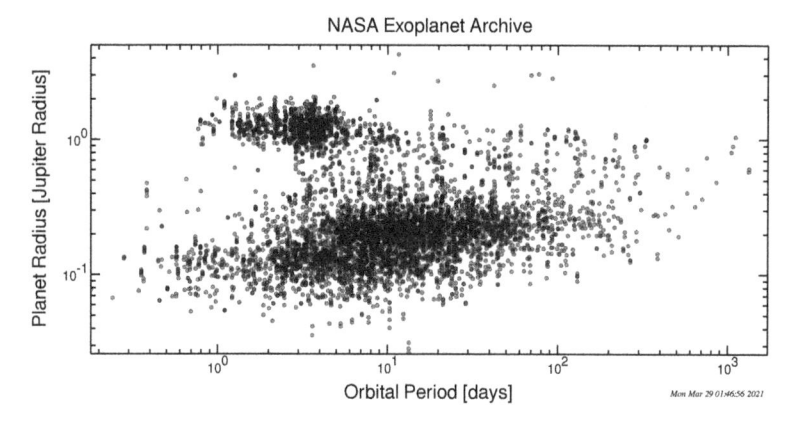

Fig. 3.13 The diagram of planet radius versus orbital period for the exoplanets belonging to the *Kepler* sample. The error bars have been suppressed for clarity. Data were taken from the NASA Exoplanet Archive in March 2021

3.3.2 Transit Surveys

The *Kepler* mission sample of thousands of transiting planet candidates (see Fig. 3.13) has enabled a large number of analyses aimed at determining the occurrence rates of planets with radii as small as 1 $R_\oplus$, and at correlating planet occurrence estimates in the *Kepler* field with stellar properties (e.g., Howard et al. 2012; Fressin et al. 2013; Petigura et al. 2013a, b; Dressing and Charbonneau 2013, 2015; Burke et al. 2015; Foreman-Mackey et al. 2015; Fulton et al. 2017; Fulton and Petigura 2018; Zhu and Wu 2018; Narang et al. 2018; Mulders et al. 2018; Petigura et al. 2018; Hardegree-Ullman et al. 2019; Hsu et al. 2019; Berger et al. 2020; Bryson et al. 2020a, b; Lu et al. 2020; Yang et al. 2020).[4]

With time, the sophistication of the approaches to the analysis of *Kepler* data for the purpose of occurrence rate calculation has increased. Most of the effort has been devoted to the improved characterization of both completeness and reliability against false alarms of the content of the *Kepler* catalog of candidates. The first studies utilized simple Gaussian cumulative distribution functions, or their linear approximations, as proxies for pipeline completeness, or identified restricted samples with an assumed high degree of completeness. These approaches evolved further into using Poisson-likelihood-based calculations and approximate Bayesian computations that took advantage of the most significant effort to characterize the reliability of the *Kepler* planet candidates catalog realized in the final DR25 catalog paper (Thompson et al. 2018). The interested reader can find additional details on the difficulties inherent to accurate occurrence rate calculations (using the *Kepler* survey as a guiding example) in Sect. 3 of Chap. 4. We will focus here on the methodological approach

---

[4] A non-exhaustive list of of planet occurrence rate papers, heavily focused on the analysis of *Kepler* mission data, can be found at exoplanetarchive.ipac.caltech.edu/docs/occurrence_rate_papers.html.

of Howard et al. (2012), who provided the first such estimate for transit candidates within 0.25 au of solar-type stars, matching the period range of Howard et al. (2010b) analysis based on Doppler data.

Howard et al. (2012) adopted a carefully crafted strategy to effectively use Eq. (3.8) in the occurrence rate calculation. At the time, the Gaia mission had yet to provide direct distance estimates for the overwhelming majority of stars in the *Kepler* field, so they made careful cuts in effective temperature (4100 K $< T_\text{eff} <$ 6100 K) and surface gravity (4.0 $< \log g <$ 4.9), and selected only bright stars with *Kepler* bandpass magnitudes $K_\text{p} <$ 15 mag, so as to ensure sufficient photometric precision for transit signal detectability. Similarly, transit candidates were selected based on a high threshold in terms of signal-to-noise ratio (S/N) of the detection:

$$\text{S/N} = \frac{\delta}{\sigma_\text{CDPP}} \sqrt{\frac{n_\text{tr} T_\text{dur}}{3 \text{ hr}}} \quad (3.10)$$

where δ is from Eq. 3.6, n_tr is the number of transits observed by *Kepler* in a 90-day quarter, T_dur is the transit duration (see Eq. 3.15), σ_CDPP is the Combined Differential Photometric Precision, i.e. the empirical rms noise in 3-hr time interval bins coming from the *Kepler* pipeline. The transit candidates where selected so as to have $R_\text{p} > 2 R_\oplus$, $P <$ 50 days, and S/N $>$ 10, substantially a regime of nearly complete detection efficiency, with minimal pollution ($\sim 5\%$) from false positives. The top panel of Fig. 3.14 shows the final selection of candidates made by Howard et al. (2012).

The occurrence rate calculation was performed on a grid of cells in logarithmically spaced intervals of orbital period and radius. If $n_\star$ is the number of stars around which that planet could have been detected with S/N $>$ 10, and $p_\text{tr} = R_\star/a$ is the geometric transit probability, then the average occurrence rate of planets within a cell is:

$$f_\text{cell} = \sum_{j=1}^{n_\text{pl,cell}} \frac{1/p_{\text{tr},j}}{n_{\star,j}}, \quad (3.11)$$

where the sum is over all detected planets within the cell. The actual planet count is thereby augmented taking into account that for each detected planet there are actually $1/p_\text{tr}$ planets in all orbital inclinations orbiting $n_\star$ stars. The bottom panel of Fig. 3.14 shows the result of the full occurrence rate calculation, color-coded in terms of occurrence per logarithmic area units ($\frac{d^2 f}{d \log P \, d \log R_\text{p}}$).

The distribution of planet occurrence provides crucial clues on planet formation, migration, and evolution processes. Summing occurrence over all periods in the various radius bins, the top panel of Fig. 3.15 shows occurrence as a function of planet radius, modeled with a power-law of the form $df/d \log R_\text{p} = k R^\alpha$, with $k = 2.9 \pm 0.5$ and $\alpha = -1.92 \pm 0.11$. Quite clearly, planet occurrence increases with decreasing planet radius. This is qualitatively in agreement with the power-law dependence of the planet mass function found by Howard et al. (2010b), albeit with a significantly steeper dependence on radius. The bottom panel of Fig. 3.15

3 The Demographics of Close-In Planets

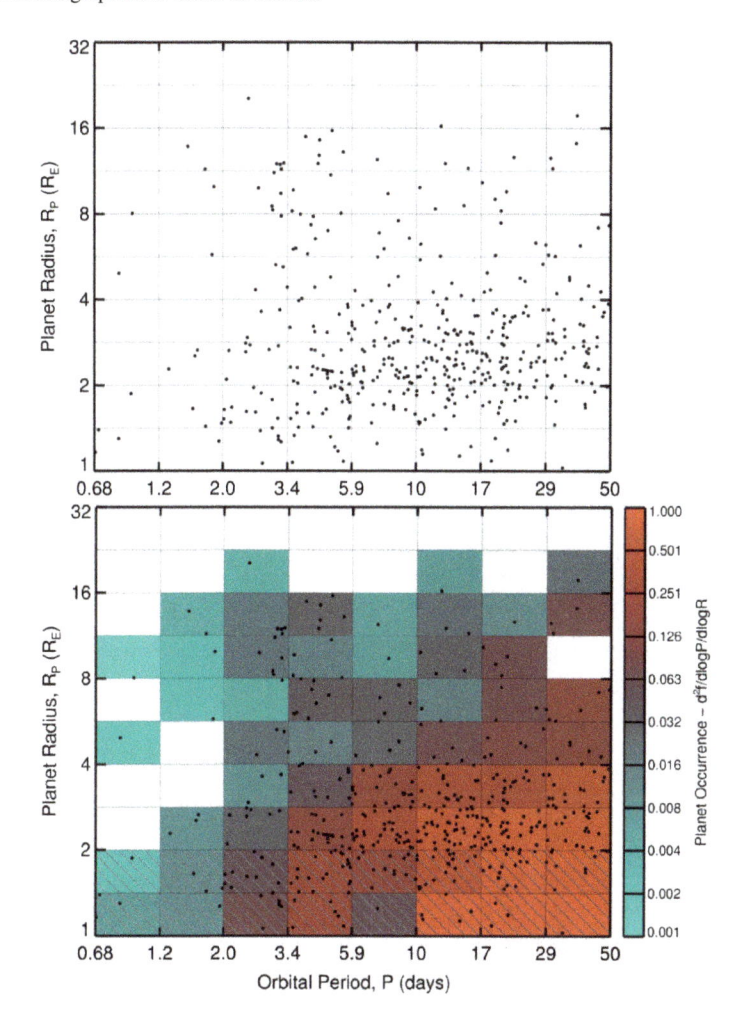

Fig. 3.14 *Top*: planet radius versus orbital period for the *Kepler* transit candidates with $P < 50$ days. it Bottom: planet occurrence in the radius-orbital period plane for solar-type stars from *Kepler*. See text for details. Figures adapted from Howard et al. (2012) and reproduced with permission from A. Howard

shows instead the measured planet occurrence as a function of orbital period using power-law fits with exponential cutoffs below a characteristic period P_0 in the form:

$$\frac{dN}{d\log P} = kP^\beta \left(1 - e^{-(P/P_0)^\gamma}\right). \tag{3.12}$$

Planet occurrence increases with $\log P$ for all planet classes considered ($\beta > 0$), while P_0 increases with decreasing planet radius, suggesting that the migration and parking mechanism that deposits planets close-in depends on planet radius.

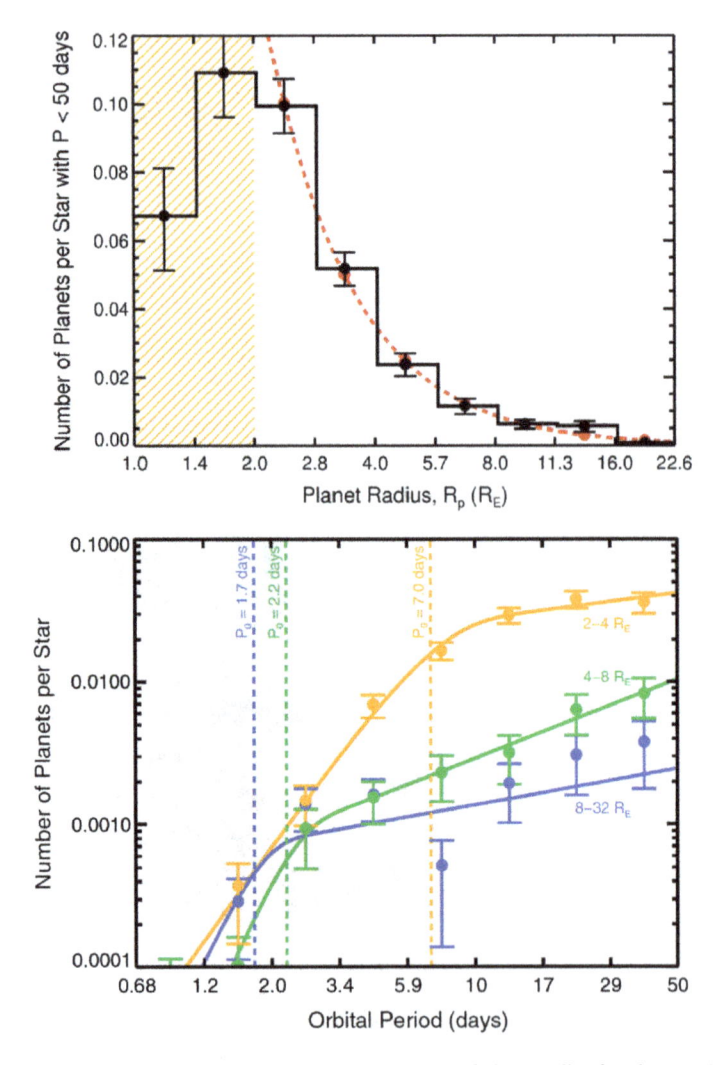

Fig. 3.15 *Top*: histogram of planet occurrence as a function of planet radius for planets with $P_{orb} <$ 50 days. The dashed line represents a power-law fit to occurrence measurements. In particular, $df/d \log R_p = k R_p^\alpha$, with $k = 2.9 \pm 0.5$ and $\alpha = -1.92 \pm 0.11$. *Bottom*: the filled circles represent the measurements of planet occurrence as a function of the orbital period with best-fitting models superimposed. Figures adapted from Howard et al. (2012) and reproduced with permission from A. Howard

Finally, Howard et al. (2012) extended the occurrence rate calculation to include cooler and hotter stars beyond the G-K dwarf sample initially selected, to unveil a clear trend of increasing close-in planet frequency within decreasing $M_\star$, which was fitted with a linear relation as a function of T_{eff}:

3 The Demographics of Close-In Planets

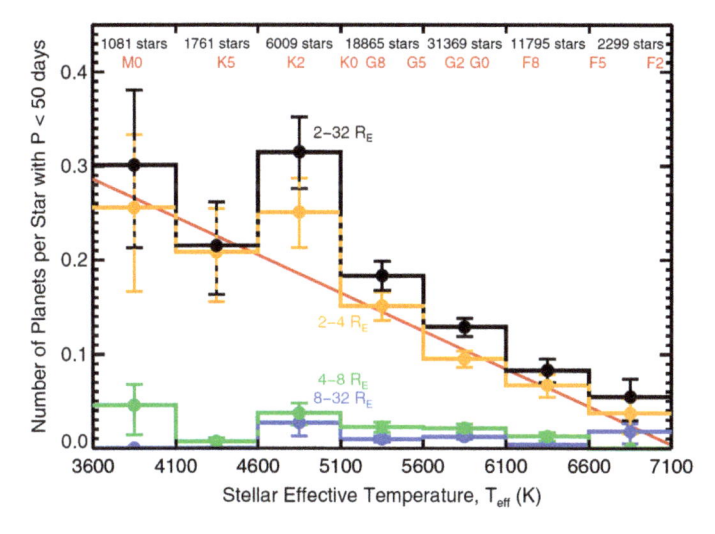

Fig. 3.16 Planet occurrence as a function of stellar effective temperature. Figure taken from Howard et al. (2012) and reproduced with permission from A. Howard

$$f = f_0 + k_T \left(\frac{T_{\text{eff}} - 5100 \text{ K}}{1000 \text{ K}} \right). \tag{3.13}$$

The trend in occurrence (shown in Fig. 3.16) is produced entirely by the population of super Earths and Neptunes with $R_p < 8 R_\oplus$, while for $R_p > 8 R_\oplus$ the opposite trend is recorded, in agreement with the findings from Doppler surveys of increasing giant planet frequency with stellar mass.

The findings by Howard et al. (2012) were extended to radii below $2 R_\oplus$ and $P > 50$ days in a follow-on study by Petigura et al. (2013b). They utilized an independent pipeline for transit candidates identification, a larger extent of *Kepler* photometry, partial reconnaissance spectroscopy of the primary stars and the same Howard et al. (2012) methodology to eventually estimate that $26 \pm 3\%$ of Sun-like stars harbor an Earth-size planet ($1 - 2 R_\oplus$) with $P = 5$–100 days (see Fig. 3.17). They produced an extrapolation of the occurrence rate of such companions to the $P = 200$–400 days range (none was directly detected), corresponding to planets that receive stellar flux comparable to Earth, and derived a first reliable estimate of the frequency of Earth-like planets: $\eta_\oplus = 5.7^{+1.7}_{-2.2}\%$.

The large body of works on global occurrence rate calculations and trends with stellar properties in the *Kepler* field, which appeared since the Howard et al. (2012) and Petigura et al. (2013b) analyses, has broadly confirmed the fact that planets of increasingly smaller size and longer period are increasingly more common (e.g., Dong and Zhu 2013; Fressin et al. 2013; Batalha 2014; Burke et al. 2015). However, differences in the assumptions underlying the baseline occurrence rates (e.g., detection and vetting completeness, characterization of false alarms and astrophysical false positives, choice of the parent stellar population) can translate into significant

Fig. 3.17 The measured distributions of planet sizes for $R_p > 1\,R_\oplus$ and $P_{\rm orb} = 5 - 100$ days. The first three bars are annotated to reflect the number of planets detected (light-gray bars) and missed (dark-gray bars). Figure adapted from Petigura et al. (2013b), ©2013—National Academy of Sciences

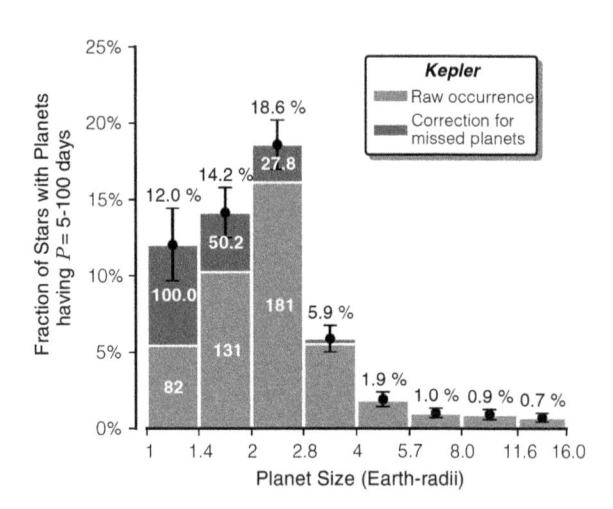

differences in terms of the specific numbers for different types of planets orbiting different types of stars (the reader can find additional details on such issues in Chap. 4, Sect. 3).

For example, while the trend of increasing planet frequency with decreasing stellar mass is not in doubt, absolute values and strength of the dependence measured by the different authors do not necessarily agree with each other (e.g., Foreman-Mackey et al. 2014; Dressing and Charbonneau 2015; Mulders et al. 2015a; Silburt et al. 2015; Mulders et al. 2018; Narang et al. 2018; Hsu et al. 2019; Hardegree-Ullman et al. 2019; Yang et al. 2020). Even more open for debate is the possible dependence of occurrence rates of small planets in the *Kepler* field with stellar metallicity (e.g., Buchhave et al. 2012, 2014; Buchhave and Latham 2015; Wang and Fischer 2015; Zhu et al. 2016; Petigura et al. 2018; Wilson et al. 2018; Zhu 2019; Lu et al. 2020), which as we have seen is still an open issue also for Doppler surveys. As for the much sought-after $\eta_\oplus$ estimate, increasingly sophisticated attempts to constrain it have been produced in recent times. When robust corrections for completeness and reliability are included, this value appears to be constrained to lie in the approximate range 5-50%, the dominant source of uncertainty being the virtual lack of true detections in the relevant parameter space. Table 3.1 summarizes the state of affairs of Earth-like planets occurrence rates.

A key aspect of the occurrence rate calculations based on *Kepler* data is that the use of a simple product of power laws in period and radius when modeling exoplanet population statistics appears as a sub-optimal choice. This was already apparent in Howard et al. (2012), Fressin et al. (2013), and Petigura et al. (2013a), where a power law was a poor fit to the observed planet population as a function of radius. In particular, additional structure in the radius distribution for $R_p \lesssim 3.0\,R_\oplus$ was smeared in those early studies due to the large uncertainties (typically 40%) in the stellar radii. The combination of spectroscopically-determined radii based on data collected as part of the California-Kepler Survey (CKS; Petigura et al. 2017) and the availability

3 The Demographics of Close-In Planets

Table 3.1 Rates of occurrence of habitable-zone Earth-like planets as measured by several authors

Planet type (planet-radius range in $R_\oplus$)	$\eta_\oplus$ (planets per star)	Reference	Notes
0.5–1.5	$0.37^{+0.48}_{-0.21}$–$0.60^{+0.90}_{-0.36}$	Bryson et al. (2021)	FGK dwarfs[a]
0.5–1.5	$0.58^{+0.73}_{-0.33}$–$0.88^{+1.28}_{-0.51}$	Bryson et al. (2021)	FGK dwarfs[b]
0.75–1.5	$0.13^{+0.09}_{-0.06}$–$0.11^{+0.07}_{-0.05}$	Kunimoto and Matthews (2020)	G dwarfs[a,b]
0.5–1.5	$0.302^{+0.181}_{-0.113}$	Bryson et al. (2020a)	GK dwarfs[a,c]
0.5–1.5	$0.126^{+0.095}_{-0.055}$	Bryson et al. (2020a)	GK dwarfs[a,d]
0.7–1.5	$0.11^{+0.06}_{-0.04}$	Pascucci et al. (2019)	FGK dwarfs[a,f]
0.7–1.5	$0.05^{+0.07}_{-0.03}$	Pascucci et al. (2019)	FGK dwarfs[a,g]
0.75–1.5	$0.33^{+0.10}_{-0.12}$	Hsu et al. (2020)	M dwarfs[a]
0.75–1.5	$0.04 - 0.40$	Hsu et al. (2019)	GK dwarfs[e]
0.85–1.4	0.33	Hsu et al. (2019)	GK dwarfs[a]
0.72–1.7	0.34 ± 0.02	Zink and Hansen (2019)	G dwarfs[a]
1.0–1.5	$0.41^{+0.29}_{-0.12}$	Hsu et al. (2018)	GK dwarfs[h]
1.0–1.5	$0.31^{+0.02}_{-0.03}$	Garrett et al. (2018)	G dwarfs[a]
0.5–1.5	$0.88^{+0.04}_{-0.03}$	Garrett et al. (2018)	G dwarfs[a]
0.5–1.0	$0.215^{+0.148}_{-0.099}$	Kopparapu et al. (2018)	G dwarfs[i]
1.0–1.75	$0.145^{+0.071}_{-0.0.061}$	Kopparapu et al. (2018)	G dwarfs[j]
0.7–1.5	0.36 ± 0.14	Mulders et al. (2018)	G stars[a]
1.0–1.5	$0.16^{+0.17}_{-0.07}$	Dressing and Charbonneau (2015)	M dwarfs[a]
1.5–2.0	$0.12^{+0.10}_{-0.05}$	Dressing and Charbonneau (2015)	M dwarfs[a]
1.0–1.5	$0.21^{+0.08}_{-0.08}$	Burke et al. (2015)	G dwarfs[k]
0.5–1.5	$0.50^{+0.40}_{-0.20}$	Burke et al. (2015)	G dwarfs[k]
1.0–2.0	$0.064^{+0.034}_{-0.011}$	Silburt et al. (2015)	FGK dwarfs[l]
0.6–1.7	$0.017^{+0.018}_{-0.009}$	Foreman-Mackey et al. (2014)	G dwarfs[a]
1.0–2.0	0.00059	Schlaufman (2014)	G stars[m,n]
1.0–2.0	$0.057^{+0.022}_{-0.017}$	Petigura et al. (2013b)	G stars[m]
0.5–1.4	$0.15^{+0.13}_{-0.06}$	Dressing and Charbonneau (2013)	M dwarfs[a]
0.5–1.4	$0.48^{+0.12}_{-0.24}$	Kopparapu (2013)	M dwarfs[a]
0.5–2.0	0.34 ± 0.14	Traub (2012)	FGK dwarfs[o]
0.8–2.0	$0.028^{+0.019}_{-0.009}$	Catanzarite and Shao (2011)	FGK dwarfs[p]
0.5–3.0	2.75 ± 0.33	Youdin (2011)	G dwarfs[(q)]

Notes: [a]conservative Habitable Zone estimate (Kopparapu 2013); [b]optimistic Habitable Zone estimate (Kopparapu 2013); [c]corrected for reliability; [d]not corrected for reliability; [e]$237 \leq P \leq 500$ days; [f]model #4; [g]model #7; [h]$237 \leq P \leq 320$ days; [i]$0.28 \leq S_{inc} \leq 1.0$; [j]$0.30 \leq S_{inc} \leq 1.12$; [k]$237 \leq P \leq 500$ days; [l]$0.99 \leq a_{HZ} \leq 1.7$ au; [m]$200 \leq P \leq 400$ days; [n]the author also required the Earth-sized planet to have a long-period giant-planet companion; [o]$228 \leq P \leq 1377$ days; [p]$0.75 \leq a_{HZ} \leq 1.8$ au; [q]$P < 1$ year

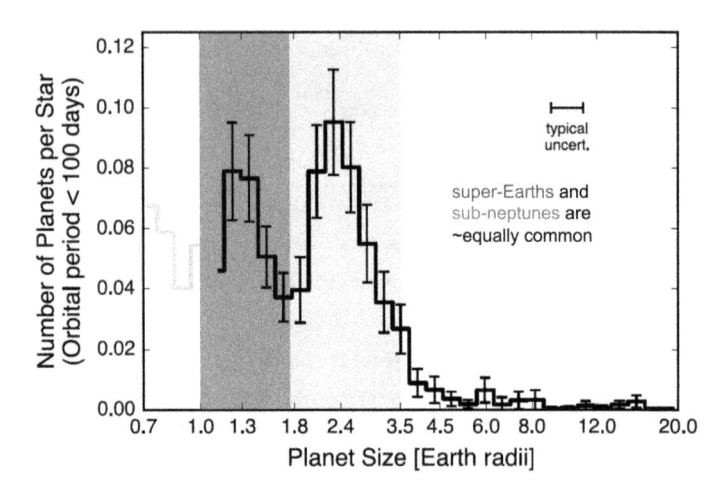

Fig. 3.18 The distribution of close-in planet sizes. The figure was showed by A. Howard in May 2019, during his lecture, and represents an update of the distribution reported by Fulton and Petigura (2018)

of extremely precise direct distance estimates provided by the second intermediate data release (DR2) of the Gaia mission (Gaia Collaboration 2016a, b, 2018) allowed to reduce typical uncertainties on the radius of stars in the *Kepler* field by close to a factor of 10. Using the improved knowledge of the stellar radii, Fulton et al. (2017) and Fulton and Petigura (2018) revisited the radius distribution of close-in ($P < 100$ days), small-size ($R_p < 4.0\,R_\oplus$) planets orbiting bright, unevolved solar-type stars, unveiling its clear bimodality (Fig. 3.18). In particular, the occurrence rate distribution at 1.5–2.0 $R_\oplus$ is suppressed by factor ~ 2. This 'gap' splits the population of close-in small planets into two size regimes of nearly identical intrinsic frequency: super Earths with $R_p < 1.5\,R_\oplus$ and sub-Neptunes with $R_p = 2.0$–$3.0\,R_\oplus$. The physical interpretation for the existence of the radius gap and in-depth studies of these two classes of small planets are the objective of much of recent research in the field, to which we now turn our attention.

3.4 Super-Earths and Sub-Neptunes

3.4.1 Key Questions

The bimodality of the small-radius planet distribution uncovered based on *Kepler* mission data suggests that super-Earths and sub-Neptunes are primarily rocky planets, which were born with primary atmospheres a few percent by mass accreted during the early stages of formation in the protoplanetary disk. Theoretical modeling suggests that terrestrial planet core sizes reach a maximum of ~ 1.6–$1.7\,R_\oplus$

(e.g., Fortney et al. 2007; Valencia et al. 2007; Rogers and Seager 2010; Mordasini et al. 2012b; Rogers 2015; Dorn et al. 2015). Planets with larger radii are mostly low-density and the inclusion of an extended atmosphere becomes a necessary ingredient. The radius gap is thus observed to occur precisely at the transition radius separating planets with and without gaseous envelopes.

Planets above the radius gap, the sub-Neptunes, were somehow able to retain their atmospheres, while planets below it, the super Earths, likely have lost their atmospheres and appear as naked cores. The mechanism that drives atmospheric loss for these planets remains an outstanding question. As with the mass-loss mechanism, the origin of the compositional properties of these two classes of planets is still a matter of debate. The two plausible formation pathways presently considered are[5]:

(a) the growth and migration of embryos from beyond the ice line (*the migration model*, e.g., Ida and Lin 2010; Morbidelli et al. 2015);
(b) inward-drifting of pebbles that coagulate to form planets close-in (*the drift model*, e.g., Ormel et al. 2017; Johansen and Lambrechts 2017).

In principle, density determination for transiting planets with measured radius and mass allows to directly infer their bulk composition, and one would therefore expect that the structural properties of exoplanets so determined would allow to distinguish between proposed scenarios for their formation and evolution. However, mass-radius relationships for small, low-mass planets from theoretical modeling (e.g., Bitsch et al. 2019; Turbet et al. 2020, and references therein) carry intrinsic degeneracies, with planets of vastly different composition (with different amounts of rock, ices, and/or gas) predicted to have observationally indistinguishable bulk densities (e.g., Adams et al. 2008; Miller-Ricci et al. 2009; Lozovsky et al. 2018), particularly for sub-Neptunes above the radius gap. Follow-up atmospheric characterization measurements are therefore necessary to help break any compositional degeneracies. However, density measurements are still of critical importance, as atmospheric analyses rely on the precise (typically, better than 20%) knowledge of mass and radius (e.g., Batalha et al. 2019).

The key questions on the true nature of close-in super Earths and sub-Neptunes can be summarized as follows: Did they form *in situ* or beyond the snowline? What causes the rock-gas transition and at what exact radius? What mechanism sculpted the radius gap? What is the diversity of planet core masses and compositions? Does it depend on stellar properties? There are the two observational channels routinely explored today to tackle these questions, as we discuss in the remainder of this Section: (1) in-depth statistical studies of the radius distribution of small *Kepler* planets, and (2) detailed investigations of the mass-radius relation for small planets.

---

[5] Excellent reviews of the subject can be found in Chap. 1 and, e.g., in Bean et al. (2021) and references therein.

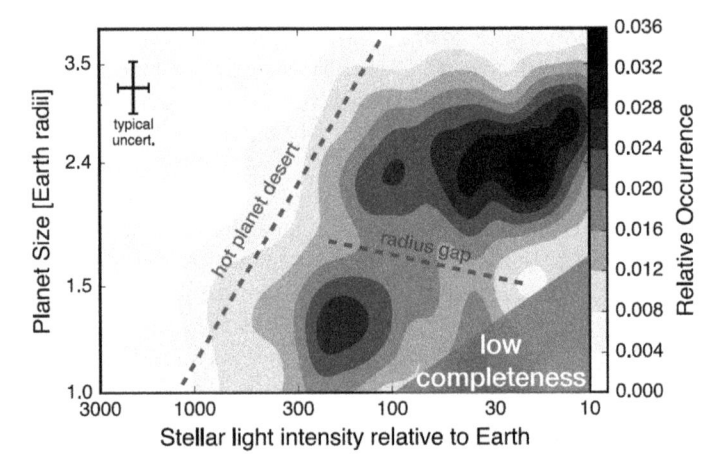

Fig. 3.19 Two-dimensional distribution of planet size and incident stellar flux. There are at least two peaks in the distribution. The figure was showed by A. Howard in May 2019, during his lecture, and represents an update of the distribution reported by Fulton et al. (2017) and Fulton and Petigura (2018)

3.4.2 Precise Radius Demographics

In the Fulton et al. (2017) and Fulton and Petigura (2018) studies, the nature of the two populations of super Earths and sub-Neptunes is best investigated when the planet radius distribution is plotted as a function of incident flux (Fig. 3.19). First of all, there is a clear dearth of $R_p > 2\,R_\oplus$ sub-Neptunes orbiting in high incident flux environments ($S_{inc} > 300\,S_\oplus$). These should be the easiest to detect, yet they do not appear in the Fulton et al. (2017) and Fulton and Petigura (2018) samples. The two studies reinforced the evidence for a scarcity of short-period Neptune-type (in size and mass) planets discussed in early analyses of exoplanet-population statistics from the *Kepler* mission, which had brought about the concept of 'Neptunian desert' (Szabó and Kiss 2011; Mazeh et al. 2016; Lundkvist et al. 2016). Second, most of the super-Earths with radii below the gap are orbiting in environments with $S_{inc} > 200\,S_\oplus$, while sub-Neptunes with radii above the gap typically receive $S_{inc} < 80\,S_\oplus$. Finally, from Fig. 3.19 it is also clear that the gap is present over a very wide range of $S_\oplus$, and there is hint of dependence of the central radius gap value on stellar insolation levels (or orbital period).

In the Fulton et al. (2017) and Fulton and Petigura (2018) studies the location of the peaks in occurrence for super Earths and sub-Neptunes is found at $R_p \sim 1.3\,R_\oplus$ and $R_p \sim 2.4\,R_\oplus$, respectively. More recent analyses with different sub-samples of *Kepler* planets all clearly resolved the radius valley, shifted the location of the peaks closer to $\sim 1.5\,R_\oplus$ and $\sim 2.7\,R_\oplus$, moved the center of the radius valley at ~ 1.9–$2.0\,R_\oplus$, and more robustly established that its position decreases with orbital period, characterizing the slope of the valley as a power law $R \propto P^\gamma$, with $\gamma \simeq -0.1$ (Van Eylen et al. 2018; Martinez et al. 2019; Petigura 2020) for the small-planet population

3 The Demographics of Close-In Planets

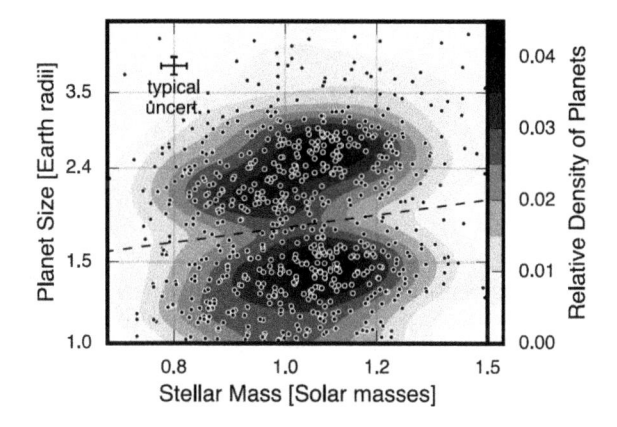

Fig. 3.20 Two-dimensional distribution of stellar mass and planet size. A dashed line is plotted at the location of the gap to guide the eye. Figure adapted from Fulton and Petigura (2018) and reproduced with permission from B. J. Fulton

around F-G-K dwarfs in the *Kepler* field. A slope with an opposite sign is instead recorded for the sample of close-in super Earths and sub-Neptunes found by *Kepler* around M dwarfs (Cloutier and Menou 2020). Fulton and Petigura (2018), Wu (2019) and Cloutier and Menou (2020) also showed that the feature locations (radius of both peaks and valley) move to smaller planet radii with decreasing mass of the F-G-K-M stellar primaries (see Fig. 3.20),

The radius gap (or valley), the 'desert' of hot sub-Neptunes, the location of the peaks in occurrence for super Earths and sub-Neptunes in different regimes of incident flux, and the trends of the feature locations with stellar mass provide some of the most formidable observational constraints for our understanding of the origin and composition of exoplanets with radii between those of Earth and Neptune. The gap has so far been explained primarily in terms of two atmospheric mass-loss mechanisms: core-powered mass-loss (e.g., Ginzburg et al. 2018; Gupta and Schlichting 2019 and photoevaporation (e.g., Owen and Wu 2016, 2017; Lopez & Fortney 2013; Jin and Mordasini 2018).

3.4.2.1 Radius Valley from Core-Powered Mass-Loss

In the core-powered mass-loss model (Ginzburg et al. 2016, 2018; Gupta and Schlichting 2019) the luminosity from a young, cooling rocky core heats a planet's envelope and drives its thermal evolution and mass loss. After a few Gyr of evolution the two end-member states are: (*i*) super-Earths, stripped rocky cores found below the radius valley, and (*ii*) sub-Neptunes, engulfed in H/He atmospheres and located above the valley (see Fig. 3.21). This mechanism is indeed capable to match the valley's location, shape and slope in planet radius–orbital period parameter space, and the relative magnitudes of the planet occurrence rate above and below the valley.

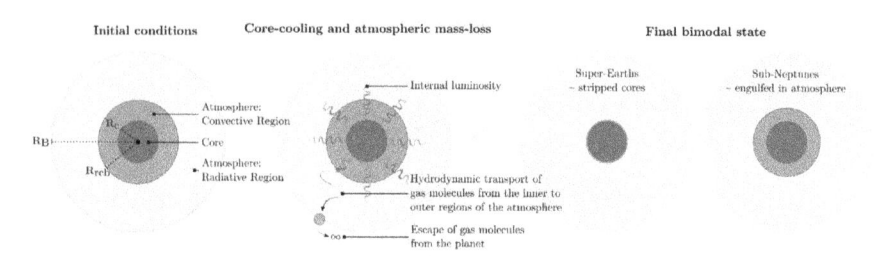

Fig. 3.21 Schematic drawing of planet evolution regulated by the core-powered mass-loss mechanism. The main components of the planet structure (left-hand panel) are the core (dark grey), the convective region (grey), and the radiative region (light grey). The thermal evolution and the atmospheric mass loss at the Bondi radius (middle panel) bring to two possible states at the end of evolution (right-hand panel). Figure taken from Gupta and Schlichting (2019) and reproduced with permission from Akash Gupta

In order to reproduce the correct position of the valley, the model requires that the observed planet population be predominantly composed of rocky cores with typical water–ice fractions of less than $\sim 20\%$.

A key element of the model is that the mass-loss mechanism is a by-product of planet formation processes, therefore it mostly depends on the properties of the planet. The studies carried out thus far have yet to show a dependence of the planet distribution parameters such as S_{inc} and R_p with stellar mass, which is instead clearly observed.

3.4.2.2 Radius Valley from Photoevaporation

The fact that small planets at short orbital periods are strongly influenced by the radiation of their host stars is generally expected within a broad range of theoretical frameworks (for a review of the subject see e.g., Owen 2019). At young ages, close-in planets, with extended atmospheres still in the cooling and contraction processes, absorb more efficiently a large fraction of the central stars' high-energy luminosity component L_{XUV}, which in turn emit the largest fraction of their total (bolometric) luminosity L_{Bol} in this channel precisely when they are young. The evolution of a star's high-energy flux is typically split into two different regimes (e.g., Jackson et al. 2012): a period lasting on the order of 100 Myr during which L_{XUV} is a constant fraction of L_{Bol} (the so-called saturated regime), followed by a clear decline in high-energy output that falls as $L_{XUV}/L_{Bol} \propto t^a$, with $a < -1.0$ (see Fig. 3.22). It is primarily in the saturated regime that the X-ray and UV radiation heats the outer layers of a close-in planet's envelope and drives mass loss, with the atmosphere undergoing severe photoevaporative effects (see cartoon of Fig. 3.23).

Theoretical models predicting that atmospheric erosion of short-period planets should result in the presence of a 'photoevaporation valley' in the radius distribution of planets around $1.7 - 2.0 \, R_\oplus$ appeared in the literature before the radius gap was

3 The Demographics of Close-In Planets 173

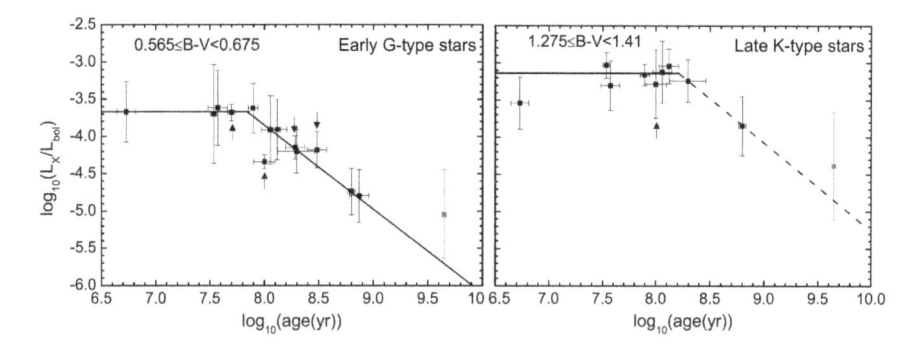

Fig. 3.22 X-ray-to-bolometric luminosity ratio versus age for earlier- (left panel) and later-type (right panel) stars in open clusters analyzed by Jackson et al. (2012). The solid lines indicate the fits to the data, while the dashed line indicates a less certain fit. Figure adapted from Jackson et al. (2012)

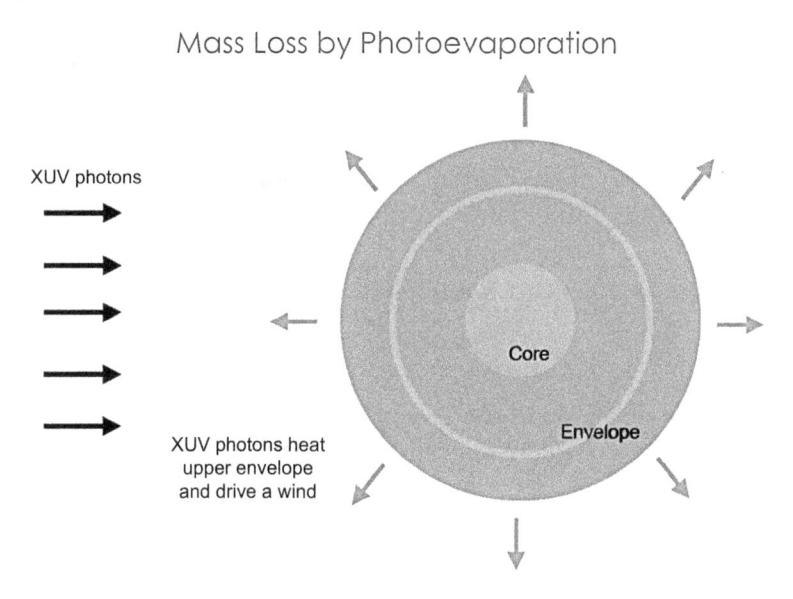

Fig. 3.23 Schematic drawing of mass loss by photoevaporation

actually observed (Owen and Wu 2013; Jin et al. 2014; Lopez and Fortney 2014; Chen and Rogers 2016). An example of the very good match between theoretical expectations and observational evidence is shown in the two panels of Fig. 3.24. Photoevaporation within ~ 100 Myr or so effectively 'herds' small planets into a population of closer-in super-Earths with ~ 1.4 $R_\oplus$, which have been mostly stripped of their envelopes and one of sub-Neptunes further out that retained a significant fraction of their extended atmospheres and have approximately twice the core's radius (~ 2.6 $R_\oplus$).

Fig. 3.24 Comparison between predicted and observed population. *Top:* Result of a planetary-evolution simulation; final radii of planets with an initial mass $M_p < 20\,M_\oplus$, as a function of separation from a Sun-like star. Different colors represent different core masses. Figure adapted from Owen and Wu (2013) and reproduced with permission from James Owen. *Bottom:* same as Fig. 3.19

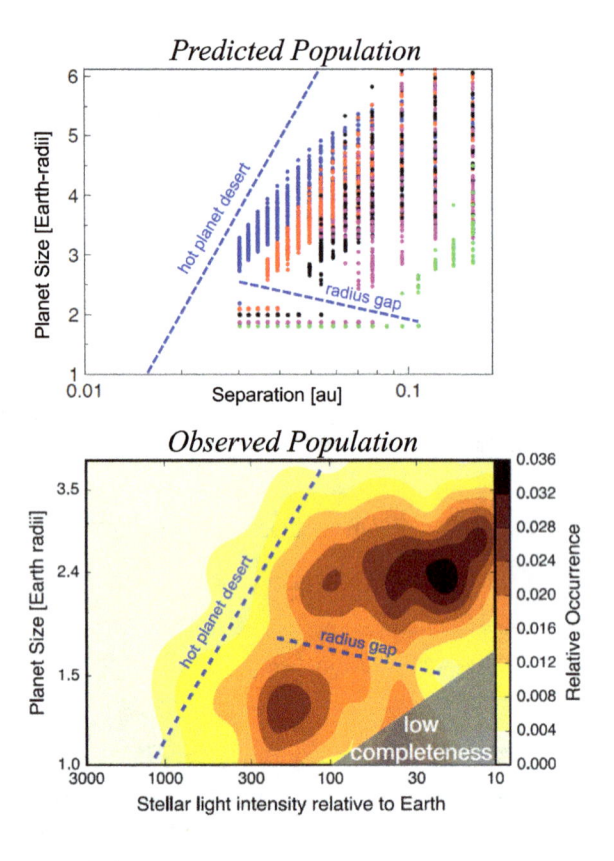

The exact details of the location, shape, and slope of the radius valley are a function of the adopted recipes in terms of planet formation processes, compositional properties of the formed planets, and the physics of evaporation (e.g., Owen and Wu 2017; Lopez & Rice 2018). In the most successful photoevaporation model to-date (Owen and Wu 2017), the valley location is found to be at:

$$R_{\text{valley}} \sim 1.85\,R_\oplus \left(\frac{\rho_{M_\oplus}}{5.5\,\text{g cm}^{-3}}\right)^{-1/3} \left(\frac{M_c}{3\,M_\oplus}\right)^{1/4}, \qquad (3.14)$$

where M_c is the core mass, and $\rho_{M_\oplus}$ is the density of a $1 - M_\oplus$ core that depends only on the core composition. For terrestrial composition, $\rho_{M_\oplus} = 5.5\,\text{g cm}^{-3}$, while it is 11.0, 4.0, and 1.4 g cm^{-3} for pure iron, silicate, and water/ice cores, respectively (Fortney et al. 2007). As shown in Fig. 3.25, the observed data exclude ice-rich cores and favor compositions that are roughly terrestrial, namely silicate–iron composites. In the Owen and Wu (2017) model, the observed features in the small-size planet radius distribution in the *Kepler* field can be explained in terms of a single population of planets initially formed with typical core masses around $3\,M_\oplus$ and terrestrial compositions, that were born with typical H/He envelopes of a few percent in mass.

3 The Demographics of Close-In Planets

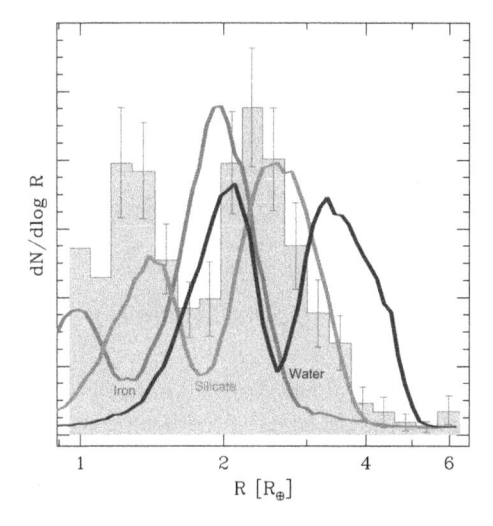

Fig. 3.25 Comparison between the observed radius distribution and models of evolving planets with cores made up of pure iron ($\rho_{core} = 11\,\mathrm{g\,cm^{-3}}$), pure silicate ($\rho_{core} = 41\,\mathrm{g\,cm^{-3}}$), and pure water ($\rho_{core} = 1.3\,\mathrm{g\,cm^{-3}}$). Figure adapted from Owen and Wu (2017) and reproduced with permission from James E. Owen

Furthermore, as the time-integrated ratio L_{XUV}/L_{Bol} is a strong function of stellar mass ($\propto M_\star^{-3}$), the photoevaporation model also predicts shifts in the S_{inc} and R_p distribution with $M_\star$ that are also observed (Fig. 3.20, see also Fulton and Petigura 2018).

3.4.2.3 Other Ways of Creating the Radius Valley

Other small planet formation mechanisms have been proposed that could potentially produce a gap in the size distribution. These include: delayed formation in a gas-poor disk (e.g., Lee et al. 2014; Lee and Chiang 2016) and sculpting by giant impacts (e.g., Lee et al. 2015; Schlichting et al. 2015; Inamdar and Schlichting 2016). However, such models fail to reproduce some of the observed features of the radius distribution, such as the changing location of the gap radius as a function of orbital period or the correct gap radius location.

A key point to consider is that both the photoevaporation and core-powered mass-loss models are able to reproduce the correct position of the radius valley only under the assumption that the remnant cores populating the first peak of the radius distribution are mostly rocky in composition. This directly implies that such ice-poor planets must have formed within the water ice line (Owen and Wu 2017; Gupta and Schlichting 2019). However, a pure dry core composition for most short-period exoplanets is not really expected from formation models (Raymond et al. 2018; Bitsch et al. 2019; Brügger et al. 2020), as the combination of prominent accretion beyond the ice line and type I migration tends to move small-mass planets with $M_p \lesssim 20\,M_\oplus$ inwards quite effectively (e.g., Tanaka et al. 2002). Recent work coupling planet formation and evolution modeling (Venturini et al. 2020) has shown that a bimodality in core mass and composition from birth naturally renders a radius valley at 1.5–2.0 $R_\oplus$. In this framework, the first peak of the *Kepler* size distribution is confirmed to be populated by bare rocky cores, as shown extensively by others (Owen and Wu 2017; Jin and Mordasini 2018; Gupta and Schlichting 2019), while the second peak can host half-rock–half-water planets with thin or non-existent H-He atmospheres, as suggested by a few previous studies (Dorn et al. 2017; Zeng et al. 2019).

3.4.3 Structural Properties of Small Planets

The mass-radius diagram for transiting exoplanets is the fundamental tool that allows to directly compare the observational data with structural models, expressed in terms of iso-density curves that describe the mass-radius relation for a fixed composition, For almost a decade since the discovery of the first transiting planet (Charbonneau et al. 2000), inferences on the structure of exoplanets were confined to the realm of hot Jupiters, to which wide-field ground-based transit surveys (e.g., TrES, XO, HAT-Net, HATSouth, Super-WASP) are mostly sensitive. Nevertheless, the growing numbers of Doppler-detected low-mass ($m_p \sin i < 30\,M_\oplus$) companions and the expectations from very high-precision space-based transit photometry missions (CoRoT and *Kepler*) motivated the development of early structural models and theoretical mass-radius relations for super Earths and sub-Neptunes (e.g., Valencia et al. 2007; Seager et al. 2007; Fortney et al. 2007).

3.4.3.1 The Dawn of Density Measurements of Small Planets

From a historical perspective, the two highly successful space-borne transit surveys, CoRoT and *Kepler*, were not the first to provide initial insights on the structural properties of small planets with $R_p \sim 1 - 4\,R_\oplus$. Indeed, the first planet with a Neptune-like radius was discovered by Gillon et al. (2007) transiting the nearby M-dwarf GJ 436, as part of a follow-up project to search for transits of known Doppler-detected planets. Soon thereafter, the first ground-based transit survey aimed specifically at finding small-size planets around mid- to late-M stars, the MEarth project, identified a transiting sub-Neptune with $R_p = 2.79 \pm 0.2\,R_\oplus$ around the M4.5 V dwarf GJ 1214 (Charbonneau et al. 2009).

Eventually, the era of space-based transit photometry unfolded. Figure 3.26 shows the mass-radius diagram for small planets as of summer 2011. The three notable additions belong to the class of ultra-short-period ($P_{orb} < 1$ day) planets (see Sect. 3.7): a) CoRoT-7 b, the first rocky super-Earth with measured radius (Léger et al. 2009; Hatzes et al. 2010), Kepler-10 b, the first rocky planet delivered by *Kepler* (Batalha et al. 2011), and 55 Cnc e, a hot super Earth that constitutes one of the two most important results from the MOST satellite (Winn et al. 2011). This handful of keystone planets is representative of the diversity of compositional properties of small planets. By comparison with more recent structural and evolutionary models (Zeng

3 The Demographics of Close-In Planets

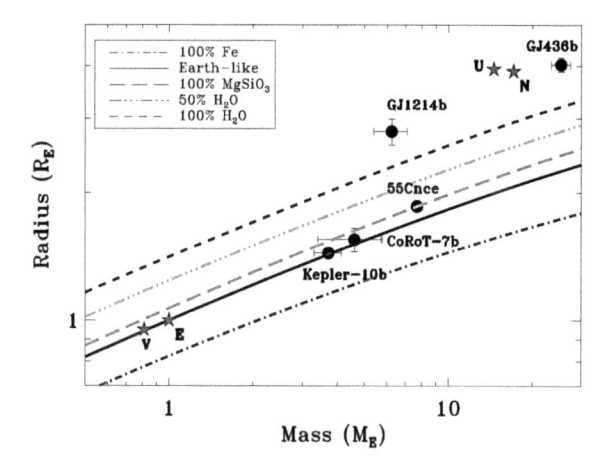

Fig. 3.26 The mass-radius diagram for small transiting exoplanets in mid-2011. The positions of the rocky and ice-giant planets of the Solar System are also shown. The different curves depict internal structure models of a variable composition from Zeng et al. ((Zeng et al. 2019)) (as reported in the legend): 100% iron, Earth-like (32.5% Fe + 67.5% MgSiO3), pure rock (100% MgSiO3), 50% Earth-like rocky core + 50% H₂O layer by mass, and 100% H₂O mass

and Sasselov 2013; Howe et al. 2014; Zeng et al. 2019) one infers that GJ 1214 b and GJ 436 b could be described as rocky worlds with a variable degree of volatiles in their interiors and substantial (2%–10% in mass), thick H₂/He gaseous envelopes of primordial origin (e.g., Lozovsky et al. 2018; Zeng et al. 2019). For Kepler-10 b and CoRoT-7 b, a truly Earth-like composition fits very well the observed mass and radius (e.g., Zeng et al. 2019), while 55 Cnc e has either a very low iron fraction or an envelope of low-density volatiles (e.g., Dai et al. 2019).

3.4.3.2 An Instructive Example

One of the relevant results of the Eta-Earth Keck survey was the discovery announced by Howard et al. (2011) of a warm ($P = 9.5$ days) low-mass planet around the bright K1 dwarf HD 97658. The planet was successively found transiting based on observations with the MOST and Spitzer satellites (Dragomir et al. 2013; Van Grootel et al. 2014). Its measured mass and radius are 7.8 $M_\oplus$ and 2.2 $R_\oplus$, respectively, making it a typical resident of the second peak in the radius distribution of small planets.

The physical properties of HD 97658 b imply a bulk composition that cannot be uniquely determined. Its mass and radius are compatible with those of a planet with a rocky core, a large silicate mantle, and a thin H₂/He envelope. However, the same mass and radius equivalently match those of a water-rich sub-Neptune, with no gaseous envelope (Fig. 3.27). Alternatively, the similar mass and radius can also be reproduced by structural models encompassing a 10%–20% super-critical water hydrosphere topping a 80%–90% mantle-like composition interior (e.g., Mousis et al.

Fig. 3.27 The mass and radius of HD 97658 b highlight the problem of compositional degeneracies for sub-Neptunes, that cannot be resolved solely based on their physical properties (see text)

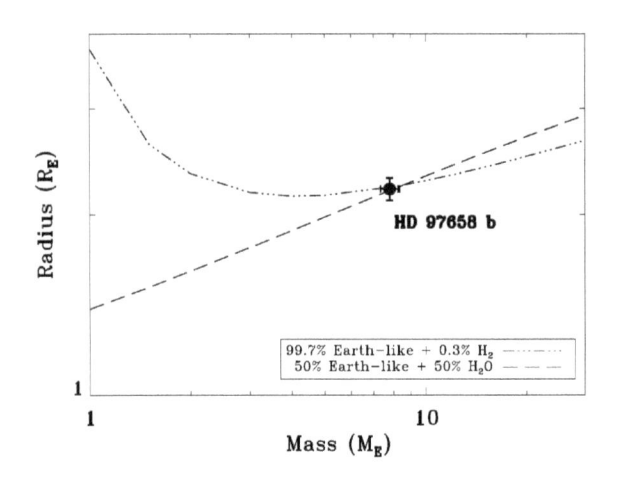

2020). As a consequence, HD 97658 b's formation history is also unclear, as a rocky super Earth with a small H_2/He envelope mass fraction could have formed *in situ*, while a water-rich sub-Neptune is expected to have formed beyond the snowline ($\gtrsim 2$ au) and then migrated inward to its present location at ~ 0.1 au.

To break compositional degeneracies for sub-Neptunes, atmospheric characterization measurements should play a decisive role. In particular, measurements of the transmission spectra allow to directly estimate the mean molecular weight of the planet's atmosphere (e.g., Miller-Ricci et al. 2008a, b; Fortney et al. 2013), which in turn provides improved constraints on its interior composition. Knutson et al. (2014a) utilized HST/WFC3 to perform near-infrared transmission spectroscopy of HD 97658 b's atmosphere, retrieving a featureless transmission spectrum. The most likely explanation is that the planet's atmosphere is dominated by clouds or hydrocarbon hazes that prevent one from detecting molecular absorption features. Similar results have been derived with HST/WFC3 transmission spectroscopy of GJ 1214 b (Kreidberg et al. 2014) and GJ 436 b (Knutson et al. 2014b). This is a potentially strong limitation towards gaining further insights on the overall structural properties of sub-Neptunes such as HD 97568 b. Fortunately, successful detections of molecular absorption have also been reported in the most recent literature for such objects (e.g., Tsiaras et al. 2019; Benneke et al. 2019; Kreidberg et al. 2020; Guilluy et al. 2021; Mikal-Evans et al. 2021). Even in the case of flat transmission spectra measured with HST, there are encouraging prospects for accessing strong molecular features in the atmospheres of sub-Neptunes at longer wavelengths with JWST (Crossfield and Kreidberg 2017; Kawashima et al. 2019).

3.4.3.3 Modern-Day Mass-Radius Relationships

The wealth of $R_p < 4\,R_\oplus$ transiting planet candidates uncovered by the *Kepler* mission refocused the investment of observing time with high-resolution spectro-

3 The Demographics of Close-In Planets

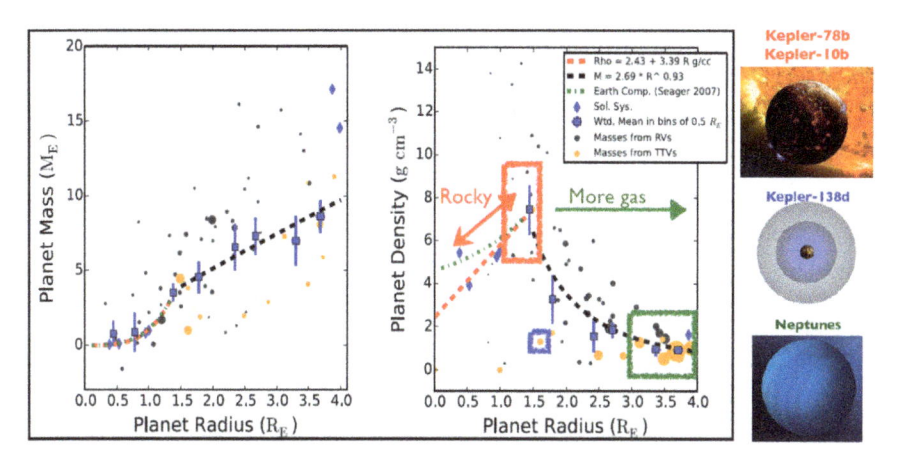

Fig. 3.28 The mass-radius diagram as of 2014. The colored boxes refer to the position of Earth, Super-Earth and Neptunes planets (see the legend on the top of the right-hand panel). Prototypes of these types of planets are shown to the right of the plots. Figure adapted from Weiss and Marcy (2014) and reproduced with permission from Lauren M. Weiss

graphs with high-precision RV measurement capabilities in follow-up programs for mass determination. Marcy et al. (2014) presented a first significant sample of mass measurements (or meaningful upper limits) for small *Kepler* candidates based on Keck/HIRES RVs. This study enabled the first empirical determination of the mass-radius relation for super Earths and sub-Neptunes. Weiss and Marcy (2014) used a sample of 65 planets with $P < 100$ days and with either measured masses or upper limits (including $\sim 10\%$ of objects with negative masses!). They fit a simple parametric model, a power-law, to the observed mass-radius distribution, obtaining $R_p \propto M_p^{0.27}$ for $R_p < 1.5 R_\oplus$ and $R_p \propto M_p^{1.07}$ for $1.5 R_\oplus \leq R_p \leq 4 R_\oplus$. They showed an increase in density with R_p, with a clear peak at $\rho_p \sim 7.6 \, \text{g cm}^{-3}$ for $R_p \sim (1.4\text{--}1.5) R_\oplus$, followed by a sharp decline at larger radii (see Fig. 3.28).

The Weiss and Marcy (2014) analysis constituted the first comparative study of the observed mass-radius relation for close-in super Earths and sub-Neptunes with theoretical expectations. It broadly confirmed the notion that objects with $R_p \lesssim 1.5 R_\oplus$ are mostly rocky in composition, while planets with larger radii must be composed of increasingly larger amounts of volatiles and/or H_2/He gaseous envelopes. Weiss and Marcy (2014) also found a large scatter around the empirical $M_p - R_p$ relation, indicating a significant diversity in planet composition at a given radius.

After the end of the original *Kepler* mission in 2013, its continuation, the K2 mission, and more recently (since 2018) the TESS mission have produced > 1000 new transiting planet candidates in the super-Earth and sub-Neptune radius regime. Most importantly, a large fraction of the candidates orbits around much brighter stars than the original *Kepler* sample. This has enabled more successful, systematic follow-up programs for mass determination with the Doppler method. As of March 2021, there are 192 planets with $R_p < 4 R_\oplus$ and a non-zero mass measurement reported with any precision. This number becomes 146, 102, and 34 with the request of mass measurements good to better than 30%, 20%, and 10% precision, respectively.[6]

With growing numbers of better characterized small-size, low-mass planets, modeling efforts have revisited the initial studies on theoretical mass-radius relations (e.g., Seager et al. 2007; Fortney et al. 2007; Rogers and Seager 2010; Valencia et al. 2010; Lopez et al. 2012). On the one hand, consensus appears to have been reached on threshold radii for a given composition, representing the radius above which a planet has a very low probability of being of that specific composition. Rogers (2015) and Lozovsky et al. (2018) converged on a threshold value of $\sim 1.6\,R_\oplus$ for purely rocky planets, and Lozovsky et al. (2018) determined that planets with radii above $\sim 2.6\,R_\oplus$ (and up to the radius of Neptune) must have retained envelopes of hydrogen and helium typically amounting to 1%–10% in terms of mass fraction. On the other hand, the details of the compositional properties of super Earths and sub-Neptunes have been fine-tuned, particularly in the latter case. Recent work has not only revisited the notion that they could be rocky worlds with small amounts of volatiles in their interiors and substantial, thick H_2/He gaseous envelopes of primordial origin (e.g., Owen and Wu 2017; Van Eylen et al. 2018; Gupta and Schlichting 2019), but it has also re-assessed the likelihood that sub-Neptunes might be 'water worlds' or 'ocean planets', containing significant amounts of water in primarily solid (ices) form, with thin or non-existent H_2/He-dominated atmospheres (e.g., Léger et al. 2004; Zeng et al. 2019; Madhusudhan et al. 2020; Venturini et al. 2020). Finally, recent investigations have also produced predictions on the mass-radius relation for strongly irradiated water-rich rocky planets that might also possess endogenic thick H_2O-dominated, steam atmospheres, in which up to 100% of the planetary water content appears in vaporized form (Dorn et al. 2018; Turbet et al. 2019; Zeng et al. 2019; Mousis et al. 2020).

From an (semi-)empirical point of view, updates to the Weiss and Marcy (2014) power-law fits have been published in the recent past by Wolfgang et al. (2016), Zeng et al. (2016), Bashi et al. (2017), Chen and Kipping (2017), and most recently by Otegi et al. (2020). These works used different cuts in the properties of the small planet population, different prescriptions on the statistical confidence with which planet radii and particularly masses are determined, and differ also in aspects of the methodological approach. Despite some degree of heterogeneity in the planet samples investigated, the above analyses provide broadly consistent results, as summarized in Table 3.2. Other determinations of the mass-radius relationship based on slightly more flexible Bayesian or non-parametric approaches (Ning et al. 2018; Ma & Ghosh 2019; Kanodia et al. 2019; Ulmer-Moll et al. 2019) capture its main features similarly well, reducing in part the spread around the relation.

The two panels of Fig. 3.29 show the state-of-the-art of the mass, radius, and density measurements of super Earths and sub-Neptunes, restricted to mass determinations to better than 20% precision. In the left panel, the power-law fits from Otegi et al. (2020) in the rocky and volatile-rich regime are overplotted. The two populations are rather clearly identified, The dispersion in the relation for rocky planets

---

[6] Data from the transiting exoplanet catalogue TEPCat: https://www.astro.keele.ac.uk/jkt/tepcat/.

3 The Demographics of Close-In Planets

Table 3.2 Parametric mass-radius relations derived for small planets by past studies

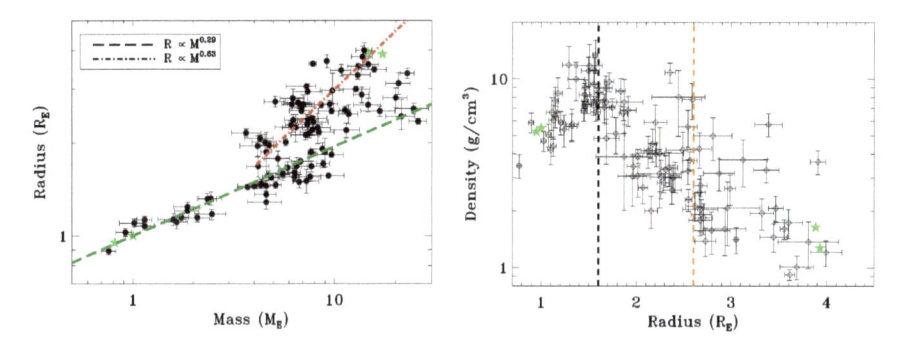

Authors	Power-law Fit	Regime of validity
Weiss and Marcy (2014)	$R_p \propto M_p^{0.27}$	$R_p < 1.5\,R_\oplus$ (rocky)
Weiss and Marcy (2014)	$R_p \propto M_p^{1.07}$	$1.5\,R_\oplus \leq R_p \leq 4.0\,R_\oplus$ (non-rocky)
Wolfgang et al. (2016)	$R_p \propto M_p^{0.66}$	$R_p < 4\,R_\oplus$ (small planets)
Zeng et al. ((Zeng et al. 2016))	$R_p \propto M_p^{0.27}$	$1\,M_\oplus < M_p < 8\,M_\oplus$ (small planets)
Bashi et al. ((Bashi et al. 2017))	$R_p \propto M_p^{0.55}$	$R_p < 12.1\,R_\oplus$, or $M_p < 124\,M_\oplus$ (small planets)
Chen and Kipping ((Chen and Kipping 2017))	$R_p \propto M_p^{0.28}$	$M_p < 2\,M_\oplus$ (Earth-like)
Chen and Kipping ((Chen and Kipping 2017))	$R_p \propto M_p^{0.59}$	$2\,M_\oplus < M_p < 120\,M_\oplus$ (Neptune-type)
Otegi et al. ((Otegi et al. 2020))	$R_p \propto M_p^{0.29}$	$\rho_p > 3.3$ g cm^{-3} (rocky)
Otegi et al. ((Otegi et al. 2020))	$R_p \propto M_p^{0.63}$	$\rho_p < 3.3$ g cm^{-3} (volatile-rich, up to $M_p \sim 120\,M_\oplus$)

Fig. 3.29 *Left*: the state-of-the-art mass-radius diagram for super-Earth and sub-Neptune exoplanets having a mass that was determined with a precision better than 20%. Dashed and dot-dashed lines are the best-fitting lines for the rocky and volatile-rich populations from Otegi et al. (2020). The green stars denote the Solar System planets as in Fig. 3.26. *Right*: the radius-density diagram for the same population. The two vertical dashed lines at 1.6 $R_\oplus$ and 2.6 $R_\oplus$ indicate the threshold radii for the rocky to volatile-rich transition and for the transition between water worlds and planets with gaseous envelopes, respectively

is rather small, with indications of a population of planet with very large iron cores with masses in the approximate range $4 - 10\,M_\oplus$. The most massive rocky planet has $M_p \sim 25\,M_\oplus$. The dispersion around the relation for volatile-rich planets is much larger. In the right panel, the threshold radius of 1.6 $R_\oplus$ for the rocky to volatile-rich transition rather clearly matches the point in which ρ_p changes its dependence with R_p (from direct to inverse). At the transition radius of 2.6 $R_\oplus$ between water worlds and planets with gaseous envelopes proposed by Lozovsky et al. (2018) nothing special seems to happen.

It is finally worth mentioning how the growing sample of super Earths and sub-Neptunes with well-determined properties can be investigated in a statistical sense in order to uncover patterns (e.g., overdensities) in parameter space that might represent

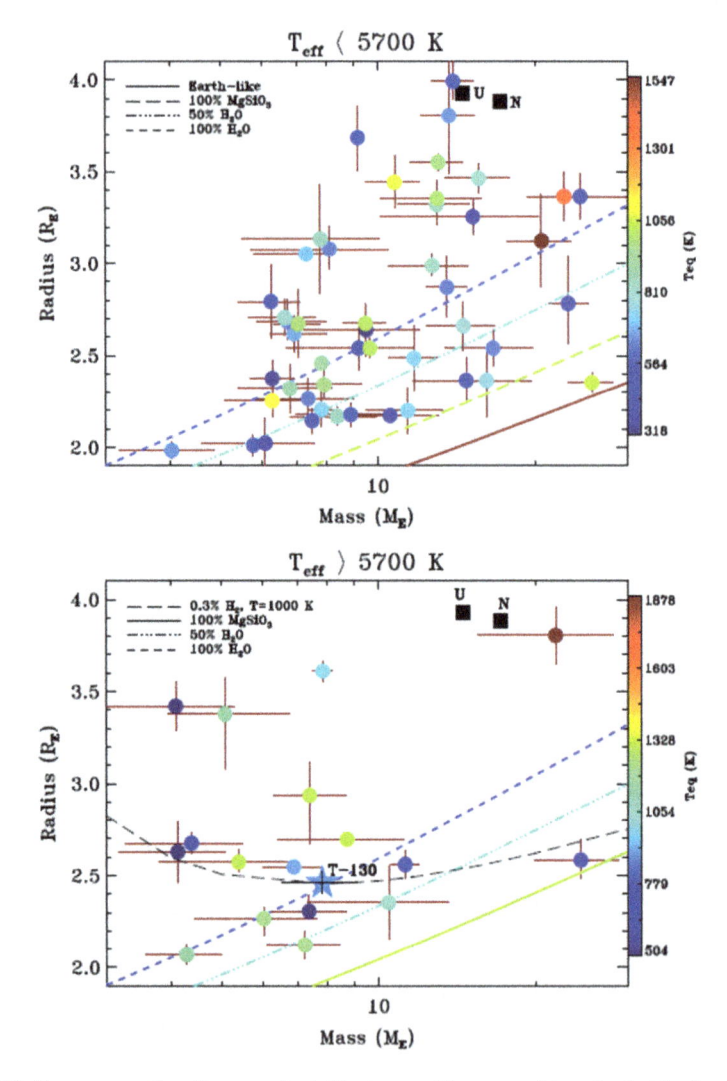

Fig. 3.30 *Top*: mass-radius diagram of sub-Neptunes with masses determined at the 3 σ level (or better), orbiting primaries with $T_{\text{eff}} < 5700$ K. The objects are color-coded by their equilibrium temperature. The different curves depict internal structure models of variable composition from Zeng et al. 2019, as reported in the legend. *Bottom*: same plot, but for sub-Neptunes around primaries with $T_{\text{eff}} < 5700$ K

fossil evidence of the planets' formation and evolution history, as it has been done using radius alone (Fulton et al. 2017; Fulton and Petigura 2018). Considering all sub-Neptunes with masses detected to better than 30% precision, Sozzetti et al. (2021) showed a possible lack of $M_p > 10\,M_\oplus$ companions around G- and F-type stars with $T_{\text{eff}} > 5700$ K, particularly for objects with $R_p \gtrsim 2.5\,R_\oplus$ (see Fig. 3.30). This can be interpreted as fossil evidence of planet formation around stars of varied mass. More massive primaries typically have higher-mass disks. Upon reaching the critical mass (10 $M_\oplus$ or so), newly formed cores have a higher chance of quickly accreting large amount of gas, ending up their formation process as giant planets.

3.5 Planet Multiplicity

The average number of planets per star is a very difficult parameter to constrain precisely, because of the variety of selection effects and variable sensitivity to areas of the parameter space inherent to different detection techniques. Our Solar System has 8 planets, which sounds like a high degree of multiplicity, and a very interesting architecture with the following relevant features: (*a*) quasi-geometric progression in orbital spacings; (*b*) clear hierarchy in mass, with inner terrestrial planets 10–100 s of times lower in mass than the outer gas giants; (*c*) almost circular orbits; (*d*) mostly prograde orbits; (*e*) almost coplanar orbits, with low mutual inclinations (Fig. 3.31).

Are we typical? This is one key question in the field of exoplanet demographics, for which we have yet to find the answer. On the one hand, by comparison with the astonishing diversity of the orbital and physical properties of known planetary systems to-date (see Fig. 3 in Chap. 1 for examples from RV and transit surveys), one would be inclined to draw the conclusion that the one we live in is *unlike* any

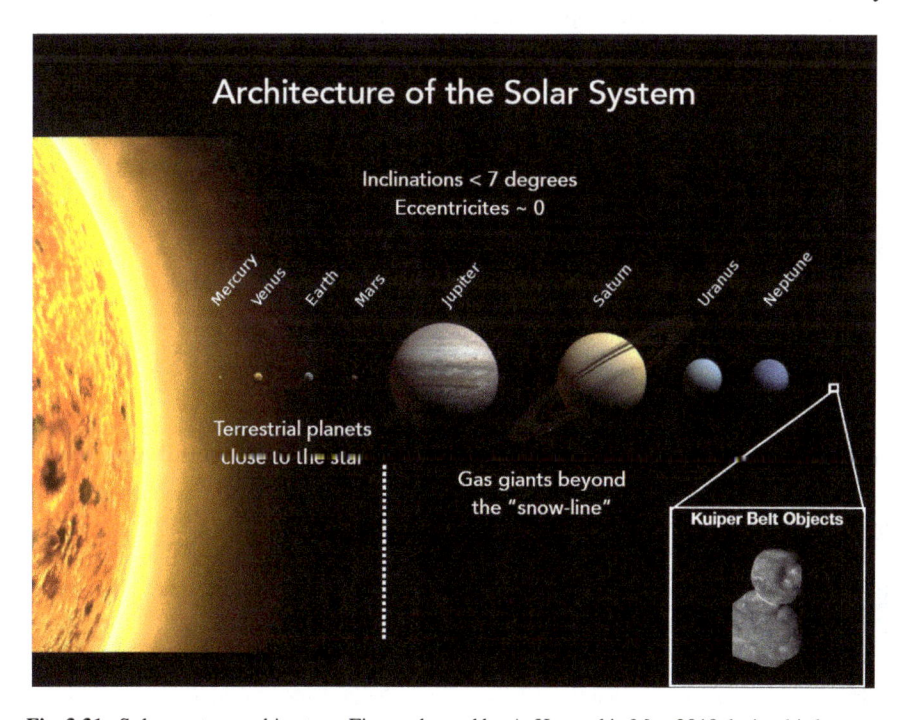

Fig. 3.31 Solar-system architecture. Figure showed by A. Howard in May 2019 during his lectures

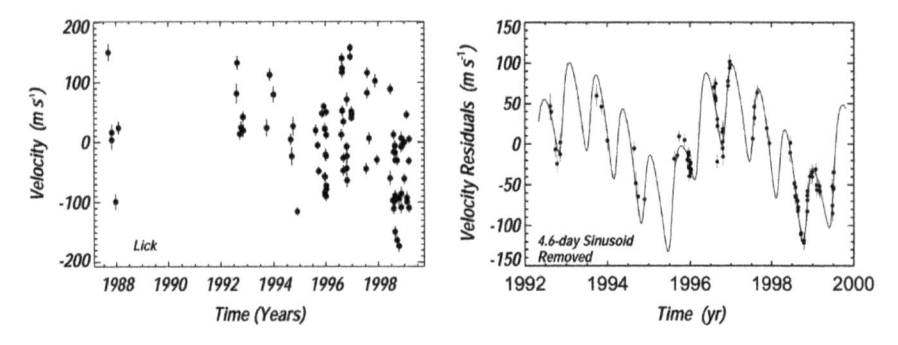

Fig. 3.32 *Left*: RV measurements for υ And from the Lick Observatory. *Right*: residual velocities for υ And after removal of the 4.6-day Keplerian signal from the data. The line shows the best-fit model describing the RV variations due to the outer two companions. Figures taken from Butler et al. (1999) and reproduced with permission from R. P. Butler

other planetary systems, and the rate of occurrence of Solar-System analogs might be (much) below 1% (e.g., Schlaufman 2014. See also Chap. 1). On the other hand, the limited sensitivity of detection techniques to Solar-System-like architectures suggests that the result is still due, at least in part, to observational biases rather than an intrinsically low frequency of true analogs of the Solar System.

The study of the global architecture of planetary systems and the identification of trends and patterns in their properties constitute some of the most formidable observational constraints for planet formation, orbital migration, and dynamical evolution models. As far as multiple-planet systems at short and intermediate separation are concerned, the Doppler and transit techniques have allowed us to make large strides toward the understanding of how different other systems can be from our own.

3.5.1 Multi-planet Systems from Doppler Surveys

3.5.1.1 Early Investigations

The detection of the first multi-planet system around a solar-type star[7] was announced four years after the discovery of 51 Peg b by Butler et al. (1999). Three giant planetary companions were found orbiting the solar-type star υ Andromedae: a hot Jupiter ($M_p \sin i = 0.7\, M_{Jup}$) with $P = 4.6$ d, and two super-Jupiters ($M_p \sin i \sim 2.0 M_{Jup}$ and $M_p \sin i \sim 4.0 M_{Jup}$) on longer periods (241 d and 1278 d, respectively).

Figure 3.32 shows the original RV time-series and the multi-Keplerian orbit fit to the RV signals induced by the two outer gas giants. The amplitudes and periods of

---

[7] Prior to this announcement, Wolszczan and Frail (1992) had published the detection, based on the pulsar timing technique, of two terrestrial-mass companions orbiting at 0.36 au and 0.47 au from the neutron star PSR 1257+12.

the RV modulations could not be explained in terms of any known star-related effects (rotation, pulsations, magnetic cycles) that could mimic a truly Keplerian signal, and their planetary nature therefore safely established. A decade later, Wright et al. (2009) presented a detailed analysis of 67 exoplanets in 28 multiple systems, at the time comprising 14% of known host stars of extrasolar planets within 200 pc. Figure 3.33 presents the chart of semi-major axes and minimum masses for the Wright et al. (2009) sample.

This remarkably diverse sample enabled the first systematic investigation of the statistical properties of multi-planet systems. In particular, Wright et al. (2009) uncovered difference in the population compared to that of systems with only one detected planet: *a*) multiple systems appear to have typically lower eccentricities, *b*) planets in multiple systems are somewhat lower in mass, and *c*) the distribution of orbital distances for multi-planet systems differs from that of single planets. Wright et al. (2009) also found marginal evidence for multi-planet hosts to be slightly more metal-rich that host stars of single planets, while no differences with stellar mass between the two populations arose.

The Wright et al. (2009) sample contained important information on orbital spacing of exoplanet systems. In particular, $\sim 20\%$ of the sample was known at the time to be in or near mean-motion resonances, characterized by period ratios that are nearly equal to ratios of small integers. At least one system (GJ 876) had been shown to undergo strong dynamical interactions, with changes in orbital elements on timescales comparable to those of the RV observations. The spacings of the other systems indicated instead the presence of secular interactions, primarily dictated by different types of apsidal motion (libration, circulation, or at the boundary between the two states), and affecting particularly the values of eccentricity and mutual inclinations on very long timescales.

Broadly speaking, the observed orbital properties (in particular the eccentricity) of the multiple-planet systems in the Wright et al. (2009) sample, as well as their dynamical features in terms of resonant and secular behavior are consistent with planet formation by core/pebble accretion followed by two possible pathways of orbital evolution: (1) smooth gas-driven migration in the protoplanetary disk and (2) a more violent history of early dynamical interactions involving planet-planet scattering and subsequent high-eccentricity migration and tidal circularization. The latter evolutionary channel appears to explain more effectively the properties of both close-in single hot-Jupiter systems and multiple systems with $a < 5$ au (more on this in Sect. 3.6.1. See also Chap. 1).

3.5.1.2 Further Characterization of the Population

As of April 2021, there are ~ 400 Doppler-detected exoplanets in ~ 160 multiple systems, including over 30 systems with at least one transiting planet.[8] The left-hand panel of Fig. 3.34 shows the distribution of period ratios for planet pairs with

---

[8] Source: exoplanet.eu and https://exoplanetarchive.ipac.caltech.edu/ .

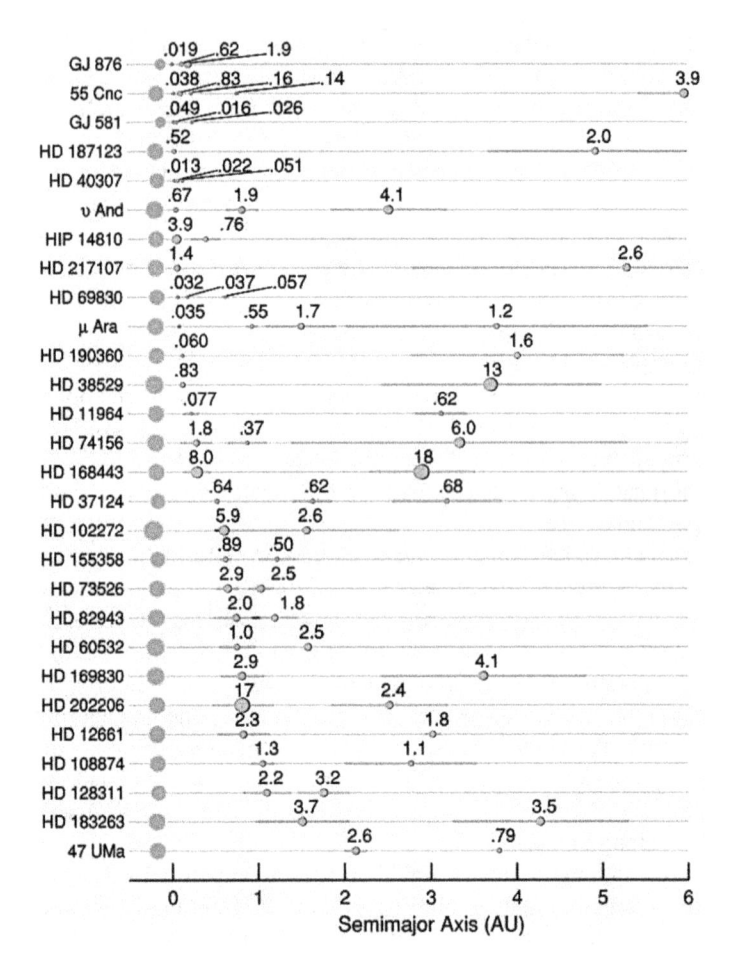

Fig. 3.33 Chart of semi-major axes and minimum masses for the 28 known multiplanet systems as of 2009. The first circle in each row represent the star, while the others represent the hosted planets. The sizes of the circles are proportional to the cube root of the stellar mass and the planetary $M_p \sin i$, respectively. The horizontal line intersecting the planet represents the periapse to apoapse excursion. Figure adapted from Wright et al. (2009) and reproduced with permission from Jason T. Wright

$P_{out}/P_{in} < 4$. There are clear peaks just wide of the 3:2 and 2:1 resonances, and another peak just interior to the 4:1 resonance. The smallest period ratio is recorded for the pair of super Earths HD 215152 b,c (Delisle et al. 2018), $\sim 1\%$ wide of the 5:4 resonance. From the right-hand panel of Fig. 3.34 we learn that the peak near the 2:1 resonance is populated by pairs of giant planets with total mass $\gtrsim 1\,M_{Jup}$, as it had already been noted by (Winn and Fabrycky 2015). The pairs in the peaks next to the 3:2 and 4:1 resonances are instead primarily composed of low-mass planets with masses $\lesssim 0.1\,M_{Jup}$. In general, most pairs in multi-planet systems detected in RV surveys do not lie near a period commensurability, however their mass correlates rather clearly with period ratio (Fig. 3.35).

3 The Demographics of Close-In Planets

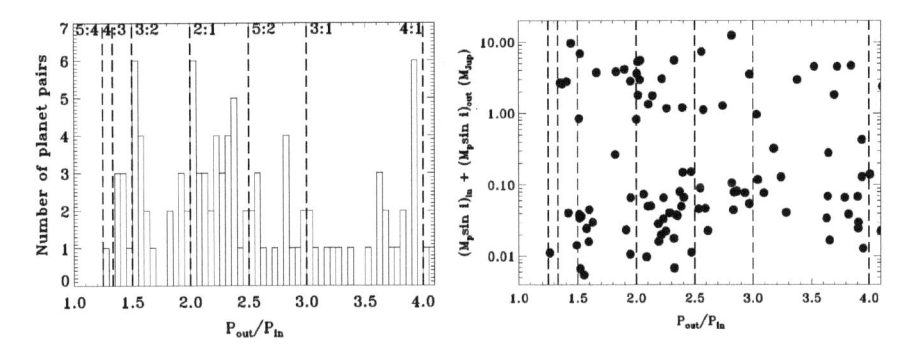

Fig. 3.34 *Left*: the distribution of period ratios for Doppler-detected planet pairs with $P_{out}/P_{in} < 4$. Vertical lines indicate first- and higher-order resonances. *Right*: total mass in a pair versus period ratio. The vertical lines are the same as in the left panel

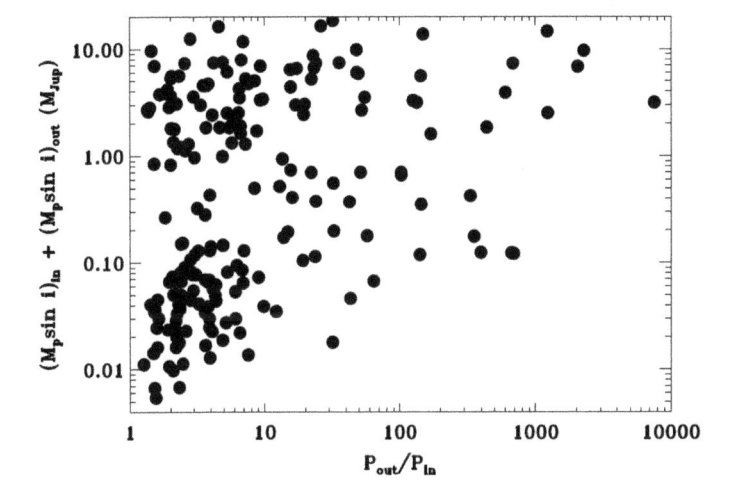

Fig. 3.35 Same as Fig. 3.34, but for all values of period ratio

The four panels of Fig. 3.36 show other interesting features of the population of multi-planet systems from RV surveys. First, the mass ratio clearly correlates with period ratio (top left panel), with only 28% of pairs in which the outer companion is the less massive one. Second, pairs with relatively low mass ratios ($\lesssim 4$) are more often composed of low-mass planets (top right panel), with a sparse population ($\sim 10\%$) of systems with high mass ratios (> 10) composed of an inner low-mass planet and an outer massive companion with a large period ratio. The masses of planet pairs are strongly correlated across all mass regimes (bottom left panel), but particularly so for companions with minimum masses below $10\,M_\oplus$. The sample of inner planets with $10\,M_\oplus \lesssim M_p \sin i \lesssim 30\,M_\oplus$ is instead often accompanied by an outer, significantly more massive companion. Finally, low-mass pairs are almost exclusively found around K- and M-dwarf primaries ($M_\star \lesssim 0.8\,M_\odot$), while the sam-

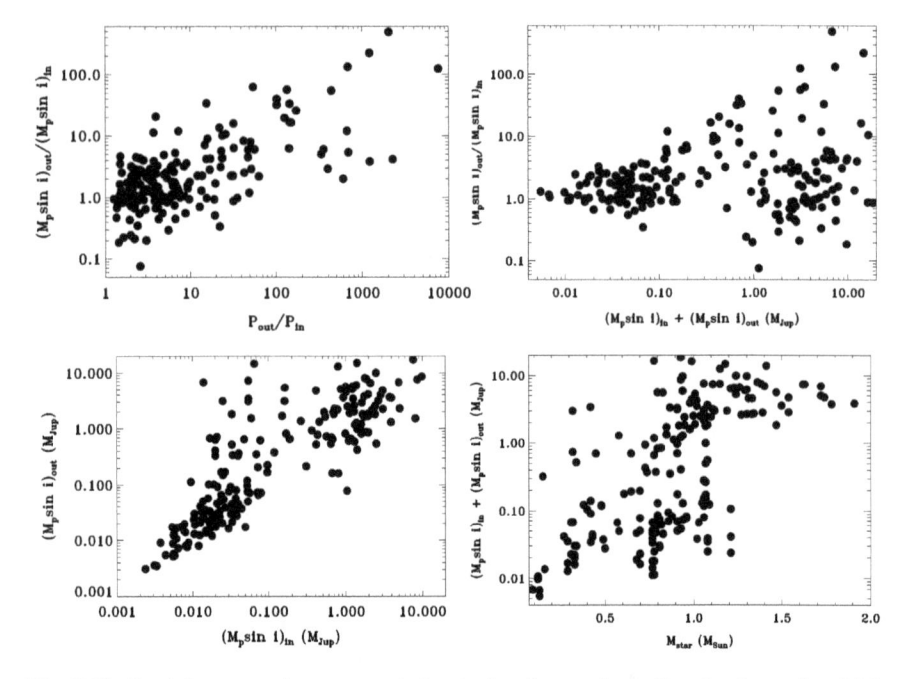

Fig. 3.36 *Top left*: mass ratio versus period ratio for planet pairs in Doppler-detected multiple systems. *Top right*: mass ratio versus total mass in pair. *Bottom left*: correlation diagram between minimum masses of the inner and outer planet in pairs. *Bottom right*: total mass in planet pairs versus mass of the stellar hosts

ple of planet pairs orbiting higher-mass stars is dominated by high-mass planets (bottom right panel).

Compact systems of low-mass planets at short orbital separations are increasingly more difficult (and very observing-time consuming) to unveil with high statistical confidence in the detections around increasingly higher-mass stars, due to their very low-amplitude RV signals (e.g., Delisle et al. 2018; Udry et al. 2019; Hara et al. 2020). It is therefore difficult for Doppler surveys to establish whether the apparently higher low-mass, close-in planet multiplicity for later spectral types is indeed real or due to observational biases. For one thing, the highest planet multiplicity has been recorded for the solar-type star HD 10180, which hosts at least six, and up to nine!, planets, 5 or 6 inside the orbit of Mercury (Lovis 2011; Tuomi 2012). The fact that adjacent pairs of close-in low-mass companions appear to have rather similar (minimum) masses and similar spacings is instead quite intriguing, and awaits a clear explanation from a theoretical viewpoint.

The structure of peaks and deficits near resonance in the period ratio distribution can be interpreted in terms of a couple of alternative scenarios of formation and subsequent orbital evolution of low-mass multi-planet systems. A viable path is *in-situ* formation in gas-poor conditions, in the late stages of the disk's lifetime, with little or no orbital migration (e.g., Lithwick et al. 2012; Ogihara et al. 2018;

3 The Demographics of Close-In Planets

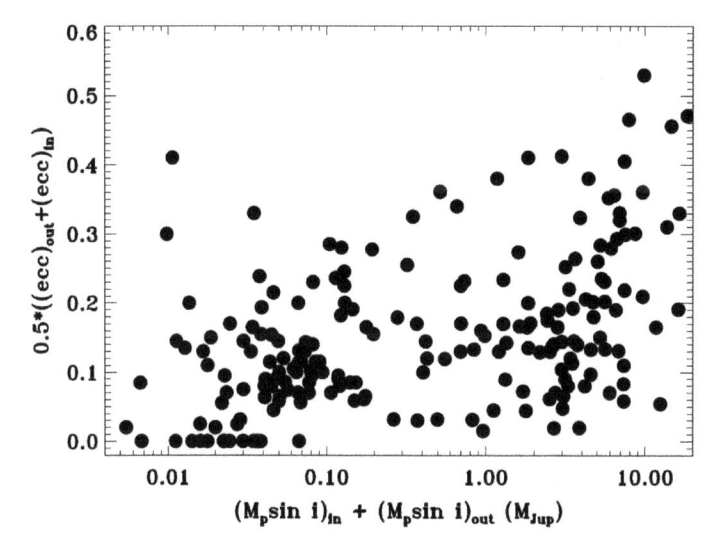

Fig. 3.37 Average eccentricity versus total mass for planet pairs in Doppler-detected multiple-planet systems

Terquem and Papaloizou 2019; Choksi and Chiang 2020, and references therein). Alternatively, such systems might have formed early in the gas-rich regions of the protoplanetary disk, undergone smooth disk-driven migration and resonance capture, and subsequently escaped resonance due to dynamical instability effects driven by a variety of other physical processes (e.g., Goldreich and Schlichting 2014; Izidoro et al. 2017; Lambrechts et al. 2019).

The average eccentricity of planet pairs is clearly correlated with the total mass (see Fig. 3.37), with low-mass multiples often compatible with having almost perfectly circular orbits.[9] As the Super Earths and Neptunes uncovered by Doppler surveys in multiple systems tend to have quite similar minimum masses (Fig. 3.36), it is tempting to conclude that they are also likely to be on nearly coplanar orbits, which could help in understanding whether their architectures have been shaped by 'clean' disk migration followed by phases of dynamical instability that retained both low eccentricities and low mutual inclinations (e.g., Esteves et al. 2020, and references therein), or they are produced by different *in-situ* formation conditions (e.g., MacDonald et al. 2020, and references therein).

Unfortunately, mutual inclination angles cannot be determined based on RV observations alone, unless planet-planet interactions produce detectable perturbations in the host star's motion beyond the simple superposition of Keplerian orbits. This has been possible in practice only for the two gas giants in a 2:1 resonance in the four-

---

[9] Note that the reported non-zero eccentricity values for low-mass companions are often compatible with zero within the error bars (not shown in the plot). It is in fact notoriously difficult to measure statistically significant small eccentricities for the low-amplitude RV signals induced by low-mass planets, as the estimated uncertainty on e scales as $\sigma_e = \frac{\sigma_K}{K} \left(\frac{2}{N_{\text{obs}}}\right)^{1/2}$.

planet system around GJ 876, which were determined to have a mutual inclination of less than 5 deg (e.g., Rivera and Lissauer 2001; Bean et al. 2009; Rivera et al. 2010). The combination of RV data and relative HST astrometry (McArthur et al. 2010) allowed to measure a significant mutual inclination (30 ± 1 deg) between the two outer gas giants orbiting υ And. Most recently, the combination of RV data and Gaia-Hipparcos absolute astrometry allowed to infer that the hot super Earth found transiting around the naked-eye solar-type star π Men (Huang et al. 2018) must be on a significantly non-coplanar configuration with the outer, eccentric, massive giant planetary companion (Damasso et al. 2020; see also De Rosa et al. 2020 and Xuan and Wyatt 2020). Smooth disk migration likely explains the architecture of the resonant pair of gas giants around GJ 876 (e.g., Dempsey and Nelson 2018 and references therein). The detailed architectural properties of the υ And and π Men systems are more naturally explained by planet-planet scattering processes (e.g., Ford et al. 2008). Any attempt at observationally determining the mutual inclination of low-mass multiples in RV surveys remains so far elusive.

3.5.2 Architecture of Kepler 'Multis': Global Patterns

Two years after theWright et al. (2009) analysis, NASA's *Kepler* mission provided a new boost to the investigations of multi-planet demographics. Among the thousands of new exoplanet candidates, it uncovered a population of $\sim$ 400 (as of April 2021) transiting multi-planet systems (Borucki et al. 2011; Lissauer et al. 2011b; Latham et al. 2011; Burke et al. 2014; Rowe et al. 2014) with orbital separations < 1 au. Based on statistical as well dynamical stability arguments, almost all of them are real ($\gtrsim$ 99%, e.g., Lissauer et al. 2012; Fabrycky et al. 2014). *Kepler* provided the first statistically robust evidence for the existence of a class of close-in, densely packed systems of small-size, low-mass planets, including record-holders Kepler-11 (six planets within Mercury's orbit; Lissauer et al. 2011a), Kepler-20 (six planets within Mercury's orbit, one of which non-transiting; Buchhave et al. 2016 and references therein), and Kepler-90 (eight planets within 1 au but 5 within Mercury's orbit; Shallue and Vanderburg 2018 and references therein). Such a large sample enables the robust identification of patterns in the intrinsic distributions of planets, which provide fundamental insights for models of the formation and evolution of planetary systems with a resolution that is not achievable by Doppler detections.

3.5.2.1 Kepler Singles Versus 'Multis'

The first question to ask is whether *Kepler*'s multi-planet systems (dubbed 'multis' thereafter) and those in which a single planet is known to transit (non-transiting planets could very well be present!) are drawn from the same parent population. The visual comparison of the population of singles and multis in the radius-orbital period space (Fig. 3.38) would make us infer this is indeed the case (except for hot Jupiters

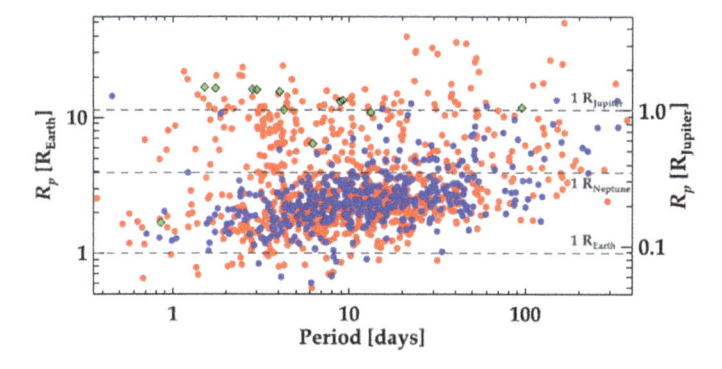

Fig. 3.38 The diagram of the planet radius versus orbital period for the planet candidates found during the first four months of the *Kepler* mission. Single candidate transiting planets are marked in red, multiple candidate transiting planets are marked in blue. The CoRoT planets known at the time are plotted in green. The figure was made by Samuel N. Quinn (Latham et al. 2011) and reproduced with permission from D. W. Latham

that tend to be 'lonely'; see, e.g., Latham et al. 2011; Steffen et al. 2012). Similarly, the radius valley seems to be present both for singles and multis (Fig. 3.39, left panel). Finally, the properties (particularly mass and metallicity) of the parent stars of *Kepler* singles and multis appear indistinguishable (an example is shown in the right panel of Fig. 3.39). This has brought many to conclude that the two populations are statistically the same (e.g., Xie et al. 2016; Munoz Romero and Kempton 2018; Zhu et al. 2018; Weiss et al. 2018a; for alternative views see, e.g., Brewer et al. 2018 and Anderson et al. 2021).

Small planet multiplicity within 1 au in the original *Kepler* field is not, however, independent of spectral type. We have already shown in Fig. 3.16 the dependence of occurrence rates with stellar mass originally uncovered by Howard et al. (2012). Several studies that followed confirmed the existence of the trend (e.g., Mulders et al. 2015a; Zhu et al. 2018). The latest work on the subject (Yang et al. 2020) has further confirmed this notion, establishing that planet hosts with $3000\,\text{K} \lesssim T_{\text{eff}} \lesssim 5500\,\text{K}$ are orbited typically by ~ 2.8 planets with $P < 400$ days, a number that drops sharply to ~ 1.8 for $T_{\text{eff}} \gtrsim 6000\,\text{K}$ (Fig. 3.40).

3.5.2.2 The *Kepler* 'Dichotomy'

With such a large number of multi-planet systems identified by *Kepler*, the answer to the question: "What is the true multiplicity distribution of exoplanetary systems out to 1 au?" appears within reach. As the transit technique is by construction prone to miss planets in systems that are not almost perfectly coplanar, in order to model the planet number distribution one needs to make an hypothesis for both the the intrinsic multiplicity distribution and the mutual inclination distribution. Lissauer et al. (2011b) modeled the former as a uniform (or Poisson) distribution and the latter

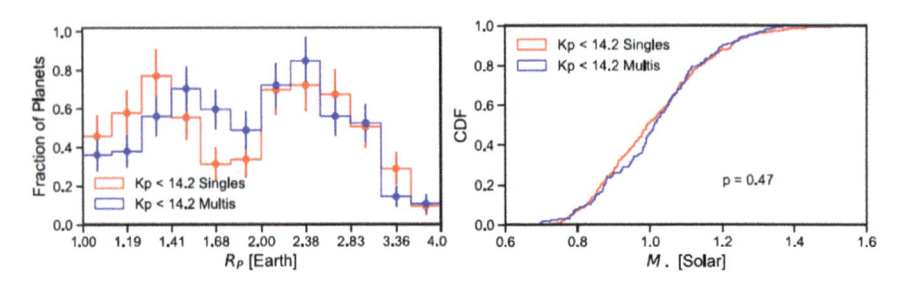

Fig. 3.39 *Left*: the distribution of planet radius for $K_p < 14.2$ mag systems with one (red) and multiple (blue) transiting planets. Only planets with $1\,R_\oplus < R_p < 4\,R_\oplus$ and $P_{\text{orb}} > 3$ days are shown. *Right*: cumulative distribution function (CDF) of the host-star masses for single and multi *Kepler* systems. Figure taken from Weiss et al. (2018a) and reproduced with permission from Lauren M. Weiss

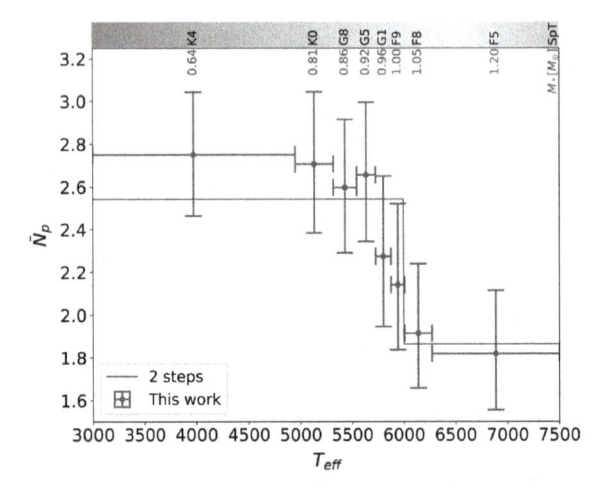

Fig. 3.40 Number of planets per star for *Kepler* planet hosts as functions of T_{eff}. Figure adapted from Yang et al. (2020) and reproduced with permission from Jia-Yi Yang

as a Rayleigh distribution, with a good match to all observed multiplicities except the singly transiting planets, whose occurrence was under-predicted by nearly 50%. The discrepancy could be reconciled if one considers the *Kepler* singles and multis as two distinct populations of planetary systems (orbiting stars with a broad range of spectral types and metallicities), a feature usually dubbed as the '*Kepler* dichotomy' (see Fig. 3.41): the first population is made up of dynamically 'cold', densely packed compact planetary systems with small mutual inclinations ($\sim 2°$), the second is composed of truly single close-in planets or dynamically 'hot' multi-planet systems (see cartoon of Fig. 3.42) with at least two companions with largely mutually inclined orbits (e.g., Johansen et al. 2012; Fang and Margot 2012; Tremaine and Dong 2012; Ballard and Johnson 2016; Munoz Romero and Kempton 2018; Mulders et al. 2018; He et al. 2019).

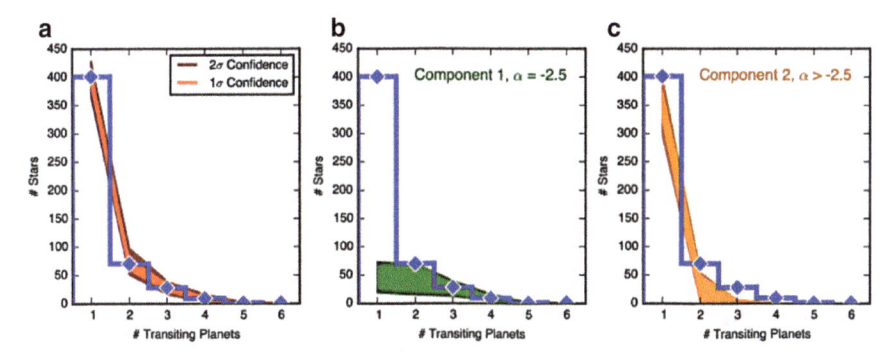

Fig. 3.41 Comparison between the *Kepler* multi-planet yield, in blue, to a best-fit power-law mixture model in multiplicity $\propto m^\alpha$ with 1σ and 2σ confidence levels. Panel **a** shows the result from the full model; panel **b** shows the contribution from a component with power-law index $\alpha = -2.5$, describing the high-multiplicity sector of the distribution; panel **c** shows the contribution from a component with power-law index $\alpha = -4$, describing the low-multiplicity sector of the distribution. Figure adapted from Ballard and Johnson (2016)

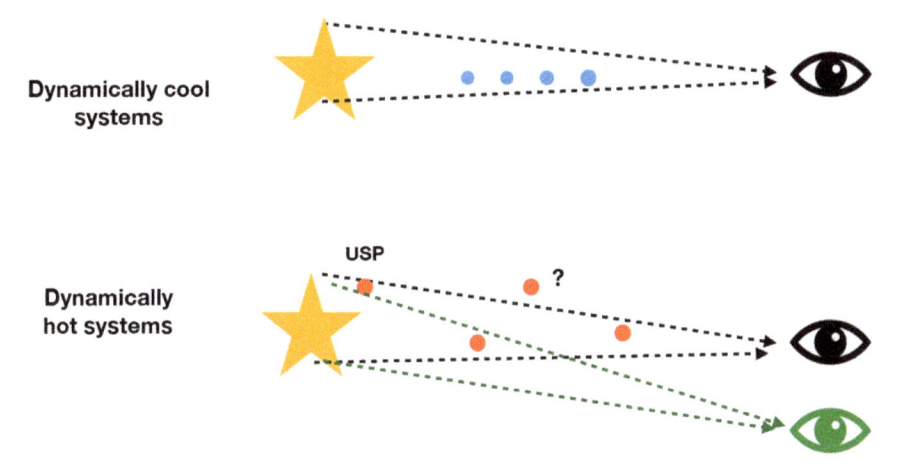

Fig. 3.42 Schematic drawing of the *Kepler* dichotomy: planetary systems densely packed and with small mutual inclinations are considered dynamically 'cold', whereas those composed of single close-in planets or several planets with large mutual inclinations are considered dynamically 'hot'

Recent work argues that the dichotomous model is not necessary, depending on the choice for the functional form of the multiplicity and mutual inclinations distributions (Gaidos et al. 2016; Bovaird and Lineweaver 2017; Sandford et al. 2019). The *Kepler* dichotomy might also artificially arise at least in part because of the increasingly lower detection efficiency of the *Kepler* pipeline with increasing planet multiplicity (Zink et al. 2019).

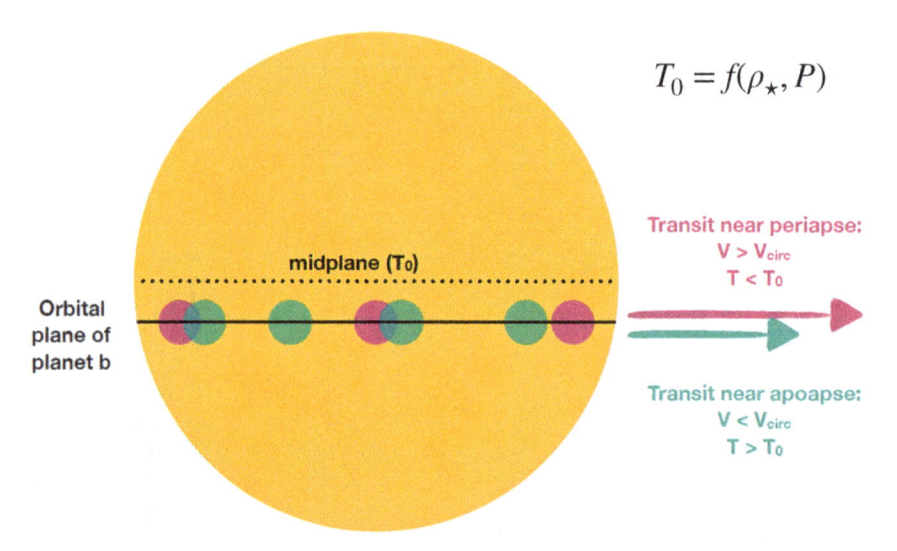

Fig. 3.43 Schematic drawing of how the transit duration, T, can change according to the orbital parameters. Here T_0 represents the transit duration corresponding to a central ($b = 0$) transit for a planet on a circular orbit ($e = 0$)

3.5.2.3 Eccentricities and Mutual Inclinations

A possibly convergent picture on the dynamical differences and similarities between *Kepler* systems with different multiplicity can be drawn when considering the combined information coming from in-depth studies of both eccentricity and mutual inclination distributions. The full duration of a transit can be expressed as:

$$T_{\text{dur}}/T_0 = \sqrt{(1 + R_p/R_\star)^2 - b^2} \frac{\sqrt{1-e^2}}{1 + e\sin\omega}, \quad (3.15)$$

with $T_0 \simeq 13\,\text{hr}\,(P/\text{year})^{1/3}\,(\rho_\star/\rho_\odot)^{-1/3}$ the transit duration corresponding to a central ($b = 0$) transit for a planet on a circular orbit ($e = 0$). See the cartoon of Fig. 3.43 for a visual description. It is therefore possible to study the distribution of T_{dur}/T_0 to infer the underlying distribution of eccentricities. Early attempts at determining the eccentricity distribution of *Kepler* multis (e.g., Moorhead et al. 2011; Kane et al. 2012; Wu and Lithwick 2013) were hampered by the pre-Gaia limited knowledge of the host stars. With more precisely determined stellar properties, the method has been applied by several authors (Van Eylen et al. 2015; Xie et al. 2016; Mills et al. 2019; Van Eylen et al. 2019). The clear, statistically robust picture portrayed by these studies is that *Kepler* multis are found on nearly-circular orbits (median $e \sim 0.05$), while *Kepler* systems with one detected transiting planet exhibit a large eccentricity dispersion, with a study-dependent median e in the range 0.15–0.3 (see Fig. 3.44 for an example). There is tentative evidence that the most eccentric *Kepler* singles might

3 The Demographics of Close-In Planets

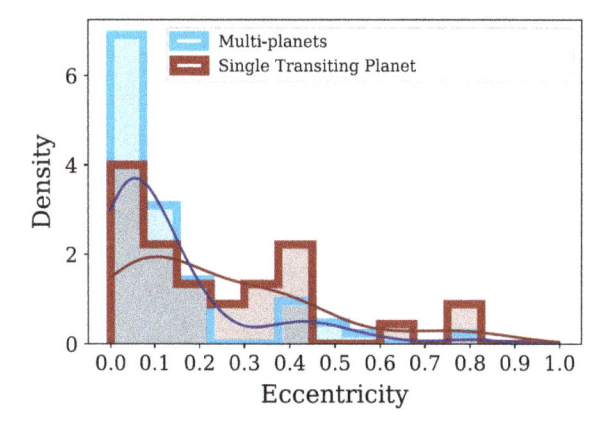

Fig. 3.44 Density histogram of the eccentricity distribution of *Kepler* systems with multiple transiting planets and systems with a single transiting planet. The bins have an arbitrarily chosen width of 0.075. Solid lines indicate kernel density estimates. Multi-planet systems clearly have a higher density of low eccentricities. Figure adapted from Van Eylen et al. (2019) and reproduced with permission from Vincent Van Eylen

preferentially orbit more metal-rich stars (Mills et al. 2019). We further elaborate on this point in Sect. 3.6.2.

The normalized transit duration ratio:

$$\xi = \frac{T_{\mathrm{dur,in}}/P_{\mathrm{in}}^{1/3}}{T_{\mathrm{dur,out}}/P_{\mathrm{out}}^{1/3}} \tag{3.16}$$

contains information on the mutual inclination of a planet pair in a multi-planet system (the subscripts in and out identify the inner and outer companion). Studies of the ξ distribution for *Kepler* multis (Fang and Margot 2012; Tremaine and Dong 2012; Fabrycky et al. 2014) agree on the fact that the mutual inclinations between *Kepler* multis can be described by a Rayleigh distribution with a dispersion of a few degrees (see Fig. 3.45).

More recent analyses (e.g., Zhu et al. 2018; He et al. 2020) have further corroborated the notion that mutual inclinations and eccentricities of *Kepler* multis are strongly correlated, and quantified the power-law dependence of the two distributions with multiplicity m ($\propto m^\gamma$, with $\gamma \sim -1.7/-2.0$ in both cases),[10] providing additional evidence for the solution of the *Kepler* dichotomy. Finally, Dai et al. (2018a) have uncovered a clear trend of increasing mutual inclination in planet pairs as a function of decreasing orbital distance of the innermost planet (for more details, see Sect. 3.7.6).

---

[10] Studies based on RV multis or a combination of transiting and RV multis have also identified an anti-correlation between orbital eccentricity and multiplicity (Limbach and Turner 2015; Zinzi and Turrini 2017; Turrini et al. 2020; Bach-Møller and Jørgensen 2021).

Fig. 3.45 Histograms of normalized transit duration ratios (Eq. 3.16). The solid histogram is the observed distribution, the dashed histogram is the distribution from simulated detections based on the best-fit model described in (Fang and Margot 2012). Figure reproduced with permission from Jean-Luc Margot

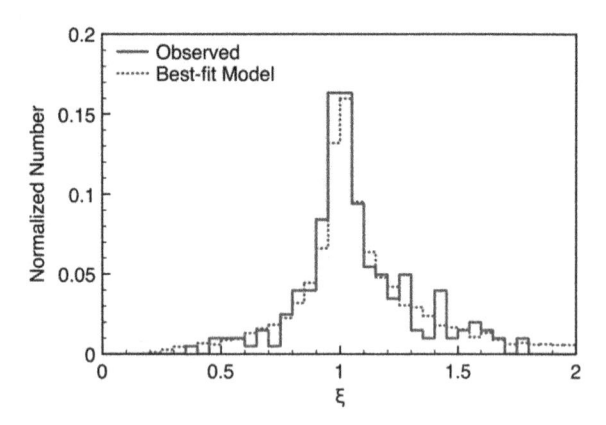

3.5.3 Architecture of Kepler 'Multis': Planet-to-Planet Patterns

Several other studies of *Kepler* exoplanetary systems have identified additional patterns in their observed architectures, particularly in relation to the intra-system variations of sizes and spacings, which are potentially very constraining for formation and evolution models.

3.5.3.1 Size Ordering and Period Ratios

The two panels of Fig. 3.46 show *Kepler* multis containing three or more, four or more planets ordered by orbital period or orbital separation of the innermost planet as well as stellar mass (Fabrycky et al. 2014; Weiss et al. 2018b). Visually, it appears that the largest planets are typically those with longer periods. It is also not difficult to get the impression that generally systems appear to have rather similar sizes and regular spacings.

Using a sample of ~ 900 planets in 365 *Kepler* multis, Ciardi et al. (2013) studied the relative sizes of the planetary radii for all planet pairs, finding that when at least one planet in a pair has radius $\gtrsim 3\, R_\oplus$ then for $\sim 70\%$ of the pairs the outer planet is larger than the inner planet. This number is almost the same as the one derived when looking at the minimum mass hierarchy in the RV sample of multis (Sect. 3.5.1.2). For pairs with both radii below $3\, R_\oplus$ no such size ordering was recorded.

The second much discussed pattern in orbital architecture of *Kepler* multis is related to the period ratio distribution. The histogram of period ratios for all pairs of planets with radii $< 4\, R_\oplus$ (Fig. 3.47) clearly shows that most of the systems are not in or close to mean motion resonances, but but there are excesses of systems just wide of the 3:2, 2:1, and 5:3 (and possibly 3:1) commensurabilities, and corresponding deficits just short of these resonances (Lissauer et al. 2011b; Delisle and Laskar 2018; Fabrycky et al. 2014; Xu and Lai 2016; Choksi and Chiang 2020). The similarity

3 The Demographics of Close-In Planets 197

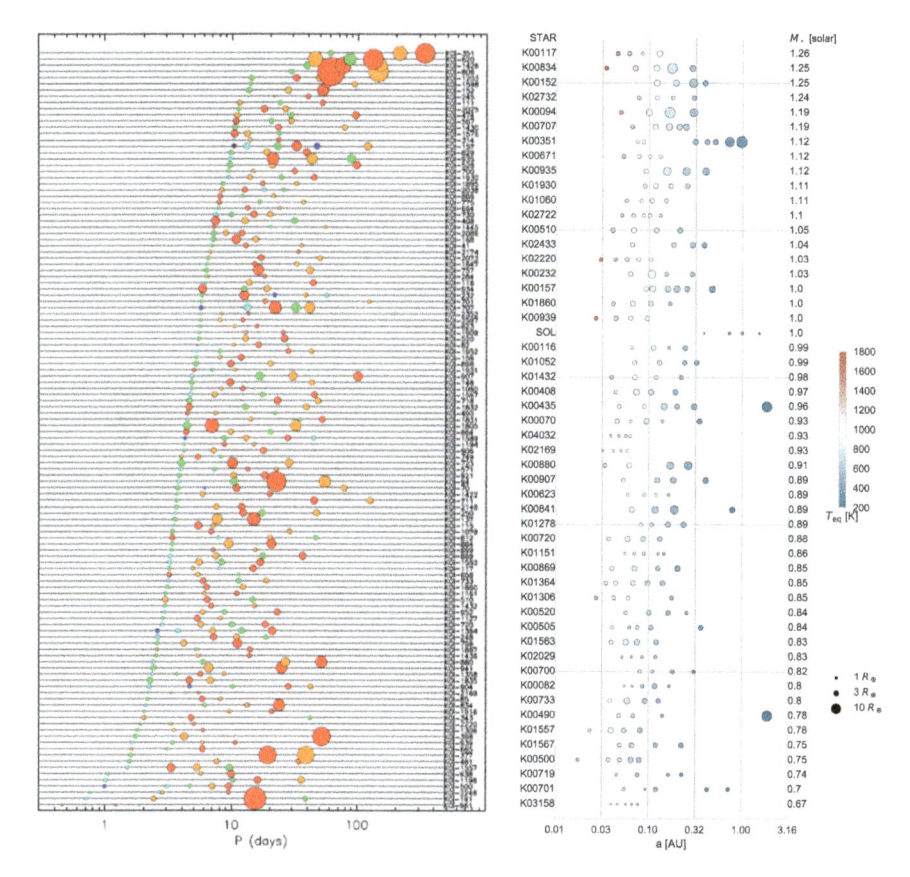

Fig. 3.46 *Left*: examples of architectures of *Kepler* multi-planet systems from Fabrycky et al. (2014). Planets are represented with circles, whose size is proportional to their radius and are colored by decreasing size within each system. The systems are ordered based on the orbital period of the inner planet. Figure reproduced with permission from Daniel C. Fabrycky. *Right*: examples of architectures of *Kepler* multi-planet systems with at least four transiting planets from Weiss et al. (2018b). Planets are represented with circles, whose size is proportional to their radius. Color indicates their equilibrium temperature. The systems are ordered based on the stellar primary mass, which is reported on the right side. The rocky planets of the Solar System are also included for comparison. Figure reproduced with permission from Lauren M. Weiss

with the histogram of Fig. 3.34 based on the RV sample of multis is compelling. The minimum period ratio is also very similar, $\simeq 1.2$ and $\simeq 1.26$ for the *Kepler* and RV multis, respectively.

3.5.3.2 The 'Peas in a Pod' Patterns

The visual appearance of similar sizes and regular spacings implies that sizes and period ratios of pairs in the same systems should be correlated. Weiss et al. (2018a) found exactly such a correlation in both parameters (Fig. 3.48). The null hypothesis

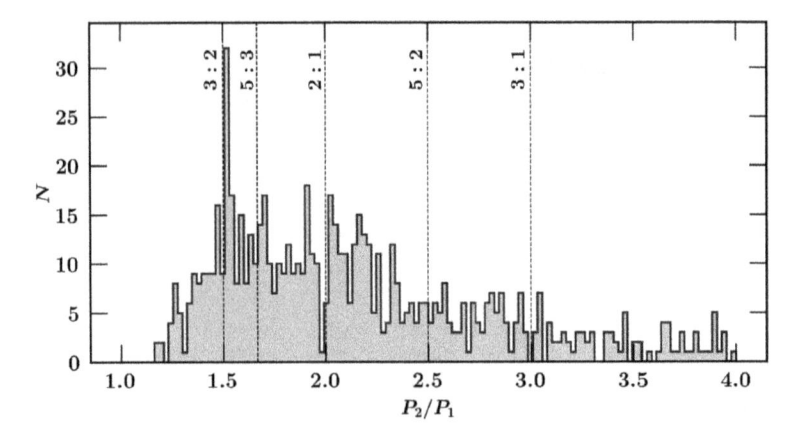

Fig. 3.47 Period ratios P_2/P_1 for all pairs of sub-Neptunes with $R_p < 4R_\oplus$ (the subscript 1 denoting the inner member of the pair and 2 the outer). N denotes the number of systems in a bin. Data taken from the NASA Exoplanet Archive on August 2019. Figure adapted from Choksi and Chiang (2020) and reproduced with permission from Nick Choksi

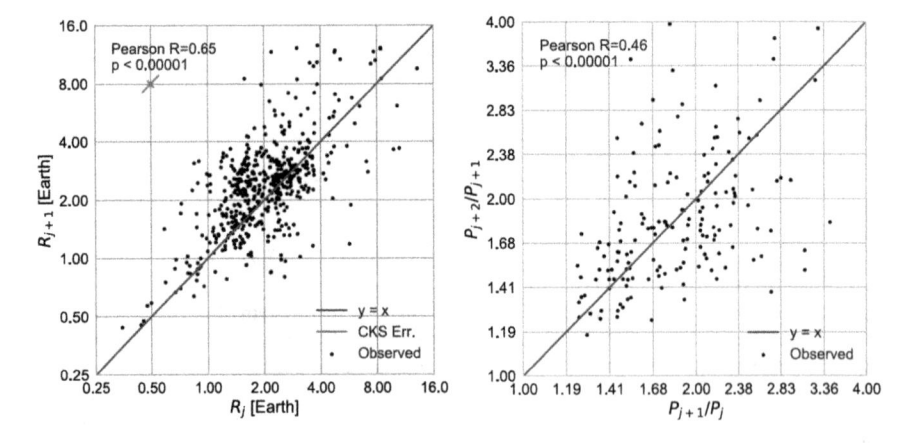

Fig. 3.48 *Left*: the radius of a planet versus the radius of the next planet out in *Kepler* multi-planet systems. *Right*: the orbital period ratio of the outer planets versus the orbital period ratio of the inner planets in *Kepler* systems with three or more planets. In both cases high Pearson correlation coefficients correspond to a very low probability of no correlation. Figures taken from Weiss et al. (2018b) and reproduced with permission from Lauren M. Weiss

of no correlation was tested in detail by Weiss et al. (2018b), based on a bootstrap procedure in which stars and number of planets were preserved, but sizes and orbital periods were randomly drawn from the observed distributions (Fig. 3.48, left-hand panel). The correlation was found to be much weaker for both parameters with respect to what was seen in real systems (Fig. 3.48, right-hand panel). The clear pattern of similar sizes and regular spacings was then termed by Weiss et al. (2018b) "peas in a pod". Strong intra-system uniformity was also shown by Millholland et al. (2017)

in mass-radius space for a sample of *Kepler* multis with mass measurements from transit-timing variations. It is very interesting to note the close similarity of the size correlation with the tight minimum-mass correlation plot of Fig. 3.36 for the RV sample of low minimum-mass multis. Finally, Weiss et al. (2018b) also found evidence for a correlation between the average planet size in an adjacent pair and the period ratio of the pair, which also closely resembles the analogous plot (in minimum mass vs. period ratio) of Fig. 3.36 for the RV sample, although on an overall different scale of period ratios.

The nature of the "peas in a pod" pattern of correlations has been the subject of a debate in the most recent literature, with evidence brought forward in support of both its astrophysical origin or it being produced by selection effects due to observational biases (He et al. 2019; Murchikova and Tremaine 2020; Weiss and Petigura 2020; Zhu 2020; Jiang et al. 2020). While the jury is still out, investigations based on state-of-the-art population synthesis models used for generating forward models of the *Kepler* survey (EPOS, Mulders et al. 2018; SysSim, Hsu et al. 2018) appear to indicate the former interpretation might be preferable (He et al. 2020; Gilbert and Fabrycky 2020).

Finally, *Kepler* multis are not only regularly, but also very closely spaced. An empirically-determined representation of the median spacing of a pair to ensure its long-term dynamical stability is (Pu and Wu 2015):

$$\Delta = 2.87 + 0.7 \log\left(\frac{t}{P_{\rm in}}\right) + 2.4\left[\left(\frac{\sigma_e}{R_H/a_{\rm in}}\right) + \left(\frac{\sigma_i}{4R_H/a_{\rm in}}\right)\right] \quad (3.17)$$

In the expression, t is the physical stability timescale, $P_{\rm in}$ the orbital period of the innermost planet, R_H is the mutual Hill radius of a planet pair (see Eq. 22 in Chap. 1), $a_{\rm in}$ is the semi-major axis of the innermost planet, and σ_e and σ_i are the multiplicity-dependent dispersions of orbital eccentricities and mutual inclinations among the planets. The spacing threshold for dynamical stability decreases from $\Delta \sim 20$ to $\Delta \sim 12$ for $n = 2$ and $n = 5$, respectively,[11] and it is rather close to the typically observed values for *Kepler* multis ($\Delta \sim 20$, see, e.g., Fang and Margot 2012; Pu and Wu 2015; Weiss et al. 2018b). One is then induced to conclude that *Kepler* multis are indeed dynamically packed, and systematically close to empirical stability limits. The major caveat in drawing this conclusion is the fact that for the overwhelming majority of *Kepler* multis dynamical mass measurements are not available, therefore the calculation of R_H relies on the use of empirical mass-radius relationships.

3.5.3.3 Possible Interpretations

The eccentricity, mutual inclination and period ratio distributions for *Kepler* small multi-planet systems (including those containing ultra-short-period companions), and the ordering of and correlations between size and spacings bear key insights on

---

[11] Delta is measured in mutual Hill radii R_H

the variety of channels for planet formation and dynamical evolution of planetary systems. A variety of mechanisms have been invoked to reproduce the mixture of dynamically hot and cold configurations of variable multiplicity in the *Kepler* sample. They all likely contribute to a higher or lesser degree to shape the details of the distributions as a function of multiplicity. These include chaotic secular interactions (Petrovich et al. 2019; Volk and Malhotra 2020), low-eccentricity migration via planet-planet scattering (Pu and Lai 2019), interactions with an outer giant planet (e.g., Johansen et al. 2012; Huang et al. 2017; Mustill et al. 2017; Becker and Adams 2017; Pu and Lai 2018; Poon and Nelson 2020; Masuda et al. 2020, effects of primordial host star obliquity and oblateness, and planetary obliquity tides (e.g., Spalding and Batygin 2016; Li and Lai 2020; Becker et al. 2020; Millholland and Laughlin 2019; Millholland and Spalding 2020; Spalding and Millholland 2020), disk migration of (not necessarily) resonant chains that subsequently disrupt due to a variety of mechanisms (e.g., Baruteau and Papaloizou 2013; Goldreich and Schlichting 2014; Pu and Wu 2015; Chatterjee and Ford 2015; Izidoro et al. 2017; Lambrechts et al. 2019; Pichierri and Morbidelli 2020), or more or less in-situ formation (e.g., Hansen and Murray 2013; Matsumoto and Kokubo 2017; Lee and Chiang 2017; Terquem and Papaloizou 2019; MacDonald et al. 2020). Additional discussion can be found in Chap. 1.

3.6 Eccentricity and Obliquity

Let us now further discuss the distributions of specific parameters, namely eccentricity and obliquity, which can be considered optimal tracers of the *dynamical temperature* of planetary systems.[12] This concept was introduced in Sect. 3.5.2.2: exoplanet systems with orbits that are almost circular and are well aligned with the host stars' equatorial planes are defined as dynamically 'cool', whereas those that present misaligned and eccentric orbits are dynamically 'hot' (see Fig. 3.42). The distributions of these parameters and how they depend on stellar and planetary properties play and important role to shed light on planet-formation mechanisms and the subsequent evolution of planetary orbits (Winn and Fabrycky 2015).

3.6.1 Eccentricities of Giant Planets

The analysis of the eccentricity as a function of the orbital period or the semi-major axis is one of the first diagnostics of the statistical properties of the exoplanet population, which has been utilized since the early days of exoplanetary science (see Fig. 3.49, where the planets of the Solar System are in the lower-right quadrant of

---

[12] This section is based on the lecture performed in May 2019 by A. Howard, who thanked Josh Winn for many slides that he presented during his lecture.

3 The Demographics of Close-In Planets   201

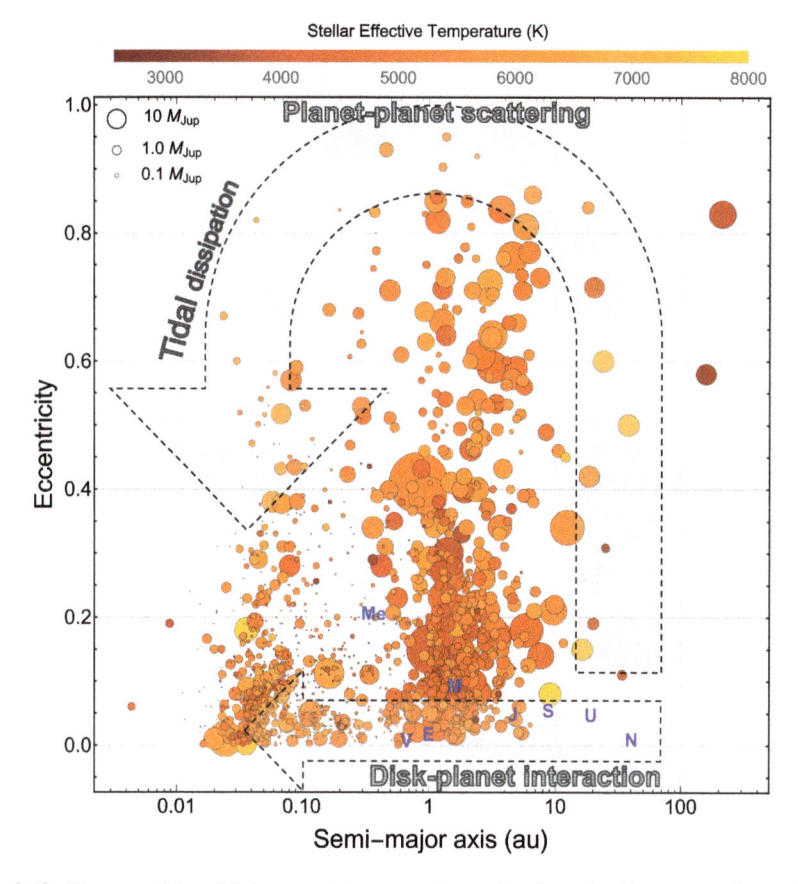

Fig. 3.49 Diagram of the orbital eccentricity versus the semi-major axis of known exoplanets. The planets are represented by circles, whose size is proportional to their mass, m_p, or $m_p \sin i$. The error bars have been suppressed for clarity. Color indicates the effective temperature of their parent stars. Data were taken from the NASA Exoplanet Archive in September 2020. The positions of the Solar-system planets are also reported by the initials of their names. Figure inspired by a similar plot from J. Winn

this semi-major axis – eccentricity diagram). All giant planets in the Solar System have very low-eccentricity orbits, and are found at at wider orbital separations than the terrestrial planets. We would expect that to be the case for exoplanetary systems as well, because, in order to become a giant planet, a core of solid material has to grow to a certain critical size, which is roughly ten times the mass of the Earth. At that critical size, the surface gravity of the planet becomes sufficient to start accreting hydrogen and helium. As we learned from Chap. 1, this phenomenon can only happen far away from the Sun, beyond the so-called *snow line*, because there are a lot more solid materials there (especially volatile chemical compounds with freezing points > 100 K, such as water, ammonia, methane, etc.) that aggregate to form large-core protoplanets, which subsequently become giant planets. Indeed, the initial expectation from theory is that the orbits of these giant planets should be roughly circular, because the effective frictional force, which exists for planets that are orbiting within a massive disk of gas and dust, would produce an effective damping of any primordial eccentricity. Instead, examining Fig. 3.49, the observational evidence tells a different story:

- first, there are many giant planets on very eccentric orbits (for example, HD 80606 b has $e \approx 0.93$; Naef et al. 2001). In our Solar system, these kind of eccentricities are typically associated with comets;
- second and even more surprising, there are giant planets existing in abundance at short orbital separations much inside the snow line, as close as $\approx 0.01 - 0.1$ au.

We already saw (Chap. 1 and Sect. 3.3) that these giant planets are known as hot and warm Jupiters and likely formed in the outer regions of the protoplanetary disk (where Jupiter did in the Solar system), but then through some processes they have *migrated* from their initial locations to where we see them today.

As already discussed in this book (see Chap. 1, Sect. 2.4 and 3.2; Chap. 2, Sect. 3.2; see also Sect. 3.5 of this Chapter), a number of physical mechanisms have been proposed that are capable of evolving significantly the primordial orbit of a giant planet, shrinking it and/or exciting its eccentricity. The two main ones are:

- early interactions between a giant planet and the protoplanetary disk in which it formed. These can lead to the loss of energy and angular momentum from this planet to the disk, resulting in the planet spiraling towards the central star, turning it into a hot Jupiter (Lin et al. 1996; Ward 1997);
- *planet-planet scattering* in multiple systems (e.g., Rasio and Ford 1996; see also Davies et al. 2014 and references therein). Gravitational interactions between planets in a system can lead to a chaotic evolution of orbital elements that can cause planet orbits to cross. When that happens, giant planets rarely collide, but instead gravitationally scatter, resulting in ejection of some of the companions while others remain on high-eccentricity orbits. Through tidal dissipation mechanisms such eccentric orbits are then shrunk and circularized.[13]

To understand which of the two models best explains the observed distributions, we can examine the orientation of the planetary orbits. If planet-planet scattering is the process that produces hot Jupiters, the same scattering encounters will randomize to some degree the orbital planes, whereas disc-planet interactions keep the system coplanar throughout the whole migration process and we will see flat architectures. These expectations are well supported by numerical simulations. For example:

- Marzari and Nelson (2009) ran hydrodynamic simulations to investigate the dynamical evolution of a Jupiter-mass planet injected into an orbit highly inclined

---

[13] Similar to this case is the Kozai-migration mechanism, in which a planetary system is dynamically perturbed by a distant third body, which could be a companion star or a massive planet (see, for example, Wu and Murray 2003). Kozai oscillations can produce a very broad range of final orientations, primarily eccentric, misaligned and even polar and retrograde orbits are possible (Fabrycky and Tremaine 2007).

3 The Demographics of Close-In Planets   203

Fig. 3.50 Simulation of the evolution of a giant planet initially in an orbit inclined with respect to a circumstellar disk. The planet started on an orbit with semi-major axis equal to 4 au, eccentricity 0.4, and inclination 20°. The exchange of angular momentum between the disk and the planet causes these three planetary orbital parameters to evolve significantly over the ≈ 1000 year timescale of the simulation. Figure adapted from Marzari and Nelson (2009)

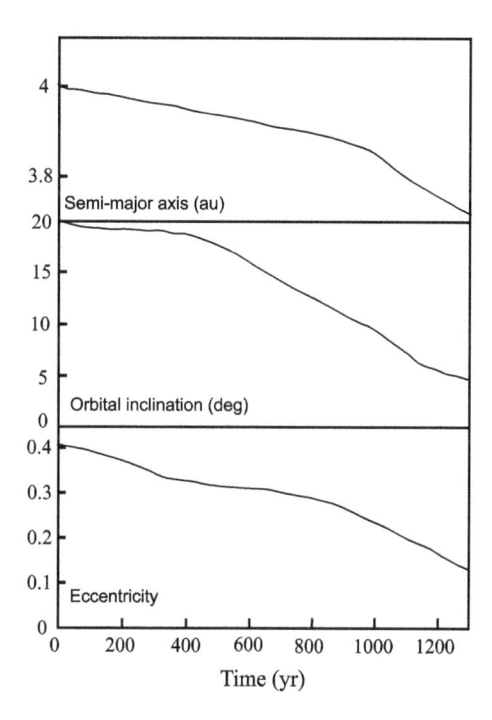

with respect to its protoplanetary disk. They found that, independently of the initial values of the semi-major axis and eccentricity of the planet, the eccentricity and inclination of its orbit are rapidly damped on a timescale of the order of 10^3 year, see Fig. 3.50.

- Chatterjee et al. (2008) performed simulations of gravitational planet-planet scattering, in systems containing three giant planets, orbiting solar-mass stars, on initially near-circular orbits and no gas disk. They found that in all of these systems, at least one planet is eventually ejected before reaching a stable configuration. In 20% of the cases, two planets are lost through ejections or collisions, leaving the system with only one giant planet, thus predicting the existence of many systems with a single eccentric giant planet. The exact distribution of eccentricities for the final remaining planets in stable orbits, after a secular evolution ($\approx 10^9$ year), depends on the choice and range of the initial mass distribution but is in general similar to what we observe in reality. The authors also found that it is possible to scatter some planets into orbits with low perihelion distances and that the inclination distribution of such planets could be significantly broadened. Therefore, these scattering events tend to amplify any initially small misalignments.

Indeed, planet–planet scattering processes can reproduce much of the observed distribution of exoplanet eccentricities for a wide range of initial conditions (Ford et al. 2008; Jurić and Tremaine 2008; Nagasawa et al. 2008; Matsumura et al. 2010; Raymond et al. 2010; Dawson and Murray-Clay 2013; Carrera et al. 2019). Furthermore,

Fig. 3.51 Diagram of the orbital eccentricity versus orbital period of known exoplanets. The error bars have been suppressed for clarity. Most of the high-eccentric, single planets are found around high-metallicity ([Fe/H] > 0) stars, as noted by Buchhave et al. (2018). Data taken from the NASA Exoplanet Archive on September 2020

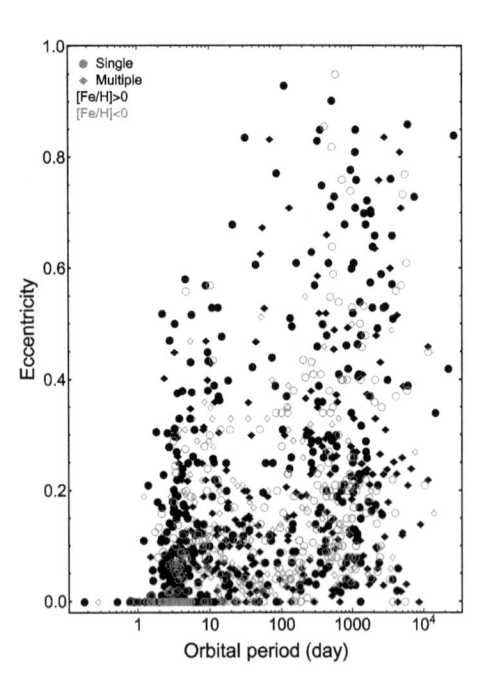

the initial metal content of protoplanetary disks also likely plays a role in the structure and architecture of planetary systems: high-metallicity systems are in fact expected to contain multiple gas giants that can perturb each other gravitationally, leading to enhanced planet–planet scattering rates. The eccentricity of cold gas giants should, therefore, increase with host-star metallicity and this is also observed (e.g., Bryan et al. 2016; Buchhave et al. 2018; see Fig. 3.51).

We would also expect that whatever mechanism perturbs the eccentricities of giant planets and/or produces hot Jupiters may also change their orbit orientation in space. Therefore, the angle between a planet's orbital axis and its host star's spin, which is also known as *spin-orbit obliquity* or simply *orbital obliquity*, ψ, represents a key parameter. Determining ψ for a statistically significant sample of exoplanetary systems, preferably spanning a range of precise parent-star ages, can allow us to figure out what the primary physical phenomenon related to the migration process of a giant planet is. We discuss in Sect. 3.6.3 the inferences obtained based on present-day statistics of ψ measurements.

3.6.2 Eccentricities of Super Earths and Sub-Neptunes

The *Kepler* space telescope (Borucki et al. 2010), with its exquisitely precise photometry, has given an enormous contribution to the study of planets with sizes smaller than that of our Jupiter. The *horn of plenty* of small exoplanets unveiled by *Kepler* has allowed theoreticians to produce robust statistical inferences on the abundances and properties of planetary systems. As we have seen in Sect. 3.3.2 and Sect. 3.4, the analysis of *Kepler* data has in particular shown how the most common planets in the Galaxy are sub-Neptunes and super-Earths.

As already discussed in Sect. 3.5.2.3, the eccentricity distribution of small-size planets in the Kepler field can be studied based on transit duration statistics (see Eq. 3.15). Using this technique, Xie et al. (2016) derived the eccentricity distributions of roughly 700 reliable *Kepler* candidates (Mullally et al. 2015), finding an interesting dichotomy. Single transiting planets have a large mean eccentricity ($\bar{e} \approx 0.3$), whereas planets in multiple systems are on nearly circular orbits (mean eccentricity $\bar{e} < 0.07$. Using a larger sample of validated *Kepler* planets (Coughlin et al. 2017) and benefiting from Gaia parallaxes to derive high-precision measurements of stellar radii (Gaia Collaboration 2016a, b, 2018), Mills et al. (2019) performed a new statistical study of the eccentricity distribution of small planets, confirming that systems with only a single transiting planet have a higher mean eccentricity ($\bar{e} \approx 0.2$), compared to those with multiple transiting planets ($\bar{e} \approx 0.05$). Mills et al. (2019) also found evidence that *Kepler* single planets with high eccentricity preferentially orbit around high-metallicity ([Fe/H] > 0) stars, extending to the regime of Super Earths and sub-Neptunes the same inference already gathered by earlier studies focused on gas giants (Dawson and Murray-Clay 2013; Buchhave et al. 2018).

The eccentricity dichotomy can be possibly understood by considering that high-metallicity environments favor the formation of giant planets. Dynamical interactions between these giant planets and inner systems can excite the eccentricities and, at the same time, can decrease the apparent multiplicity of close-in planets, which are also subjected to mutual inclination excitation (e.g., Mustill et al. 2017; Pu and Lai 2018; Masuda et al. 2020; Poon and Nelson 2020, and references therein). On the other hand, close-in multi-planet systems which are not dynamically heated by cold gas giants may not reach sufficiently high eccentricities and mutual inclinations through self-excitation mechanisms (e.g., dynamical instabilities) during the late stages of formation (e.g., Becker and Adams 2016; Poon et al. 2020).

3.6.3 Measuring the Orbital Obliquity

As mentioned in Sect. 3.6.1, the relative role of the two main mechanisms that can act to shrink the orbits of giant planets, planet-planet scattering and disk-planet interactions, can be probed using measurements of stellar obliquities, as proxies of the orbital (mis)alignment of the planets. The reader may wonder what hope do we have of measuring the obliquity ψ of the parent stars and therefore the orientations of the planets since we cannot resolve the star and the planet and directly see in which way the star is rotating and in which way the planet is orbiting. The class of transiting planets comes crucially to our aid in this case.

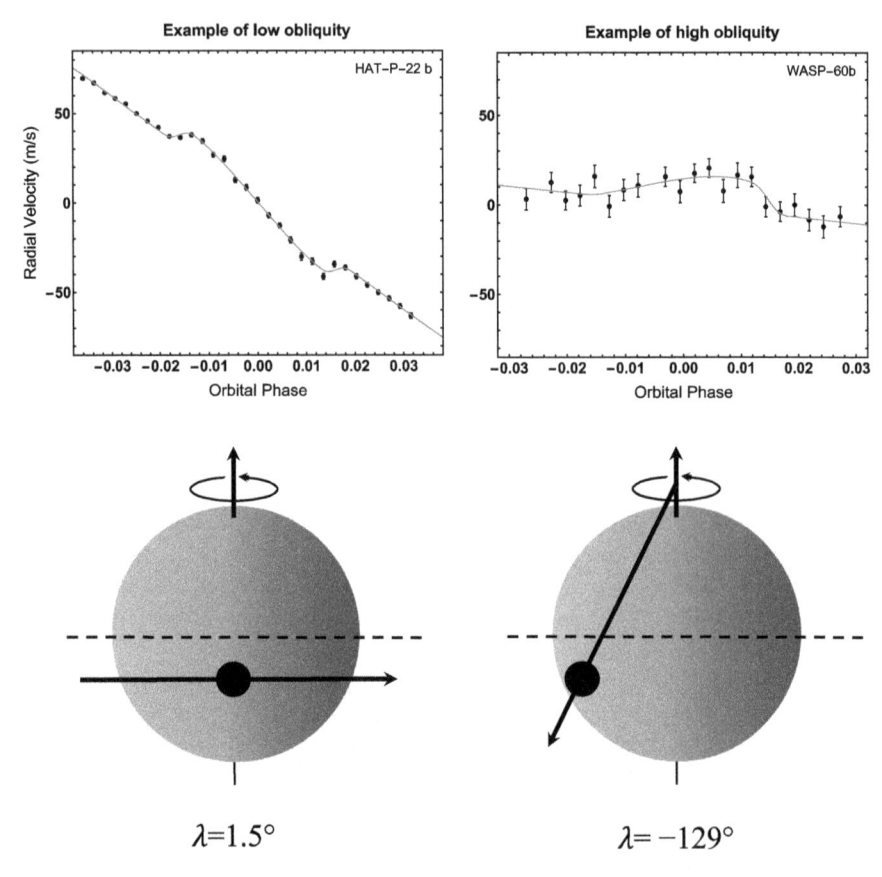

Fig. 3.52 The Rossiter-McLaughlin effect produced by two different transiting hot Jupiters, HAT-P-22 b and WASP-60 b. The phase-folded radial-velocity data (black points with error bars) were measured with the high-resolution spectrograph HARPS-N during planetary-transit events; superimposed are the best-fitting radial-velocity models (gray curves) (Mancini et al. 2018). The two planets have roughly the same impact parameter, but very different transit trajectories, which highlight the dependence of the Rossiter-McLaughlin radial velocity signature on λ

3.6.3.1 The Rossiter-McLaughlin Effect

The spin-orbit obliquity, ψ, which we learned is the angle between a planet's orbital axis and its host star's spin, is, unfortunately, not easy to determine. However, its sky projection, λ, is a quantity that is commonly measured for stars hosting transiting exoplanets through the observation of the Rossiter-McLaughlin (RML) effect (a well-known phenomenon in the context of eclipsing-binary stars), by using high-precision RV instruments. Usually, we see a transit as the apparent drop in brightness of the parent star as the planet gets in front of it (see Sect. 3.2.2.1), but if we monitor the transit event with a high-resolution spectrograph, which is sensitive to red or blue shifts of a hundred parts per million or better, then the RV time-series during transit would look similar to those in the top panels of Fig. 3.52.

3 The Demographics of Close-In Planets

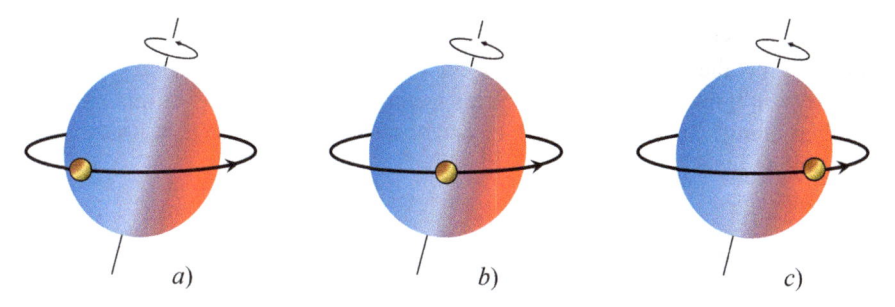

Fig. 3.53 A sketch of the Rossiter-McLaughlin effect, which can be measured during a planetary-transit event. During its rotation around the parent star, a transiting planet will block out more of the star's rotationally blue-shifted light immediately after the transit ingress (panel *a*) and more of its red-shifted light before the complete egress (panel *c*). This phenomenon changes the radial velocity profile of the parent star according to the planet's orbital phase. Figure inspired by a similar figure from Perryman (2018)

The distortions in the RV measurements on top of the Doppler shifts induced by the stellar orbital motion (see Sect. 3.2.1) occur because the parent star is rotating and so the half of the star approaching looks slightly blue-shifted, whereas the receding half looks slightly red-shifted. While transiting, a planet blocks first a small portion of the stellar surface approaching us, and then a fraction of the receding one. The stellar spectrum therefore becomes first anomalously red-shifted and then blue-shifted. This anomaly is null when the planet crosses the symmetry axis of the star. When the transit is over the effect disappears (see Fig. 3.53). By tracking the Doppler shift of the star throughout a complete planetary-transit event and by accurately measuring the shape of the RML effect, we can measure the projected stellar obliquity. If we are facing a system for which the planet's orbital plane is well aligned with the equatorial plane of the host star, i.e. a system with a low obliquity, we will measure a redshift and then a blueshift as in the top left panel of Fig. 3.52. However, in case of strong spin-orbit misalignment the planet could be spending, for example, all of its time covering the blue-shifted surface of the star and then we would only observe an anomalous redshift as in the case shown in the top right-hand panel of Fig. 3.52. Several cases of even very high-obliquity configurations have been found, including retrograde orbits (e.g., Winn et al. 2009; Narita et al. 2009).

In order to obtain an accurate modeling of the RM effect of a given planetary system and, therefore, a correct estimation of λ, it is crucial to obtain a stable out-of-transit baseline. This means that, similar to the case of a photometric transit light curve, one has to start observing the target at least ~ 60 min before ingress and ~ 60 min after egress.

The expected RV amplitude of the RML effect is approximately equal to (Winn 2010)

$$\Delta K \approx (R_p/R_\star)^2 \sqrt{(1-b^2)} v \sin i_\star \stackrel{b \simeq 0}{\approx} 52.8 \, \text{m s}^{-1} \left(\frac{R_p}{R_{\text{Jup}}}\right)^2 \left(\frac{R_\star}{R_\odot}\right)^{-2} \left(\frac{v \sin i_\star}{5 \, \text{km s}^{-1}}\right), \tag{3.18}$$

where $v \sin i_\star$ is the projected rotational velocity of the parent star and b is the impact parameter, i.e. the sky-projected distance at the conjunction between the planet and the star ($b = 0$ for a central transit). This means that giant planets orbiting around rapidly rotating stars are the best objects to observe to get precise measurements of λ, exploiting the RML effect.

3.6.3.2 Doppler Tomography

The detection of the RML effect in transiting-planetary systems as an anomalous Doppler shift of the stellar atmospheric lines is routinely obtained as far as stellar rotation is not the dominant line broadening mechanism. However, if the parent star rotates quickly ($\gtrsim 10\,\mathrm{km\,s^{-1}}$), there will be a distortion of the stellar line profiles that change as the planet blocks parts of the rotating stellar surface during the transit. In this case, as noted by Collier Cameron et al. (2010a), the parameters derived from modeling the RML effect can be severely affected by systematic errors, which arise from this time-variable asymmetry of the spectral lines. By monitoring a complete planetary transit (including the out-of-transit baseline) with high-cadence and high-resolution spectroscopy, we can detect such a distortion, also known as 'Doppler shadow', moving across the line profiles. By mapping these line profile variations as a function of stellar rotational velocity and orbital phase, the Doppler tomography technique allows isolating and tracking the component of the starlight blocked by a planet as it transits the host star, which can be self-consistently expressed as a function of the projected spin-orbit misalignment angle.

In practice, the method of Doppler tomography entails subtracting the mean, out-of-transit, line shape (i.e. the representation of the unobscured starlight) from all the in-transit spectra, so that any dark inhomogeneity on the stellar surface (corresponding to the missing starlight blocked by the transiting planet) will pop up as a positive bump. In this way, we will see the bump appearing at ingress, moving along the transit chord and disappearing at egress. As a result, in this mean line profile residual map we will detect a trail that corresponds to the Doppler shadow of the transiting planet (see Fig. 3.54). We can, then, model the shadow with a Gaussian having a width of $(R_p/R_\star)v \sin i_\star$ and an area of $1 - f(t)$, where $f(t)$ is the flux blocked by the planet, centered on $v_p(t)$. The latter quantity is the projected rotational velocity for the region of the star occulted by the planet (Cegla et al. 2016; Zhou et al. 2016), which depends on $v \sin i_\star$ and λ.

Doppler imaging techniques have been effectively applied to measure the spin–orbit angles of planetary systems with hot, fast-rotating primaries (e.g., Collier Cameron et al. 2010a,b; Miller et al. 2010; Brown et al. 2012; Gandolfi et al. 2012; Hartman et al. 2015; Gaudi et al. 2017). In the case of ultra-hot Jupiters ($T_{eq} \gtrsim 2000\,\mathrm{K}$), the *atmospheric* RML effect can be detected (Borsa et al. 2019). It corresponds to a deviation of the in-transit RVs from the classical RML effect that occurs when the atmosphere of the planet is intercepted by the mask used to create the stellar cross-correlation functions or mean line profile. The atmospheric trace will appear in the tomographic map of the line profile residuals after removal of

3 The Demographics of Close-In Planets

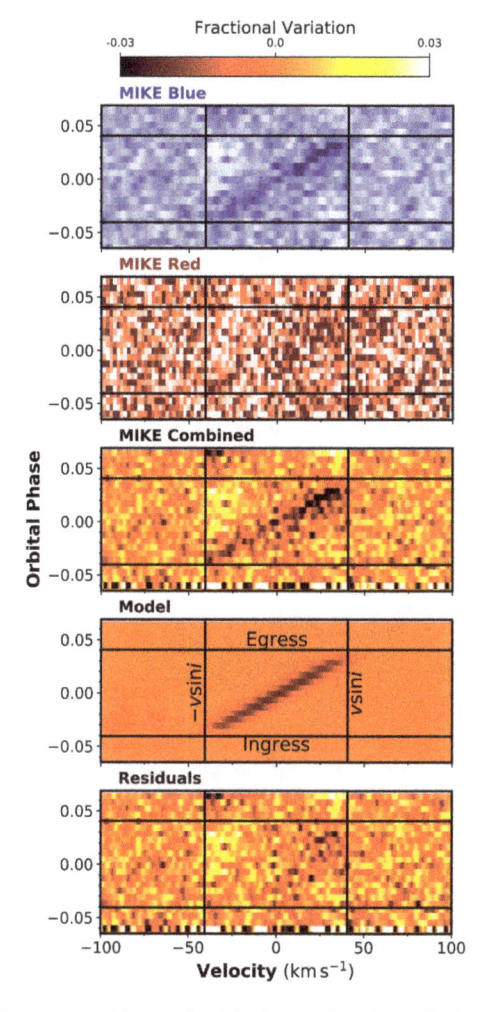

Fig. 3.54 The Doppler tomographic transit of the brown dwarf HATS-70 b, as measured by Zhou et al. (2019a) with the two cameras of the MIKE instrument on the 6.5 m Magellan Clay telescope. HATS-70 b successively blocks parts of the rotating stellar surface, inducing a 'Doppler shadow' in the stellar line profile. These line profile variations are mapped as a function of stellar rotational velocity and orbital phase. The vertical lines mark the $v \sin i_\star$ of the star, while the horizontal lines mark the timings of ingress and egress. The planetary transit signal is the dark trail crossing from bottom left to top right. The top two panels show the transit as seen with the blue and red cameras. The middle panel shows the combined data set. The bottom two panels show the best-fit model and the residuals after model subtraction. The figure is taken from Zhou et al. (2019a) and reproduced with permission from George Zhou

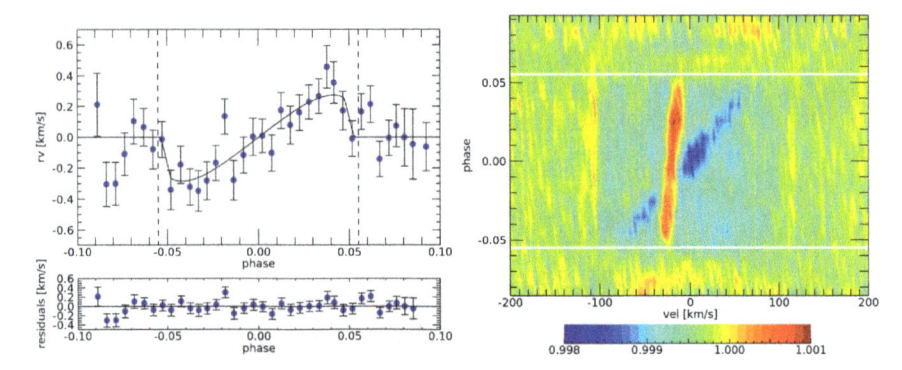

Fig. 3.55 Atmospheric RML effect of the ultra-hot-Jupiter KELT-9b. *Left*: the phase-folded radial-velocity data (points with error bars) were measured with the high-resolution spectrograph HARPS-N during four planetary-transit events of KELT-9b (the RVs of the four transits were averaged in bins of 0.005 in phase). *Right*: The Doppler tomographic map of the transit of KELT-9b as in Fig. 3.54. The line profile was averaged over four transit observations, with the horizontal white lines indicating the transit ingress and egress. Both the Doppler shadow of the planet (red) and the planetary atmospheric trace (blue) are clearly visible. The figure is taken from Borsa et al. (2019) and reproduced with permission © ESO

planetary Doppler shadow, see Fig. 3.55. By studying the shape of this deviation one can measure the extension of the planetary atmosphere that correlates with the stellar mask (Borsa et al. 2021; Rainer et al. 2021). The method is applicable only in the case of extremely hot planetary atmospheres, which show a chemical composition similar to that of late-type stars (in particular due to the presence of neutral or ionized iron).

Finally, if the star is a very fast rotator and the mass of the secondary is non-negligible (i.e. it is a brown dwarf or low-mass star), more accurate measurements of λ can be obtained by performing global modeling of the Doppler tomographic data, the RV measurements, the photometric transit light curves and the gravity-darkening effect (e.g., Zhou et al. 2019a,b).

3.6.3.3 Star-Spot Occultations

There is another fascinating technique that we can use to measure the spin-orbit angle of a planetary system, which is based on a simultaneous observation of a planetary transit and a *star-spot crossing* event. Instead of being spectroscopic, this technique is purely photometric and consists in measuring the brightness variations of an active, spotted star[14] during a transit of its planet. If there is a single star spot or a star-spot complex and the transiting planet occults it, then the transit depth will temporarily decrease as the star will appear brighter (star spots have lower effective temperature

---

[14] Just like the Sun can present sunspots, also the other stars that we observe in the Galaxy can have star spots on their photosphere.

3 The Demographics of Close-In Planets

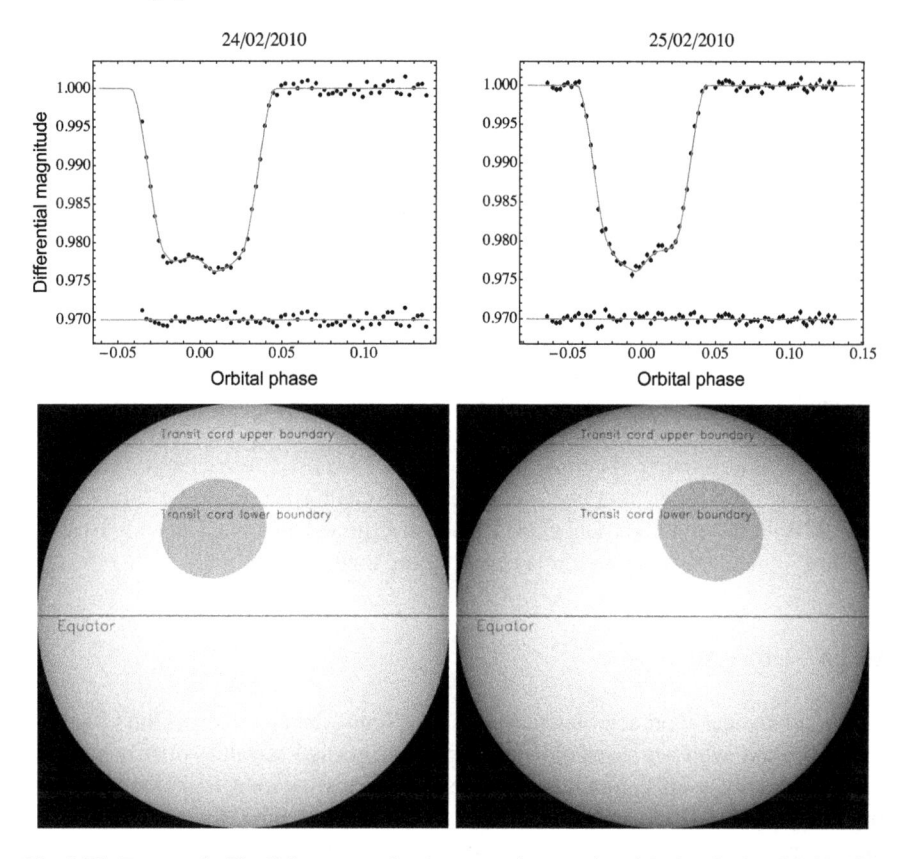

Fig. 3.56 *Top panels*: The light curves of two consecutive transits of the hot Jupiter WASP-19 b, which has an orbital period of ≈ 0.8 days, observed with the 3.6 m New Technology Telescope (Tregloan-Reed et al. 2013). *Bottom panels*: Representations of the WASP-19 stellar disc, star-spot positions, and transit chords for the two transit events with star-spot crossings. The gray-scale of each star-spot is related to its contrast. The two horizontal lines on each panel represent the upper and lower parts of the planet pass. The authors estimated $\lambda = 1.0 \pm 1.2$ deg. Data and bottom figures kindly provided by Jeremy Tregloan-Reed

than the surrounding photosphere). A small bump will then be recorded in the transit light curve, similar to those in Fig. 3.56. Now, if the planetary system is well aligned, if the star has a plausible rotation period of roughly a month and the planet has, instead, an orbital period of few days, the next time the planet transits in front of its star, we will see that the star spot has only moved slightly across the stellar surface, so that we will notice the same anomaly but occurring at a later transit phase. The phase shift of the anomalous feature will continue until the spots disappear from the visible hemisphere due to rotation. However, if the spin-orbit angle is significant, the planet will never encounter again the same star spot at the subsequent transit, we will no longer see the anomaly and we can put a lower limit on the stellar obliquity (e.g., Dai and Winn 2017).

In general, the occultation of the same star-spot complex in two or more transit events indicates that there is good spin-orbit alignment, allowing the measurement of λ with very good precision (e.g., Mancini et al. 2017). On the other hand, if in two consecutive transits we observed two different star spots, their latitude difference is completely degenerate with λ. Discriminating between the two cases is not trivial. Crucial parameters that one has to take into account are the rotational period of the host star, the flux modulation produced by the rotation of the starspots and the difference in position and time among the starspot-crossing events, which should be precisely modeled with sophisticated procedures (e.g., Désert et al. 2011; Sanchis-Ojeda et al. 2011; Sanchis-Ojeda and Winn 2011; Sanchis-Ojeda et al. 2013b; Tregloan-Reed et al. 2013; Mazeh et al. 2014; Tregloan-Reed et al. 2015; Juvan et al. 2018). The projected obliquity can also be efficiently determined by a statistical method, correlating the starspot anomalies observed in a sequence of transits (Dai et al. 2018b). In this case, a large number of consecutive transits are required, restricting this method only to long time-series photometric data collected with space missions, like *Kepler* and K2. With this method, (Dai et al. 2018b) determined the obliquities for 10 stars with hot Jupiters studying a sample of a bit more than 60 stars.

3.6.3.4 Other Methods

Another statistical approach for determining the spin-orbit alignment is the so-called $v \sin i_\star$ *method*, which is based on searching anomalously low values of the projected rotation velocity, $v \sin i_\star$, in a large sample of stars hosting transiting planets. Since low obliquity for a transiting-planet system implies that $\sin i \approx \sin i_\star \approx 1$, a low value of $v \sin i_\star$ can be the sign of high-obliquity systems (Schlaufman 2010; Winn et al. 2017a).

Other methods for measuring the obliquity have not found many applications so far, because they require specific situations, like years of ultra-accurate photometry with ≈ 1 minute cadence for studying the stellar pulsation modes (the *asteroseismic method*; Gizon and Solanki 2003), or rapidly-rotating stars and high obliquity (the *gravity-darkening method*; Barnes 2009), or very bright and rapidly rotating stars (the *interferometric method*; Kraus et al. 2020).

3.6.4 Obliquity of Transiting Planetary Systems

Having determined λ, we can also estimate the real obliquity, ψ. For this purpose, we need to know the planetary orbital inclination, i, and the stellar-spin inclination angle, $i_\star$.[15] Then, by using the following formula (Winn et al. 2007)

---

[15] Frequency analysis of the time-series light curves and the study of the oscillation modes induced by stellar rotation (*Asteroseismology*) can be used for determining $i_\star$.

$$\cos\psi = \cos i_\star \cos i + \sin i_\star \sin i \cos\lambda\,, \tag{3.19}$$

we can get the true orbital obliquity of a given transiting planetary system. To-date (September 2020), λ has been measured for roughly 160 transiting exoplanets, while ψ has been constrained for only roughly 30 ones.[16] Except for the measurements of the RML effect of few brown dwarfs, Neptune-, superEarth-size planets and, recently, three rocky planets in the TRAPPIST-1 system (Hirano et al. 2020), the degree of spin-orbit alignment has been mostly determined for hot Jupiters, representing the only class for which we have a good statistical sample.

What this large collection of λ measurements for hot Jupiters is telling us is still a matter of debate, as reviewed by (Triaud 2018). No convincing patterns or trends have emerged from plotting the projected obliquity versus other parameters such as, for example, planetary mass and radius, orbital separation, or stellar age. Interestingly, however, early investigations based on smaller samples of λ measurements tentatively identified two populations of more or less aligned hot Jupiters based on the effective temperature T_{eff} of their parent stars, the dividing line being related to the Kraft break (Kraft 1967) and falling somewhere between 6090 and 6300 K (Winn et al. 2010; Albrecht et al. 2012; Dawson 2014). This is marked with a gray zone in Fig. 3.57, where the absolute values of λ of hot Jupiters ($0.1\,M_{\text{Jup}} < M_{\text{p}} < 13\,M_{\text{Jup}}$) have been plotted versus T_{eff}.

On the left side of this plot we find planets orbiting stars with mostly convective outer envelopes, whereas on the right side host stars have mostly radiative outer envelopes. Therefore, the two populations of planet hosts are distinctively different in terms of e.g., strength of the magnetic field, rotation rate and, especially, the tidal dissipation rate (a star's ability to convert tidal oscillations into heat). Mostly convective stars are cooler and supposed to have much more rapid tidal dissipation because the convective cells are producing the turbulent cascades that lead to energy/heat loss, whereas the radiative stars are thought to have much weaker tidal dissipation. Therefore, hot Jupiters hosted by relatively cool stars ($T_{\text{eff}} < 6100\,\text{K}$), should be much more aligned than those hosted by hotter stars because in the former case tides effectively damp any possible obliquity on timescales much shorter than those in connection to the planet's orbital decay (Lai 2012; Valsecchi and Rasio 2014). However, although high-obliquity hot Jupiters were found regularly above the Kraft break, many exceptions challenge this theory, as we can see from Fig. 3.57.

It was also noted that those planetary systems, in which the hot Jupiter is massive ($M_{\text{p}} > 3\,M_{\text{Jup}}$) tend to have lower spin-orbit angles, because the parent star is much more affected by planet's tidal forces (Hébrard et al. 2011). We plot in Fig. 3.58 the measured sky-projected obliquities of all known systems hosting $0.1\,M_{\text{Jup}} < M_{\text{p}} < 80\,M_{\text{Jup}}$ companions, including brown dwarfs such as WASP-30 b, KELT-1 b and CoRoT-3 b, as a function of scaled orbital distance from the host star $a/R_\star < 12.5$. The larger circles in this plot, which are associated with companions with larger

---

[16] Data taken from TEPC at:http://www.astro.keele.ac.uk/jkt/tepcat/rossiter.html (Southworth 2011).

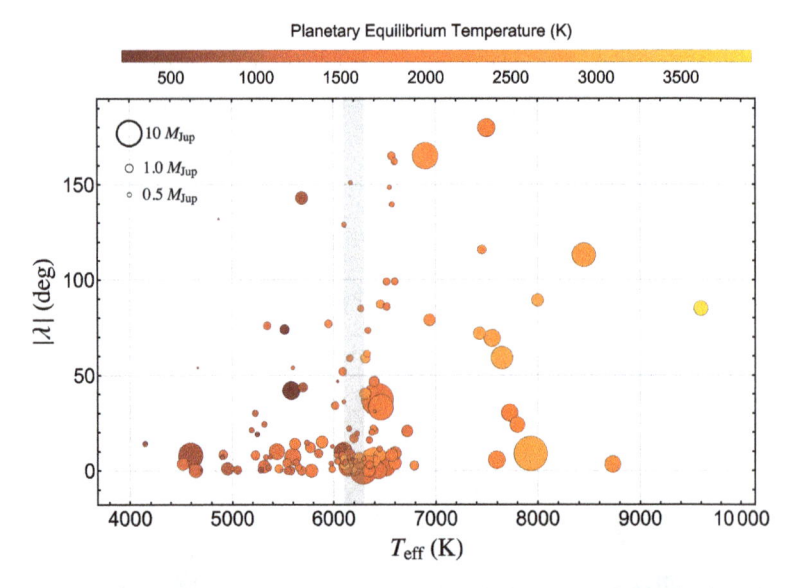

Fig. 3.57 Absolute values of the sky projected orbital obliquity angles of close-in giant planets as a function of the host star's effective temperature. The planets are represented by circles, whose size is proportional to their mass. The error bars have been suppressed for clarity. Color indicates their equilibrium temperature. The gray zone should discriminate two different populations of hot Jupiters, according to several authors (e.g., Winn et al. 2010; Albrecht et al. 2012). Data were taken from TEPCat in September 2020

masses, show that most of these systems present a good alignment (Zhou et al. 2019a).

Using the $v \sin i$ method (see Sect. 3.6.3.4), Winn et al. (2017a), Muñoz and Perets (2018) and Louden et al. (2021) selected samples of well-studied *Kepler* planet hosting stars with reliably measured photometric periods and projected rotation velocities. The aim was to investigate in a statistical sense the obliquities of *Kepler* stars that, as we know (cf. Sect 3.6.2), host planets spanning a wide range of sizes, most of which are smaller than Neptune. The above studies helped drawing a picture in which Kepler small-size planet hosts generally have higher values of projected rotational velocities, and therefore low obliquities, than control samples of field stars. The result is thus a broad indication of spin-orbit alignment for Kepler planetary systems, except for those orbiting hotter, late-F-type stars ($T_{\text{eff}} \gtrsim 6250$ K), that exhibit compatible to random orientations and therefore typically high obliquities (see Fig. 3.59). The tentatively identified trend of increasing λ (or ψ) with T_{eff} for hot Jupiters thus appears to extend to other types of planets.

Important as the statistical inferences based on the $v \sin i$ method may be, it would be highly desirable to expand the domain of true obliquity measurements for other types of planets besides giant planets and brown dwarfs. Observations of the RML effect for Neptunes and Super Earths are especially challenging given that the amplitudes of the RV anomalies can be typically on the order of a few ms^{-1},

3 The Demographics of Close-In Planets

Obliquity angle λ

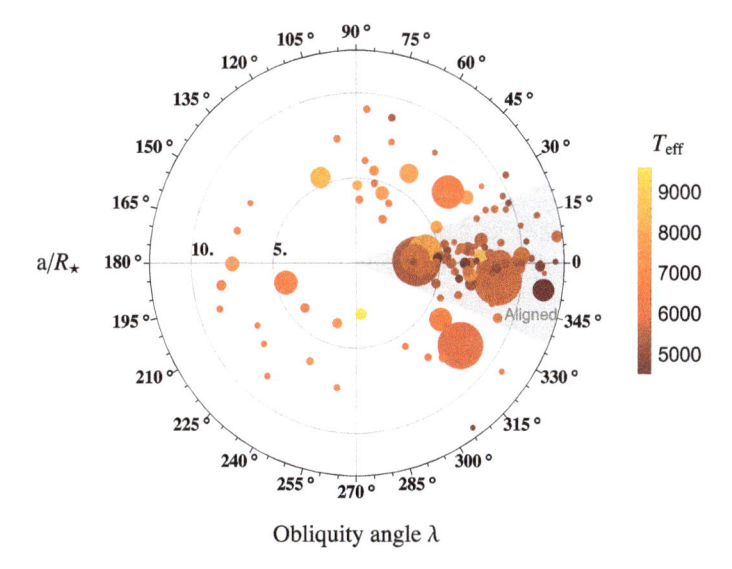

Fig. 3.58 Sky projected orbital obliquity of massive planets and brown dwarfs (0.1 $M_{\rm Jup}$ < $M_{\rm p}$ < 30 $M_{\rm Jup}$) as a function of their scaled orbital distance from the host star. The planets are represented by circles, whose size is proportional to their mass. The error bars have been suppressed for clarity. Color indicates the effective temperature of their parent stars. Data were taken from TEPCat in September 2020. Figure inspired by similar plots from J. Winn; see also Zhou et al. (2019a)

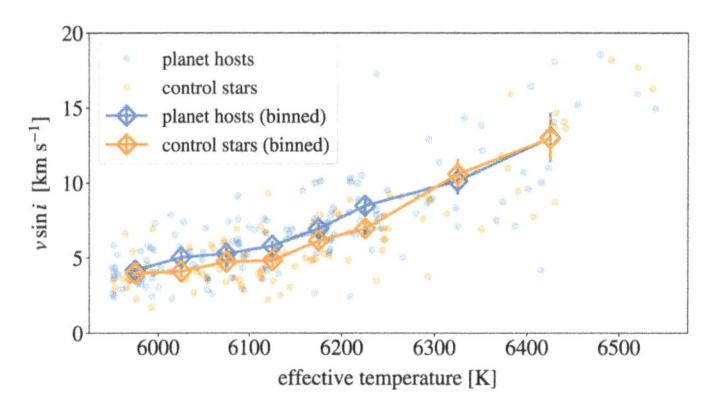

Fig. 3.59 Measured sky-projected rotation velocity versus effective temperatures for a sample of *Kepler* stars (blue points) with $T_{\rm eff}$ between 5950 and 6550 K. Orange points represent a control sample of stars with matching spectroscopic properties and random orientations. Figure taken from Louden et al. (2021) and reproduced with permission from Emma M. Louden

or smaller (see Eq. 3.18), unless the primary is a fast rotator. A particular focus has been devoted recently to obliquity measurements for transiting systems at young ages (e.g., DS Tuc A b: Zhou et al. 2020; K2-25 b: Stefansson et al. 2020; AU Mic b: Addison et al. 2020; TOI-1726 c: Dai et al. 2020), as in this case dynamical masses are very difficult to determine due to the very high levels of activity and fast rotation of the primaries, therefore detection of the RML effect also corresponds to a direct planet confirmation. So far, these systems all appear well aligned, but the regime of small-number statistics still prevents any clear conclusions to be drawn.

The emerging picture is that in the case of giant planets the wide range of measured obliquity values implies that dynamical processes such as planet-planet scattering and secular perturbations are responsible for tilting their orbits. The situation for smaller planets is less clear, as obliquities might be excited or damped based on a number of possible mechanisms, either environmental in nature (originally misaligned protoplanetary disks, dynamical interactions with outer giant planets or stellar companions; e.g. Bate et al. 2010; Spalding and Batygin 2015; Anderson and Lai 2018; Takaishi et al. 2020), or star-specific (stochasticity of internal gravity waves in hotter stars; e.g. Rogers et al. 2012).

3.7 Ultra-Short Period Planets

3.7.1 Introduction

One of the most intriguing findings in the field of exoplanetary science was the discovery of small planets with ultra-short orbital periods, which are operationally defined as having $P \lesssim 1$ day. A key step forward from space-based surveys came in 2009 during the *CoRoT* mission, when Corot-7b was announced, a small ($\sim 1.5 - 1.6\, R_\oplus$) transiting planet with a very short period of 0.85 days (Léger et al. 2009). Two years later, the two super-Earth planets Kepler-10b and 55 Cnc e were found with periods smaller than 0.8 days (Batalha et al. 2011; Winn et al. 2011). Later on, many other surveys, both in space and from the ground, continued finding new very short-period planets. For instance, by early 2021, 98 giant planets ($R > 8 R_\oplus$) had been discovered with periods between 2 and 3 days, 65 with periods between 1 and 2, and 8 with periods shorter than one day (namely, NGTS-10b, WASP-19b, WASP-43b, HATS-18b, NGTS-6b, WASP-103b, KELT-16b, HIP 65 A b; see exoplanetarchive.ipac.caltech.edu). The *Kepler* spacecraft has unveiled the presence of over 100 ultra-short period (USP) small planets with $R_p \lesssim 2\, R_\oplus$. It is clear that, at least for solar-type stars, super Earths with periods shorter than one day occur as frequently as hot Jupiters with periods of ~ 1-10 days and that their discovery was previously hampered mainly because of their small signals.

The study of short-period planets is important for many reasons. They are so close to their host stars that the geometric probability for transits can be as large as 40%. Moreover, the expected surface temperatures of USP planets can reach 2000 −

3000 K, allowing the detection of thermal emission from the planetary surface (e.g., Sanchis-Ojeda et al. 2013a), and the induced RV signals can be as high as a few m/s, allowing the planet masses to be measured even for moderately faint stars (Howard et al. 2013; Pepe et al. 2013). Therefore, this class of objects is important to understand the formation and evolution of short-period planets in general, but also to study star-planet interactions, atmospheric erosion, photoevaporation, and other processes arising from strong irradiation and tidal forces (Winn et al. 2018, and references therein).

3.7.2 The Case of Kepler-78b and the USP Planets

One object that emerged from the 4-year *Kepler* survey of about 200,000 stars was the Earth-sized planet Kepler-78b (Sanchis-Ojeda et al. 2013a). This planet, with an orbital period of 8.5 hours, is important because is one of the very smallest planets for which both mass and radius have been measured to better than 20% precision (Howard et al. 2013; Pepe et al. 2013). Based on spectroscopic monitoring performed with HIRES@Keck, Howard et al. (2013) reported a radius of $1.20 \pm 0.09\ R_\oplus$ and a mass of $1.69 \pm 0.41\ M_\oplus$, implying a planet's mean density of 5.3 ± 1.8 g/cm^3. Similar findings were also found by Pepe et al. (2013) from independent measurements using the HARPS-N@TNG spectrograph. All these results are similar to the Earth values, suggesting for Kepler-78b a rock/iron composition and negligible atmosphere (see Fig. 3.60).

Howard et al. (2013) explored some possibilities for the interior structure of Kepler-78b using a simplified two-component model (Fortney et al. 2007) consisting of an iron core surrounded by a silicate mantle, which is a model reproducing correctly the masses of Earth and Venus. Applied to Kepler-78b, the model gives an iron fraction of $\sim 20\%$, similar to that of the Earth and Venus and smaller than that of Mercury. Moreover, with a star-planet separation of 0.01 au, the dayside of Kepler-78b is heated to temperatures of 2300-3100 K, therefore any gaseous atmosphere around the planet would probably have been lost due to photoevaporation by intense starlight. Kepler-78b represents a 'prototype' for the class of USP planets (the interested reader can find additional details in the recent review by Winn et al. (2018) on Kepler-78 and the ultra-short period planets in general.)

3.7.3 The 1-Day Cutoff and Planet Occurrence

The boundary at one day for the orbital periods of USP planets is an arbitrary choice, mainly due to the fact that such $P_{\text{orb}} < 1$ day regime was previously unexplored. Therefore, there is no sharp astrophysical distinction between the USP planets with periods less than one day and the short-period planets with periods of 1–10 days. Following the reasoning of Winn et al. (2018), it is possible to verify that the angular

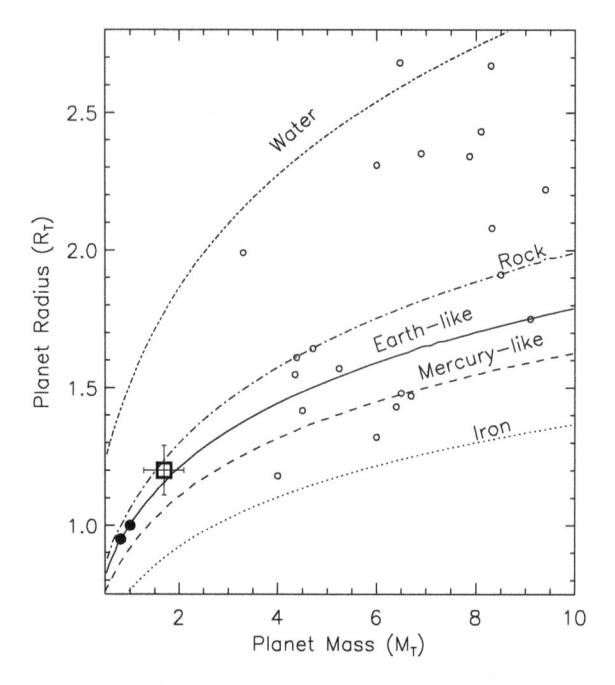

Fig. 3.60 The planet Kepler-78b (open square) placed on a planetary-mass diagram. Earth and Venus are represented by filled circles, while the other extrasolar planets with well characterized mass and radius are plotted as open circles. Model mass-radius relationships for idealized planets consisting of 100% iron, rock (Mg_2SiO_4), and pure water are shown as dotted, dot-dashed, and dot-dot-dot dashed lines, respectively (Seager et al. 2007; Fortney et al. 2007). Solid and dashed lines denote Earth-like composition (67% rock, 33% iron) and Mercury-like composition (40% rock, 60% iron). Exoplanet masses and radii (without restrictions on uncertainties) were taken from the Exoplanet Orbit Database (Wright et al. 2011; September 2020). Data for Kepler-78 were taken from (Howard et al. 2013). Figure inspired by a similar plot by Howard et al. (2013)

diameter of a main-sequence star in the sky of the USP planet is around fifty times wider than the Sun in the sky of the Earth. This means that tidal interactions lead to relatively rapid orbital and spin evolution. Therefore, when there is a USP planet around a main-sequence star it is reasonable to assume that the planet has a circular orbit and is tidally locked (with a permanent dayside and nightside). However, the orbit of an Earth-like USP planet does not have enough angular momentum to spin up the star and achieve a stable double-synchronous state, leading to its spiraling towards the star (see, e.g., Patra et al. 2017). Moreover, a one-day planet around a solar-like star intercepts around 2500 times the flux of the Sun impinging on the Earth and should be bathed in strong UV and X-ray radiation. If all incident energy is re-radiated locally, the planet's surface can easily reach temperatures as high as $\gtrsim 3000$ K (as in fact determined for Kepler-78b). This would lead to a complete loss of any hydrogen-helium atmospheric envelope (Owen and Wu 2016) and the melting of silicates and iron, resulting in the so-called "lava worlds" or "hot Earths" (Léger et al. 2011; Rouan et al. 2011).

3 The Demographics of Close-In Planets

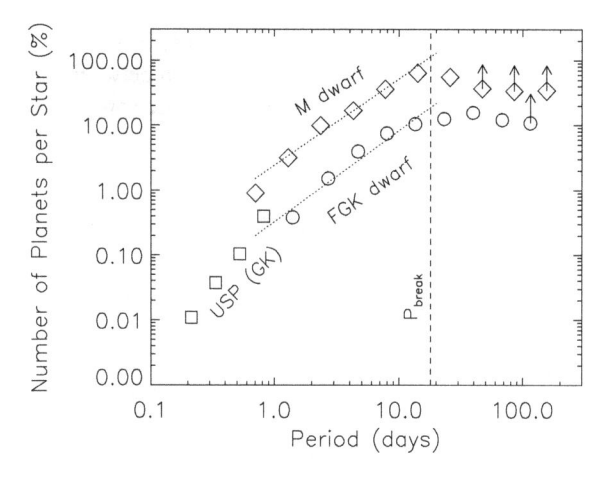

Fig. 3.61 Occurrence rates of sub-Neptunes orbiting FGKM dwarfs (Fressin et al. 2013; Dressing and Charbonneau 2015). The P_{break} at ~ 10 days is shown as vertical dashed line. Short-period planets at $P < P_{break}$ appear distributed according to a $P^{1.4-1.5}$ power law (dotted lines). USP planets at $P < 1$ days using data from Sanchis-Ojeda et al. (2014) are shown as open squares. Points without arrows correspond to sub-Neptunes larger than $0.5\, R_\oplus$ for M-dwarf hosts, and larger than $0.8\, R_\oplus$ for FGK stars. Points with arrows represent sub-Neptunes larger than $1\, R_\oplus$ (M dwarfs) and larger than $1.25\, R_\oplus$ (FGK dwarfs). Figure adapted from Lee and Chiang (2017)

From *Kepler* data, Sanchis-Ojeda et al. (2014) found that the occurrence rate of USP planets shows a strong dependence on the host star's mass. They measured an occurrence rate in the range $\sim 1.10\%$ and $\sim 0.83\%$ for M and K dwarfs, respectively, and $\sim 0.51\%$ and $\sim 0.15\%$ for G and F dwarfs, respectively. This represents the evidence that cooler stars are more likely to host USP planets. Occurrence rates of sub-Neptunes ($R < R_\oplus$) orbiting FGKM dwarfs as obtained by the *Kepler* mission in combination with RV surveys are shown in Fig. 3.61 (Fressin et al. 2013; Dressing and Charbonneau 2015), where USP planets around GK stars are marked with open squares (Sanchis-Ojeda et al. 2014). At periods $P > 100$ days sub-Neptunes appear to be evenly distributed, while at shorter periods they are less common. The occurrence rate as a function of orbital period follows a broken power law, with a break at $P_{break} \sim 10$ days (e.g., Youdin 2011; Mulders et al. 2015a). Within ~ 10 days, the occurrence rate scales approximately as $P^{1.4-1.5}$ depending on the spectral type, while beyond ~ 10 days the occurrence rate shows a plateau. The lower occurrence rate of planets at $P < P_{break}$ may reflect a truncation of their disks, perhaps due to the magnetosphere of their host stars (e.g., Mulders et al. 2015a). In fact, disk-locking theory posits that the inner disk edge corotates with the host star in equilibrium (e.g., Camenzind 1990; Königl 1991; Romanova and Owocki 2016). The disk-locking scenario is supported observationally by "dippers" in the light curves of young low-mass stars with relatively evolved disks (see, e.g., Ansdell et al. 2020; Frasca et al. 2020, and references therein). These stars exhibit material lifted out of the disk midplane, near the corotation radius (see, e.g., Stauffer et al. 2015; Ansdell et al. 2016). The distribution of rotation periods of low-mass stars younger than ~ 5 Myr typically ranges from 0.2 to 20 days with peaks around ∼ 10 days and falls towards shorter periods. This result has been interpreted as a possible connection between disk truncation at corotation and occurrence rate of short-period planets (Lee and Chiang 2017).

3.7.4 Host Star Iron Abundance and Planet Formation

The connection between stellar iron abundance and the presence of planets offers an observational link between conditions during the epoch of planet formation and mature planetary systems. This is because the star iron abundance (usually expressed as [Fe/H], the iron to hydrogen abundance ratio measured with respect to that of the Sun) is thought to reflect the metallicity of the protostellar nebula and the protoplanetary disk from which planets form. Since metal-rich protoplanetary disks are thought to have enhanced surface densities of solids, one expects that metal-rich disks form cores of gas giant planets and terrestrial planets with higher efficiency than metal-poor disks (core-accretion theory; Lissauer et al. 1995; Pollack et al. 1996). As we have already noticed in Sect. 3.3.1, the increase in giant planet occurrence with iron abundance is well established, while in the regime of lower planetary masses this trend appears less solid.

Figure 3.62 shows the distribution of [Fe/H] for stars in three samples of *Kepler* stars analyzed through Keck spectroscopy (Winn et al. 2017b):

1. USP planets selected from Sanchis-Ojeda et al. (2014);
2. hot Jupiters with radii in the range $4 - 20\,R_\oplus$ and orbital periods shorter than 10 days;
3. hot small planets with radii $< 4\,R_\oplus$ and orbital periods between 1 and 10 days.

Even at a glance, it is clear that the hot Jupiters appear to be weighted toward higher [Fe/H] than either the USP planets or the hot small planets. The distributions for the USP and the small planets appear similar to one another. From a two-sample Kolmogorov-Smirnov test, Winn et al. (2017b) concluded that it is very unlikely that the USP planets and hot Jupiters are drawn from the same distribution, while the USP planets and the hot smaller planets have distributions that are indistinguishable. The association between high metallicity and occurrence of USP planets is of particular interest because it was postulated that USP planets may be the remnants of disrupted hot Jupiters (Jackson et al. 2013; Valsecchi et al. 2015). Since this correlation was not found, it is unlikely that USP planets are the evaporated cores of hot Jupiters that reached so close to their host stars that they completely lost their gas through photoevaporation, or Roche lobe overflow, or other processes. It remains quite possible that they are solid cores of formerly gaseous planets smaller than Neptunes.

Recently, Petigura et al. (2018), based on California-*Kepler* Survey results, modeled the metallicity (M) distribution of planets as $df \propto 10^{\beta M} dM$, where β is related to the strength of any metallicity correlation. The authors found that the correlation gets steeper with decreasing orbital period and increasing planet size, from $\beta = -0.3 \pm 0.2$ for warm super-Earths to $\beta = 3.4 \pm 0.9$ for hot Jupiters. This result

3 The Demographics of Close-In Planets   221

Fig. 3.62 Metallicity distribution of three statistical samples: UPS planets (solid line; [Fe/H]$_{mean}$ ~ 0.06 dex), hot Jupiters (dashed line; [Fe/H]$_{mean}$ ~ 0.21 dex), and hot small planets (dotted line; [Fe/H]$_{mean}$ ~ 0.05 dex). Figure adapted from Winn et al. (2017b)

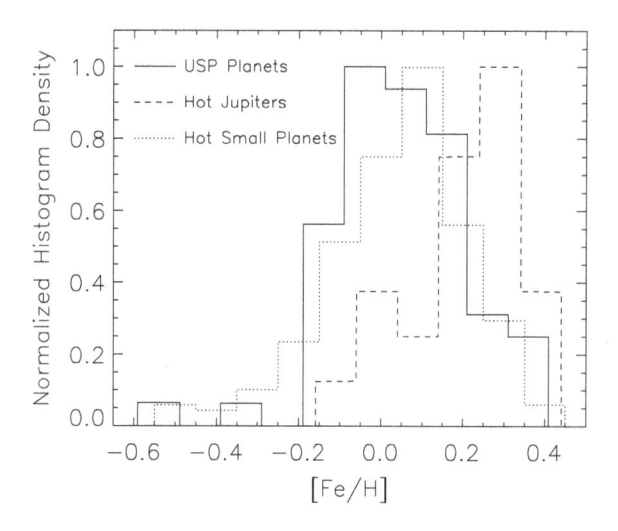

supports the idea that high metallicities in protoplanetary disks increase the mass of the largest rocky cores or the speed at which they are assembled, enhancing the production of planets with higher masses. They conclude that the association between high metallicity and short-period planets may reflect disk density profiles that facilitate the inward migration of solids or higher rates of planet-planet scattering.

3.7.5 Composition of Hot Earths

High-cadence RV monitoring of USP planets allows to fully sample their orbits in just a few nights of observation. Furthermore, since the RV semi-amplitude scales with $P^{-1/3}$, the shorter the orbital period, the higher the amplitude of the RV signal, increasing the likelihood of swift detection. However, because the planets tend to be small, the Doppler signals typically have K_p not exceeding a few m/s, making their detection challenging nonetheless. Accurate masses have been measured for a few USP planets, as shown in Fig. 3.63, where the data points are color-coded according to the level of insolation by the host star. Indeed, one might have expected the more strongly irradiated planets to have a higher density, as a consequence of photoevaporation. However, no clear correlation between planetary mean density and level of irradiation appears. This could be due to the fact that all USP planets are so strongly irradiated that they have been entirely stripped of any preexisting hydrogen/helium atmospheres, as all of them show irradiation values much higher than the threshold of 650 $F_\oplus$ above which close-in sub-Neptunes have undergone photoevaporation (Lundkvist et al. 2016). In the mass-radius diagram for USP planets with well-characterized parameters, radii smaller than 2.2 $R_\oplus$ and orbital periods smaller than 1 day, most planets are placed between the theoretical relations for pure

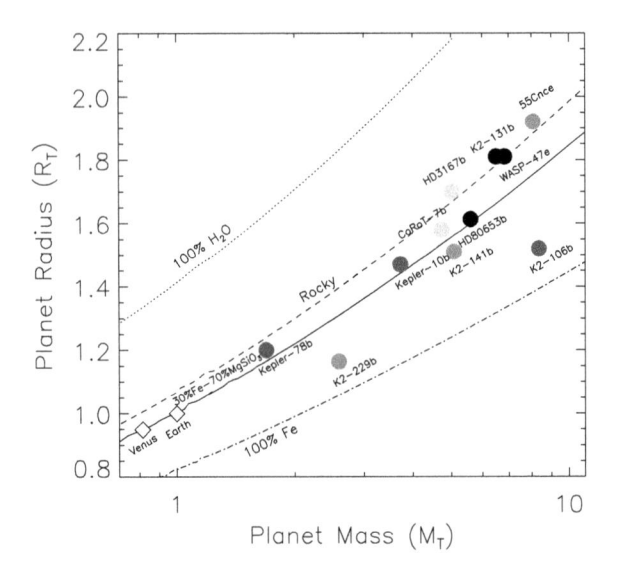

Fig. 3.63 Masses and radii of USP planets for which very accurate Doppler mass measurements have been reported (filled circles). The circles are color-coded according to four levels of irradiation by the star in units of $F_\oplus$, the insolation level received by Earth: light grey (< 2000 $F_\oplus$), dark gray (2000 – 3000 $F_\oplus$), very dark gray (3000 – 4000 $F_\oplus$), black (> 4000 $F_\oplus$). The diamonds represent Venus and Earth. The curves are theoretical mass-radius relationships for planets of different compositions: pure iron, iron core and silicate mantle, pure rock, and pure water (Zeng et al. 2016). References for planetary masses and radii are: Howard et al. (2013); Haywood et al. (2014); Weiss et al. (2016); Demory et al. (2016); Dai et al. (2017); Gunther et al. (2017); Christiansen et al. (2017); Vanderburg et al. (2017); Malavolta et al. (2018); Santerne et al. (2018); Frustagli et al. (2020). Figure inspired by a similar plot by (Dai et al. 2017)

rock (100% $MgSiO_3$) and Earth-like (30% Fe and 70% $MgSiO_3$) compositions (see Fig. 3.63). K2-229b has high density suggesting a massive iron core compatible with that of Mercury, although it was expected to be similar to that of Earth based on host-star chemistry (Santerne et al. 2018).

Dressing et al. (2015) claimed that planets heavier than 6 $M_\oplus$ should have gaseous H/He envelopes, while Rogers (2015) found that planets with orbital periods shorter than 50 days and radii smaller than 1.6 $R_\oplus$ are predominantly rocky. Later, Lopez (2017) found that planets on USP orbits could often retain their envelopes even at high irradiation levels if they formed with very high-metallicity water dominated envelopes. This would imply that planets larger than 1.6 $R_\oplus$ to be more massive than 6 $M_\oplus$ and with densities compatible with water envelopes. Of the eleven USP planets for which mass and radius have been both measured with high accuracy, six are larger than $\sim$ 1.5-1.6 $R_\oplus$ (HD3167b, K2-131b, WASP-47e, 55 Cnc e, HD 80653b, and CoRoT-7b). Three of them (namely, K2-131b, WASP-47e, 55 Cnc e) have masses heavier than 6 $M_\oplus$ and a low mean density ($\sim$ 6.0–6.4 g/cm^3) compatible with a layer of volatiles (like water) surrounding a rocky-iron body. For K2-106b, Sinukoff et al. (2017) reported a planetary mass and radius of 9.0 $M_\oplus$ and 1.82 $R_\oplus$, compatible with a rocky composition and a water mass fraction of $\sim 2\%$, while the values by Gunther et al. (2017) shown in Fig. 3.63 pointed to an iron-rich composition and a water mass fraction $< 20\%$ (Dai et al. 2017). For HD 3167b, Christiansen et al. (2017) reported a planetary mass and radius of $5.02\,M_\oplus$ and $1.70\,R_\oplus$ (consistent with a rocky composition and a water mass fraction $\sim 15\%$, as shown in Fig. 3.63), while Gandolfi et al. (2017) reported $5.69\,M_\oplus$ and $1.575\,R_\oplus$, suggesting a predominantly rocky composition (with a water mass fraction $< 10\%$). The position of HD 80653b and its density is consistent with an Earth-like composition of rock and iron with no significant envelope of volatiles or H/He surrounding the planet (Frustagli et al. 2020).

3.7.6 Formation and Early Evolution of USP Planets

The formation and early evolution of USP planets and their relationship to wider-orbit planets are not fully understood. Dai et al. (2018b) considered a sample of *Kepler* and *K2* multiple-planet systems with the innermost planet of radius smaller than $4\,R_\oplus$ (to avoid giant planets) and $a/R_\star < 12$ (to consider inclinations of 85–90°). They measured the minimum difference $\Delta I = |I_1 - I_2|$ between the fitted orbital inclinations of the innermost two planets as a function of the orbital distance of the innermost planet (see Fig. 3.64). ΔI is equal to the mutual inclinations only if the trajectories of the two planets across the stellar disk are parallel and on the same hemisphere of the star. Among the systems with the closest-orbiting planets ($a/R_\star < 5$), there are about 10 systems for which $\Delta I = 5° - 10°$, larger than the typical mutual inclinations inferred for planets in wider orbits. The planets with $a/R_\star > 5$ almost all have $\Delta I < 5°$. This means that the planets with smaller values of $a/R_\star$ have a broader distribution of ΔI, nearly filling the full range of inclinations compatible with transits. Similar results were obtained by, e.g., Tremaine and Dong (2012) and Fabrycky et al. (2014). Dai et al. (2018b) also noted that higher values of ΔI are associated with larger period ratios. They found that period ratios tend to be higher when the inner planet's period is shorter than about one day (Steffen and Farr 2013). This means that the innermost planets have experienced both inclination excitation and orbital shrinkage, and therefore that they are dynamically more separated.

The results by Dai et al. (2018b) can be interpreted as follows: the shortest-period planets tend to have larger mutual inclinations, and thus are likely to be observed to transit even when the wider-orbiting companions do not transit. These results indicate that the shortest-period planets have a different orbital architecture, with higher mutual inclinations and larger period ratios, meaning that whatever processes led to the extremely tight orbits of these planets were also responsible for tilting the orbit to a higher inclination.

Several theories have been proposed for the formation of very short-period sub-Neptune planets:

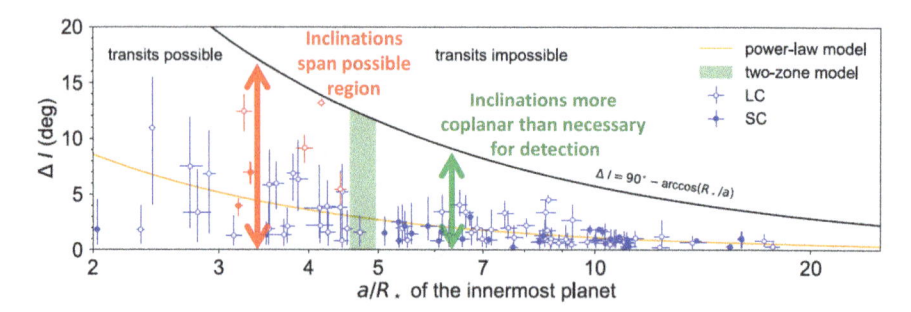

Fig. 3.64 Inclination difference ΔI versus $a/R_\star$ of the innermost planet of a sample of *Kepler/K2* multiple-planet systems. The orange solid line represents a power law model by Dai et al. (2018b), while the black solid line is the boundary above which the inner planet would not transit. Figure adapted from Dai et al. (2018b)

- Spalding and Batygin (2016) proposed that if the host star was initially rotating rapidly, with a non-zero obliquity, the planets' orbits would undergo nodal precession at different rates becoming misaligned, with the innermost planet being most strongly affected.
- Lee and Chiang (2017) proposed that planets form from material collected near the innermost edge of the protoplanetary disk, begin with nearly circular well-aligned orbits, and then the innermost planet undergoes tidal orbital decay.
- In the "secular dynamic chaos" scenario, the innermost planet of a multi-planet system is launched into a high-eccentricity orbit via chaotic secular interactions with its companion planet. If the period is short and the eccentricity becomes high enough, tidal interactions with the host star shrink the orbit. Since eccentricity and inclination are excited together, this theory predicts that the shortest-period planets should have larger mutual inclinations (Petrovich et al. 2019).
- In the "forced-eccentricity migration" scenario, in which the interaction with outer companions continually excites the eccentricity of the innermost planet, eccentricity tides dissipate energy and shrink the orbit (Hansen and Murray 2015).

3.7.7 Brief Summary on USP Planets

Even though the origin of USP planets still remains a matter of debate, we can summarize their main properties as follows:

- USP planets tend to be in multi-planet systems (with longer-period companions), and very few of them have known transiting companions;
- the metallicity distribution of USP-planet-hosting stars does not resemble that of hot-Jupiter hosts and it is indistinguishable from that of close-in sub-Neptunes hosts;

- only very few USP planets have densities compatible with the presence of substantial water envelopes;
- the shortest-period planets likely migrated inwards via a dynamical process that excited orbital eccentricity and mutual inclinations simultaneously.

In conclusion, USP planets seem to represent a subset of sub-Neptunes, most of which have lost their atmospheres because of photoevaporation, or Roche lobe overflow, or other processes, with the exception of a few larger planets that have retained a water (or other volatiles) envelope.

References

Adams, E.R., Seager, S., Elkins-Tanton, L.: Astrophy. J. **673**, 1160 (2008)
Addison, B. C., Horner, J., Wittenmyer, R. A., et al.: (2020). arXiv:2006.13675
Albrecht, S., Winn, J.N., Johnson, J.A., et al.: Astrophys. J. **757**, 18 (2012)
Alonso, R., Brown, T.M., Torres, G., et al.: Astrophys. J. Lett. **613**, L153 (2004)
Alsubai, K.A., Parley, N.R., Bramich, D.M., et al.: Acta Astron. **63**, 465 (2013)
Anderson, K.R., Lai, D.: Mon. Noti. R. Astrono. Soc. **480**, 1402 (2018)
Anderson, S.G., Dittmann, J.A., Ballard, S., et al.: Astron. J. **161**, 203 (2021)
Ansdell, M., Gaidos, E., Hedges, C., et al.: Mon. Not. R. Astron. Soc. **492**, 572 (2020)
Ansdell, M., Gaidos, E., Williams, J.P., et al.: Mon. Not. R. Astron. Soc. **462**, L101 (2016)
Auvergne, M., et al.: Astron. Astrophys. **506**, 411 (2009)
Bach-Møller, N., Jørgensen, U.G.: Mont. Not. R. Astron. Soc. **500**, 1313 (2021)
Bakos, G.Á., Noyes, R.W., Kovács, G., et al.: Proc. Astron. Soc. Pac. **116**, 266 (2004)
Bakos, G.Á., Csubry, Z., Penev, K., et al.: Proc. Astrono. Soc. Pac. **125**, 154 (2013)
Ballard, S., Johnson, J.A.: Astrophy. J. **816**, 66 (2016)
Barnes, J.W.: Astrophys. J. **705**, 683 (2009)
Baruteau, C., Papaloizou, J.C.B.: Astrophy. J. **778**, 7 (2013)
Bashi, D., Helled, R., Zucker, S., Mordasini, C.: Astron. Astrophys. **604**, A83 (2017)
Bashi, D., Zucker, S., Adibekyan, V.: Astron. Astrophys. **643**, A106 (2020)
Batalha, N.M., et al.: Astrophy. J. **729**, 27 (2011)
Batalha, N.M.: Proc. Natal Acad. Sci. **111**, 12647 (2014)
Batalha, N.M., Lewis, T., Fortney, J.J., et al.: Astrophy. J. **885**, L25 (2019)
Bate, M.R., Lodato, G., Pringle, J.E.: Mon. Not. R. Astron. Soc. **401**, 1505 (2010)
Bean, J.L., Seifahrt, A.: Astron. Astrophys. **496**, 249 (2009)
Bean, J.L., Raymond, S.N., Owen, J.E.: J. Geophys. Res. **126**, 06639 (2021)
Becker, J.C., Adams, F.C.: Mon. Not. R. Astron. Soc. **455**, 2980 (2016)
Becker, J.C., Adams, F.C.: Mon. Not. R. Astron. Soc. **468**, 549 (2017)
Becker, J., Batygin, K., Fabrycky, D., et al.: Astron. J. **160**, 254 (2020)
Benneke, B., Wong, I., Piaulet, C., et al.: Astrophy. J. **887**, L14 (2019)
Berger, T.A., Huber, D., Gaidos, E., et al.: Astrono. J. **160**, 108 (2020)
Bitsch, B., Raymond, S.N., Izidoro, A.: Astron. Astrophys. **624**, A109 (2019)
Bonfils, X., Delfosse, X., Udry, S., et al.: Astron. Astrophys. **549**, A109 (2013)
Bonfils, X., Almenara, J.M., Jocou, L., et al.: Proc. SPIE **9605**, 96051L (2015)
Borsa, F., Rainer, M., Bonomo, A.S., et al.: Astron. Astrophys. **631**, A34 (2019)
Borsa, F., Allart, R., Casasayas-Barris, N., et al.: Astron. Astrophys. **645**, A24 (2021)
Borucki, W.J.: Rep. Prog. Phys. **79**, 036901 (2016)
Borucki, W.J., Summers, A.L.: Icarus **58**, 121 (1984)
Borucki, W.J., Koch, D., Basri, G., et al.: Science **327**, 977 (2010)

Borucki, W.J., Koch, D., Basri, G., et al.: Astrophy. J. **736**, 19 (2011)
Bouchy, F., Pepe, F., Queloz, D.: Astron. Astrophys. **374**, 733 (2001)
Bovaird, T., Lineweaver, C.H.: Mon. Not. R. Astron. Soc. **468**, 1493 (2017)
Brewer, J.M., Wang, S., Fischer, D.A., Foreman-Mackey, D.: Astrophy. J. **867**, L3 (2018)
Broeg, C., et al.: EPJ Web of Conf. **47**, 03005 (2013)
Brown, D.J.A., Collier Cameron, A., Díaz, R.F., et al.: Astrophy. J. **760**, 139 (2012)
Brügger, N., Burn, R., Coleman, G.-A.L., et al.: Astron. Astrophys. **640**, A21 (2020)
Bryan, M.L., Knutson, H.A., Howard, A.W., et al.: Astrophy. J. **821**, 89 (2016)
Bryson, S., Coughlin, J., Batalha, N.M., et al.: Astrono. J. **159**, 279 (2020a)
Bryson, S., Coughlin, J., Kunimoto, M., Mullally, S.E.: Astron. J. **160**, 200 (2020b)
Bryson, S., Kunimoto, M., Kopparapu, R.K., et al.: Astrophy. J. **161**, 32 (2021)
Buchhave, L.A., Latham, D.W., Johansen, A., et al.: Nature **486**, 375 (2012)
Buchhave, L.A., Bizzarro, M., Latham, D.W., et al.: Nature **509**, 593 (2014)
Buchhave, L.A., Latham, D.W.: Astrophy. J. **808**, 187 (2015)
Buchhave, L.A., Dressing, C.D., Dumusque, X., et al.: Astron. J. **152**, 160 (2016)
Buchhave, L.A., Bitsch, B., Johansen, A., et al.: Astrophy. J. **856**, 37 (2018)
Burke, C.J., Bryson, S.T., Mullally, F., et al.: Astrophy. J. Suppl. Ser. **210**, 19 (2014)
Burke, C.J., Christiansen, J.L., Mullally, F., et al.: Astrophy. J. **809**, 8 (2015)
Butler, R.P., et al.: Astrophy. J. **526**, 916 (1999)
Butler, R.P., et al.: Astrophy. J. **617**, 580 (2004)
Camenzind, M.: Rev. Modern Astron. **3**, 234 (1990)
Carrera, D., Raymond, S.N., Davies, M.B.: Astron. Astrophys. **629**, L7 (2019)
Carter, J.A., Winn, J.N.: Astrophy. J. **709**, 1219 (2010)
Catanzarite, J., Shao, M.: Astrophy. J. **738**, 151 (2011)
Cegla, H.M., Lovis, C., Bourrier, V., et al.: Astron. Astrophys. **588**, A127 (2016)
Charbonneau, D.: Research on extrasolar planets. ASP Conf. Ser. **294**, 449 (2003)
Charbonneau, D., et al.: Astrophy. J. **529**, L45 (2000)
Charbonneau, D., et al.: Nature **462**, 891 (2009)
Chatterjee, S., Ford, E.B.: Astrophy. J. **803**, 33 (2015)
Chatterjee, S., Ford, E.B., Matsumura, S., Rasio, F.A.: Astrophy. J. **686**, 580 (2008)
Chen, H., Rogers, L.A.: Astrophy. J. **831**, 180 (2016)
Chen, J., Kipping, D.: Astrophy. J. **834**, 17 (2017)
Choksi, N., Chiang, E.: Mon. Not. R. Astron. Soc. **495**, 4192 (2020)
Christiansen, J.L., Vanderburg, A., Burt, J., et al.: Astrophy. J. **154**, 17 (2017)
Ciardi, D.R., Fabrycky, D.C., Ford, E.B.: Astrophy. J. **763**, 41 (2013)
Cloutier, R., Menou, K.: Astron. J. **159**, 211 (2020)
Collier Cameron, A., Bruce, V.A., Miller, G.R.M., et al.: Mon. Not. R. Astron. Soc. **403**, 151 (2010a)
Collier Cameron, A., Guenther, E., Smalley, B., et al.: Mon. Not. R. Astron. Soc. **407**, 507 (2010b)
Collier Cameron, A.: Methods of Detecting Exoplanets, Springer, Berlin, p. 89 (2016)
Coughlin, J., Thompson, S.E. and Kepler Team: Astron. J. 152, 158 (2017)
Courcol, B., Bouchy, F., Deleuil, M.: Mon. Not. R. Astron. Soc. **461**, 1841 (2016)
Crossfield, I.J.M., Kreidberg, L.: Astron. J. **154**, 261 (2017)
Cumming, A., Butler, R.P., Marcy, G.W., et al.: Proc. Astron. Soc. Pac. **120**, 531 (2008)
Dai, F., Winn, J.N.: Astron. J. **153**, 205 (2017a)
Dai, F., Winn, J.N., Gandolfi, D., et al.: Astron. J. **154**, 226 (2017b)
Dai, F., Masuda, K., Winn, J.N.: Astrophy. J. Lett. **864**, L38 (2018a)
Dai, F., Winn, J.N., Berta-Thompson, Z., et al.: Astron. J. **155**, 177 (2018b)
Dai, F., Masuda, K., Winn, J.N., Zeng, L.: Astrophy. J. **883**, 79 (2019)
Dai, F., Roy, A., Fulton, B., et al.: (2020). arXiv:2008.12397
Damasso, M., Sozzetti, A., Lovis, C., et al.: Astron. Astrophys. **642**, A31 (2020)
Davies, M.B., Adams, F.C., Armitage, P., et al.: Protostars and Planets VI, 787 (2014)
Dawson, R.I., Murray-Clay, R.A.: Astrophy. J. **767**, L24 (2013)
Dawson, R.I.: Astrophy. J. **790**, 31 (2014)

Delisle, J.-B., Laskar, J.: Astron. Astrophys. **570**, L7 (2014)
Delisle, J.-B., Ségransan, D., Dumusque, X., et al.: Astron. Astrophys. **614**, 133 (2018)
Delrez, L., Gillon, M., Queloz, D., et al.: ıProc. SPIE **10700**, 107001I (2018)
Deming, D., et al.: Nature **434**, 740 (2005)
Demory, B.-O., Gillon, M., Madhusudhan, N., Queloz, D.: Mon. Not. R. Astron. Soc. **455**, 2018 (2016)
Dempsey, A.M., Nelson, B.E.: Astrophys. J. **867**, 75 (2018)
De Rosa, R.J., Dawson, R., Nielsen, E.L.: Astrono. Astrophys. **640**, A73 (2020)
Désert, J.M., Charbonneau, D., Demory, B.O., et al.: Astrophy. J. Suppl. Ser. **197**, 14 (2011)
Dong, S., Zhu, Z.: Astrophy. J. **778**, 53 (2013)
Dorn, C., Khan, A., Heng, K., et al.: Astron. Astrophys. **577**, 83 (2015)
Dorn, C., Venturini, J., Khan, A., et al.: Astron. Astrophys. **597**, 37 (2017)
Dorn, C., Noack, L., Rozel, A.B.: Astron. Astrophys. **614**, A18 (2018)
Doyle, L.R., et al.: Science **333**, 1602 (2011)
Dragomir, D., Matthews, J.M., Eastman, J.D., et al.: Astrophy. J. **772**, L2 (2013)
Dressing, C.D., Charbonneau, D.: Astrophy. J. **767**, 95 (2013)
Dressing, C.D., Charbonneau, D.: Astrophy. J. **807**, 45 (2015)
Dressing, C.D., Charbonneau, D., Dumusque, X., et al.: Astrophy. J. **800**, 135 (2017)
Dumusque, X., et al.: Astron. Astrophys. **598**, A133 (2017)
Eastman, J., Gaudi, B.S., Agol, E.: Proc. Astron. Soc. Pac. **125**, 83 (2013)
Endl, M., Cochran, W.D., Kürster, M.: Astrophy. J. **649**, 436 (2006)
Fabrycky, D., Tremaine, S.: Astrophy. J. **669**, 1298 (2007)
Fabrycky, D.C., Lissauer, J.J., Ragozzine, D., et al.: Astrophy. J. **790**, 146 (2014)
Fang, J., Margot, J.-L.: Astrophy. J. **761**, 92 (2012)
Fellgett, P.: Optical Acta **2**, 9 (1955)
Fernandes, R.B., Mulders, G.D., Pascucci, I., et al.: Astrophy. J. **874**, 81 (2019)
Fischer, D.A., Valenti, J.: Astrophy. J. **622**, 1102 (2005)
Fischer, D.A., et al.: Astrophy. J. **675**, 790 (2008)
Fischer, D.A., et al.: Proc. Astron. Soc. Pac. **128**, 066001 (2016)
Ford, E.B., Rasio, F.A.: Astrophy. J. **686**, 621 (2008)
Foreman-Mackey, D., Hogg, D.W., Morton, T.D.: Astrophy. J. **795**, 64 (2014)
Foreman-Mackey, D., Montet, B.T., Hogg, D.W., et al.: Astrophy. J. **806**, 215 (2015)
Fortney, J.J., Marley, M.S., Barnes, J.W.: Astrophy. J. **659**, 1661 (2007)
Fortney, J.J., Mordasini, C., Nettelmann, N., et al.: Astrophy. J. **775**, 80 (2013)
Fressin, F., Torres, G., Charbonneau, D., et al.: Astrophy. J. **766**, 81 (2013)
Frasca, A., Manara, C.F., Alcalá, J.M., et al.: Astron. Astrophys. **639**, L8 (2020)
Frustagli, G., Poretti, E., Milbourne, T., et al.: Astron. Astrophys. **633**, 133 (2020)
Fulton, B.J., et al.: Astron. J. **142**, 84 (2011)
Fulton, B.J., Petigura, E.A., Howard, A.W., et al.: Astron. J. **154**, 109 (2017)
Fulton, B.J., Petigura, E.A.: Astron. J. **156**, 264 (2018)
Fulton, B.J., Rosenthal, L.J., Hirsch, L.A., et al.: Astrophy. J. Suppl. Ser. (2021) in press (arXiv:2105.11584)
Gaidos, E., Mann, A.W., Kraus, A.L., Ireland, M.: Mon. Not. R. Astron. Soc. **457**, 2877 (2016)
Gandolfi, D., Barragán, O., Hatzes, A.P., et al.: Astron. J. **154**, 123 (2017)
Gaia Collaboration, Prusti, T., de Bruijne, J.H.J., et al.: Astron. Astrophys. **595**, A1 (2016a)
Gaia Collaboration, Brown, A.G.A., Vallenari, A., et al.: Astron. Astrophys. **595**, A2 (2016b)
Collaboration, Gaia, Brown, A.G.A., Vallenari, A., et al.: Astron. Astrophys. **616**, 22 (2018)
Gandolfi, D., Collier Cameron, A., Endl, M., et al.: Astron. Astrophys. **543**, L5 (2012)
Garrett, D., Savransky, D., Belikov, Rus, 2018, Proc. Astron. Soc. Pac. **130**, 4403
Gaudi, B.S., Stassun, K.G., Collins, K.A., et al.: Nature **546**, 514 (2017)
Ghezzi, L., Cuñha, K., Smith, V. V., et al.: Astrophy. J. **720**, 1290 (2010)
Ghezzi, L., Montet, B.T., Johnson, J.A.: Astrophy. J. **860**, 109 (2018)
Gibbs, A., Bixel, A., Rackham, B.V., et al.: Astron. J. **159**, 169 (2020)

Gilbert, G.J., Fabrycky, D.C.: Astron. J. **159**, 281 (2020)
Gillon, M., Demory, B.-O., Barman, T., et al.: Astron. Astrophys. **471**, L51 (2007)
Gillon, M., Triaud, A.H.M.J., Fortney, J.J., et al.: Astron. Astrophys. **542**, A4 (2012)
Gillon, M., et al.: Nature **542**, 23 (2017)
Ginzburg, S., Schlichting, H.E., Sari, R.: Astrophy. J. **825**, 29 (2016)
Ginzburg, S., Schlichting, H.E., Sari, R.: Mon. Not. R. Astron. Soc. **476**, 759 (2018)
Gizon, L., Solanki, S.K.: **589**, 1009 (2003)
Goldreich, P., Schlichting, H.E.: Astron. J. **147**, 32 (2014)
Greene, T.P., et al.: Astrophy. J. **817**, 17 (2016)
Griffin, R.F.: Astrophy. J. **148**, 465 (1967)
Guilluy, G., Gressier, A., Wright, S., et al.: Astron. J. **161**, 19 (2021)
Guenther, E.W., Barragán, O., Dai, F., et al.: Astron. J. **608**, 93 (2017)
Gupta, A., Schlichting, H.E.: Mon. Not. R. Astron. Soc. **487**, 24 (2019)
Hansen, B.M.S., Murray, N.: Astrophy. J. **775**, 53 (2013)
Hansen, B.M.S., Murray, N.: Mon. Not. R. Astron. Soc. **448**, 1044 (2015)
Hara, N.C., Bouchy, F., Stalport, M., et al.: Astron. Astrophys. **636**, L6 (2020)
Hardegree-Ullman, K.K., Cushing, M.C., Muirhead, P.S., Christiansen, J.L.: Astron. J. **158**, 75 (2019)
Hartman, J.D., Bakos, G.A., Buchhave, L.A., et al.: Astron. J. **150**, 197 (2015)
Hatzes, A.P., Dvorak, R., Wuchterl, G., et al.: Astron. Astrophys. **520**, A93 (2010)
Hatzes, A.P.: Methods of Detecting Exoplanets, p. 3 . Springer, Berlin (2016)
Haywood, R.D., Cameron, A.C., Queloz, D., et al.: Mon. Not. R. Astron. Soc. **443**, 2517 (2014)
He, M.Y., Ford, E.B., Ragozzine, D.: Mon. Not. R. Astron. Soc. **490**, 4575 (2019)
He, M.Y., Ford, E.B., Ragozzine, D., Carrera, D.: Astron. J. **160**, 276 (2020)
Hébrard, G., Ehrenreich, D., Bouchy, F., et al.: Astron. Astrophys. **527**, L11 (2011)
Henry, G.W., Marcy, G.W., Butler, R.P., Vogt, S.S.: Astrophy. J. **529**, L41 (1999)
Hirano, T., Gaidos, E., Winn, J.N., et al.: Astrophy. J. Lett. **890**, L27 (2020)
Horne, J.H., Baliunas, S.L.: Astrophy. J. **302**, 757 (1986)
Howard, A.W., et al.: Astrophy. J. **721**, 1467 (2010a)
Howard, A.W., Marcy, G.W., Johnson, J.A., et al.: Science **330**, 653 (2010b)
Howard, A.W., Johnson, J.A., Marcy, G.W., et al.: Astrophy. J. **730**, 10 (2011)
Howard, A.W., Marcy, Marcy, G. W., Bryson, S. T., et al.: Astrophy. J. Supplement Series **201**, 15 (2012)
Howard, A.W., Sanchis-Ojeda, R., Marcy, G.W., et al.: Nature **503**, 381 (2013)
Howard, A.W., Fulton, B.J.: Proc. Astron. Soc. Pac. **128**, 114401 (2016)
Howe, A.R., Burrows, A., Verne, W.: Astrophy. J. **787**, 173 (2014)
Hsu, D.C., Ford, E.B., Ragozzine, D., Morehead, R.C.: Astron. J. **155**, 205 (2018)
Hsu, D.C., Ford, E.B., Ragozzine, D., Ashby, K.: Astron. J. **158**, 109 (2019)
Hsu, D.C., Ford, E.B., Terrien, R.: Mon. Not. R. Astron. Soc. **498**, 2249 (2020)
Huang, C.X., Petrovich, C., Deibert, E.: Astron. J. **153**, 210 (2017)
Huang, C.X., Burt, J., Vanderburg, A., et al.: Astrophy. J. **868**, L39 (2018)
Ida, S., Lin, D.N.C.: Astrophy. J. **626**, 1045 (2005)
Ida, S., Lin, D.N.C.: Astrophy. J. **685**, 584 (2008)
Ida, S., Lin, D.N.C.: Astrophy. J. **719**, 810 (2010)
Ida, S., Lin, D.N.C.: Astrophy. J. **775**, 42 (2013)
Inamdar, N.K., Schlichting, H.E.: Astrophy. J. **817**, 13 (2016)
Izidoro, A., Ogihara, M., Raymond, S.N., et al.: Mon. Not. R. Astron. Soc. **470**, 1750 (2017)
Jackson, A.P., Davis, T.A., Wheatley, P.J.: Mon. Not. R. Astron. Soc. **422**, 2024 (2012)
Jackson, B., Stark, C.C., Adams, E.R., et al.: Astrophy. J. **779**, 2 (2013)
Jiang, C.-F., Xie, J.-W., Zhou, J.-L.: Astron. J. **160**, 180 (2020)
Jin, S., Mordasini, C., Parmentier, V., et al.: Astrophy. J. **795**, 65 (2014)
Jin, S., Mordasini, C.: Astrophy. J. **853**, 163 (2018)
Johansen, A., Davies, M.B., Church, R.P., Holmelin, V.: Astrophy. J. **758**, 39 (2012)

Johansen, A., Lambrechts, M.: Ann. Rev. Earth Planet. Sci. **45**, 359 (2017)
Johnson, J.A., Aller, K.M., Howard, A.W., Crepp, J.R.: Proc. Astron. Soc. Pac. **122**, 905 (2010)
Jurgenson, C., Fischer, D., McCracken, T., et al.: Proc. SPIE **9908**, 99086T (2016)
Jurić, M., Tremaine, S.: Astrophy. J. **686**, 603 (2008)
Juvan, I.G., Lendl, M., Cubillos, P.E., et al.: Astron. Astrophys. **610**, A15 (2018)
Kane, S.R., Ciardi, D.R., Gelino, D.M., von Braun, K.: Mon. Not. R. Astron. Soc. **425**, 757 (2012)
Kawashima, Y., Hu, R., Ikoma, M.: Astrophy. J. **876**, L5 (2019)
Knutson, H.A., Dragomir, D., Kreidberg, L., et al.: Astrophy. J. **794**, 155 (2014a)
Knutson, H.A., Benneke, B., Deming, D., et al.: Nature **505**, 66 (2014b)
König, K.: Astrophy. J. **370**, L37 (1991)
Köingl, A., Giacalone, S., Matsakos, T.: Astrophy. J. Lett. **846**, L13 (2017)
Kopparapu, R.K.: Astrophy. J. **767**, L8 (2013)
Kopparapu, R.K., Hébrard, E., Belikov, R., et al.: Astrophy. J. **856**, 122 (2018)
Kovács, G., Zucker, S., Mazeh, T.: Astron. Astrophys. **391**, 369 (2002)
Kovács, G., Hodgkin, S., Sipőcz, B., et al.: Mon. Not. R. Astron. Soc. **433**, 889 (2013)
Kraft, R.P.: Astrophy. J. **150**, 551 (1967)
Kraus, S., Le Bouquin, J.-B., Kreplin, A., et al.: Astrophy. J. **897**, L8 (2020)
Kreidberg, L., Bean, J.L., Désert, J.-M., et al.: Nature **505**, 69 (2014)
Kreidberg, L., Molliére, P., Crossfield, I. J. M., et al.: (2020). arXiv:2006.07444
Kunimoto, M., Matthews, J.M.: Astron. J. **159**, 248 (2020)
Lai, D.: Mon. Not. R. Astron. Soc. **423**, 486 (2012)
Lambrechts, M., Morbidelli, A., Jacobson, S.A., et al.: Astron. Astrophys. **627**, A83 (2019)
Latham, D.W., Rowe, J.F., Quinn, S.N.: Astrophy. J. **732**, L24 (2011)
Lee, E.J., Chiang, E., Ormel, C.W.: Astrophy. J. **797**, 95 (2014)
Lee, E.J., Chang, Ph., Murray, N.: Astrophy. J. **800**, 49 (2015)
Lee, E.J., Chiang, E.: Astrophy. J. **817**, 90 (2016)
Lee, E.J., Chiang, E.: Astrophy. J. **842**, 40 (2017)
Léger, A., Selsis, F., Sotin, C., et al.: Icarus **169**, 499 (2004)
Léger, A., Rouan, D., Schneider, J., et al.: Astron. Astrophys. **506**, 287 (2009)
Léger, A., Grasset, O., Fegley, B., et al.: Icarus **213**, 1 (2011)
Li, J., Lai, D.: Astrophy. J. **898**, L20 (2020)
Limbach, M.A., Turner, E.L.: Proc. Natl. Acad. Sci. **112**, 20 (2015)
Lin, D.N.C., Bodenheimer, P., Richardson, D.C.: Nature **380**, 606 (1996)
Lissauer, J.J., Pollack, J.B., Wetherill, G.W., Stevenson, D.J.: Neptun and Triton, vol. 37. Cruikshank Univ. of Arizona Press, Tucson (1995)
Lissauer, J.J., Fabrycky, D.C., Ford, E.B., et al.: Nature **470**, 53 (2011a)
Lissauer, J.J., Ragozzine, D., Fabrycky, D.C., et al.: Astrophy. J. Suppl. Ser. **197**, 8 (2011b)
Lissauer, J.J., Marcy, G.W., Rowe, J.F., et al.: Astrophy. J. **750**, 112 (2012)
Lithwick, Y., Xie, J., Wu, Y.: Astrophy. J. **761**, 122 (2012)
Lomb, N.R.: Astrophys. Space Sci. **39**, 447 (1976)
Lopez, E.D., Fortney, J.J., Miller, N.: Astrophy. J. **761**, 59 (2012)
Lopez, E.D., Fortney, J.J.: Astrophy. J. **776**, 2 (2013)
Lopez, E.D., Fortney, J.J.: Astrophy. J. **792**, 1 (2014)
Lopez, E.D.: Mon. Not. R. Astron. Soc. **472**, 245 (2017)
Lopez, E.D., Rice, K.: Mon. Not. R. Astron. Soc. **479**, 5303 (2018)
Louden, E.M., Winn, J.N., Petigura, E.A., et al.: Astron. J. **161**, 68 (2021)
Lovis, C., Mayor, M., Pepe, F., et al.: Nature **441**, 305 (2006)
Lovis, C., Ségransan, D., Mayor, M., et al.: Astron. Astrophys. **528**, A112 (2011)
Lozovsky, M., Helled, R., Dorn, C., Venturini, J.: Astrophy. J. **866**, 49 (2018)
Lu, C.X., Schlaufman, K.C., Cheng, S.: Astron. J. **160**, 253 (2020)
Lundkvist, M.S., Kjeldsen, H., Albrecht, S., et al.: Nat. Commun. **7**, 11201 (2016)
MacDonald, M.G., Dawson, R.I., Morrison, S.J., et al.: Astrophy. J. **891**, 20 (2020)
Madhusudhan, N., Nixon, M.C., Welbanks, L., et al.: Astrophy. J. **891**, L7 (2020)

Malavolta, L., Mayo, A.W., Louden, T., et al.: Astron. J. **155**, 107 (2018)
Maldonado, J., Micela, G., Baratella, M., et al.: Astron. Astrophys. **644**, A68 (2020)
Mancini, L., Southworth, J., Raia, G., et al.: Mon. Not. R. Astron. Soc. **465**, 843 (2017)
Mancini, L., Esposito, M., Covino, E., et al.: Astron. Astrophys. **613**, A41 (2018)
Mandel, K., Agol, E.: Astrophy. J. **580**, L171 (2002)
Marcy, G.W., Isaacson, H., Howard, A.W., et al.: Astrophy. J. Suppl. Ser. **210**, 20 (2014)
Martinez, C.F., Cunha, K., Ghezzi, L., et al.: Astrophy. J. **875**, 29 (2019)
Marzari, F., Nelson, A.F.: Astrophy. J. **705**, 1575 (2009)
Masuda, K., Winn, J.N., Kawahara, H.: Astron. J. **159**, 38 (2020)
Matsumoto, Y., Kokubo, E.: Astron. J. **154**, 27 (2017)
Matsumura, S., Peale, S.J., Rasio, F.A.: Astrophy. J. **725**, 1995 (2010)
Mayor, M., Udry, S., Lovis, C., et al.: Astron. Astrophys. **493**, 639 (2009)
Mayor, M., Marmier, M., Lovis, C., et al.: (2011). arXiv:1109.2497
Mayor, M., Queloz, D.: Nature **378**, 355 (1995)
Mazeh, T., Perets, H.B., McQuillan, A., Goldstein, E.S.: Astrophy. J. **801**, 3 (2014)
Mazeh, T., Holczer, T., Faigler, S.: Astron. Astrophys. **589**, A75 (2016)
McArthur, B.E., Benedict, G.F., Barnes, R., et al.: Astrophy. J. **715**, 1203 (2010)
McCullough, P.R., Stys, J.E., Valenti, J.A., et al.: Proc. Astron. Soc. Pac. **117**, 783 (2005)
Mikal-Evans, T., Crossfield, I.J.M., Benneke, B., et al.: Astron. J. **161**, 18 (2021)
Miller, G.R.M., Collier Cameron, A., Simpson, E.K., et al.: Astron. Astrophys. **523**, A52 (2010)
Miller-Ricci, E., Rowe, J.F., Sasselov, D., et al.: Astrophy. J. **682**, 586 (2008a)
Miller-Ricci, E., Rowe, J.F., Sasselov, D., et al.: Astrophy. J. **682**, 593 (2008b)
Miller-Ricci, E., Seager, S., Sasselov, D.: Astrophy. J. **690**, 1056 (2009)
Millholland, S., Wang, S., Laughlin, G.: Astrophy. J. **849**, L33 (2017)
Millholland, S., Laughlin, G.: Nature Astronomy **3**, 424 (2019)
Millholland, S., Spalding, C.: Astrophy. J. **905**, 71 (2020)
Mills, S.M., Howard, A.W., Petigura, E.A., et al.: Astron. J. **157**, 198 (2019)
Morbidelli, A., Lambrechts, M., Jacobson, S., Bitsch, B., Icarus 258, 418 (2015)
Morbidelli, A., Bitsch, B., Crida, A., et al.: Icarus **267**, 368 (2016)
Mordasini, C., Alibert, Y., Benz, W.: Astron. Astrophys. **501**, 1139 (2009a)
Mordasini, C., Alibert, Y., Benz, W., Naef, D.: Astron. Astrophys. **501**, 1161 (2009b)
Mordasini, C., Alibert, Y., Benz, W., et al.: Astron. Astrophys. **541**, A97 (2012a)
Mordasini, C., Alibert, Y., Klahr, H., Henning, T.: Astron. Astrophys. **547**, A111 (2012b)
Mordasini, C., Alibert, Y., Georgy, C., et al.: Astron. Astrophys. **547**, A112 (2012c)
Mordasini, C., Klahr, H., Alibert, Y., et al.: Astron. Astrophys. **566**, A141 (2014)
Moorhead, A.V., Ford, E.B., Morehead, R.C., et al.: Astrophy. J. Suppl. Ser. **197**, 1 (2011)
Mortier, A., Santos, N.C., Sozzetti, A., et al.: Astron. Astrophys. **543**, A45 (2012)
Mousis, O., Deleuil, M., Aguichine, A., et al.: Astrophy. J. **896**, L22 (2020)
Mulders, G.D., Pascucci, I., Apai, D.: Astrophy. J. **798**, 112 (2015a)
Mulders, G.D., Ciesla, F.J., Min, M., et al.: Astrophy. J. **807**, 9 (2015b)
Mulders, G.D., Pascucci, I., Apai, D., Ciesla, F.J.: Astron. J. **156**, 24 (2018)
Mullally, F., Coughlin, J.L., Thompson, S.E., et al.: Astrophy. J. Suppl. Ser. **217**, 31 (2015)
Muñoz, D.J., Perets, H.B.: Astrophy. J. **156**, 253 (2018)
Munoz Romero, C.E., Kempton, E.M.-R.: Astron. J. **155**, 134 (2018)
Murchikova, L., Tremaine, S.: Astron. J. **160**, 160 (2020)
Mustill, A.J., Davies, M.B., Johansen, A.: Mon. Not. R. Astron. Soc. **468**, 3000 (2017)
Naef, D., Latham, D.W., Mayor, M., et al.: Astron. Astrophys. **375**, L27 (2001)
Nagasawa, M., Ida, S., Bessho, T.: Astrophy. J. **678**, 498 (2008)
Narang, M., Manoj, P., Furlan, E., et al.: Astron. J. **156**, 221 (2018)
Narita, N., Sato, B., Hirano, T., Tamura, M.: Proc. Astron. Soc. Jpn. **61**, L35 (2009)
Ogihara, M., Kokubo, E., Suzuki, T.K., Morbidelli, A.: Astron. Astrophys. **615**, A63 (2018)
Ormel, C.W., Liu, B., Schoonenberg, D.: Astron. Astrophys. **604**, A1O (2017)
Otegi, J.F., Bouchy, F., Helled, R.: Astron. Astrophys. **634**, A43O (2020)

Owen, J.E., Wu, Y.: Astrophy. J. **775**, 105O (2013)
Owen, J.E., Wu, Y.: Astrophy. J. **817**, 107 (2016)
Owen, J.E., Wu, Y.: Astrophy. J. **847**, 29O (2017)
Owen, J.E.: Ann. Rev. Earth Planet. Sci. **47**, 67 (2019)
Paardekooper, S.-J., Baruteau, C., Meru, F.: Mon. Not. R. Astron. Soc. **416**, 65 (2011)
Paardekooper, S.-J., Rein, H., Kley, W.: Mon. Not. R. Astron. Soc. **434**, 3018 (2013)
Pascucci, I., Mulders, G.D., Lopez, E.: Astrophy. J. **883**, 15 (2019)
Patra, K.C., Winn, J.N., Holman, M.J., et al.: Astrophy. J. **154**, 4 (2017)
Pepe, F., Cameron, A.C., Latham, D.W., et al.: Nature **503**, 377 (2013)
Pepe, F., Cristiani, S., Rebolo, R., et al.: Astron. Astrophys. **645**, A96 (2021)
Pepper, J., et al.: Proc. Astron. Soc. Pac. **119**, 923 (2007)
Perryman, M.: The Exoplanet Handbook, 2nd edn. Cambridge University Press, Cambridge (2018)
Petigura, E.A., Marcy, G.W., Howard, A.W., et al.: Astrophy. J. **770**, 69 (2013a)
Petigura, E.A., Howard, A.W., Marcy, G. W., et al.: Proc. Natl Acad. Sci. **110**, 19273 (2013b)
Petigura, E.A., Howard, A.W., Marcy, G.W., et al.: (2017), **154**, 107
Petigura, E.A., Marcy, G.W., Winn, J.N., et al.: Astron. J. **155**, 89 (2018)
Petigura, E.A.: Astron. J. **160**, 89 (2020)
Petrovich, C., Deibert, E., We, Y.: Astrophy. J. **157**, 180 (2019)
Pichierri, G., Morbidelli, A.: Mon. Not. R. Astron. Soc. **494**, 4950 (2020)
Pickering, E.C.: Mon. Not. R. Astron. Soc. **50**, 296 (1890)
Pollacco, D.L., Skillen, I., Cameron, A.C., et al.: Proc. Astron. Soc. Pac. **118**, 1407 (2006)
Pollack, J.B., Hubickyj, O., Bodenheimer, P., et al.: Icarus **124**, 62 (1996)
Poon, S.T.S., Nelson, R.P.: Mon. Not. R. Astron. Soc. **498**, 5166 (2020)
Poon, S.T.S., Nelson, R.P., Jacobson, S.A.: Mon. Not. R. Astron. Soc. **491**, 5595 (2020)
Pu, B., Wu, Y.: Astrophy. J. **807**, 44 (2015)
Pu, B., Lai, D.: Mon. Not. R. Astron. Soc. **478**, 197 (2018)
Pu, B., Lai, D.: Mon. Not. R. Astron. Soc. **488**, 3568 (2019)
Queloz, D.: New developments in array technology and applications. In: IAU Symposium. vol. 167, 221 (1995)
Queloz, D.: Astron. Astrophys. **379**, 279 (2001)
Rainer, M., Borsa, F., Pino, L., et al.: (2021). arXiv:2103.10395
Rasio, F.A., Ford, E.B.: Science **274**, 954 (1996)
Rauer, H., et al.: Exp. Astr. **38**, 249 (2014)
Raymond, S.N., Armitage, P.J., Gorelick, N.: Astrophy. J. **711**, 772 (2010)
Raymond, S.N., Boulet, T., Izidoro, A., et al.: Mon. Not. R. Astron. Soc. **479**, 81 (2018)
Reffert, S., Bergmann, C., Quirrenbach, A., et al.: Astron. Astrophys. **574**, A116 (2015)
Ricker, G.R., Winn, J.N., Vanderspek, R., et al.: JATIS **1**, 014003 (2015)
Rivera, E.J., Lissauer, J.J.: Astrophy. J. **558**, 392 (2001)
Rivera, E.J., Laughlin, G., Butler, R.P., et al.: Astrophy. J. **719**, 890 (2010)
Rogers, T.M., Lin, D.N.C., Lau, H.H.B.: Astrophy. J. **758**, L6 (2012)
Rogers, L.A.: Astrophy. J. **801**, 41 (2015)
Rogers, L.A., Seager, S.: Astrophy. J. **712**, 974 (2010)
Romanova, M.M., Owocki, S.P.: The strongest magnetic fields in the universe. In: Beskin et al. (eds.) Space Science Series of ISSI, vol. 54, , p. 347. Springer, New York (2016)
Rouan, D., Deeg, H.J., Demangeon, O., et al.: Astrophy. J. Lett. **741**, L30 (2011)
Rowe, J.F., Bryson, S.T., Marcy, G.W., et al.: Astrophy. J. **784**, 45 (2014)
Sanchis-Ojeda, R., Winn, J.N., Holman, M.J.: Astrophy. J. **733**, 127 (2011a)
Sanchis-Ojeda, R., Winn, J.N.: Astrophy. J. **743**, 61 (2011b)
Sanchis-Ojeda, R., Rappaport, S., Winn, J.N., et al.: Astrophy. J. **774**, 54 (2013a)
Sanchis-Ojeda, R., Winn, J.N., Marcy, G.W.: Astrophy. J. **775**, 54 (2013b)
Sanchis-Ojeda, R., Rappaport, S., Winn, J.N., Kotson, M.C.: Astrophy. J. **787**, 47 (2014)
Sandford, E., Kipping, D., Collins, M.: Mon. Not. R. Astron. Soc. **489**, 3162 (2019)
Santerne, A., Brugger, B., Armstrong, D.J., et al.: Nat. Astron. **2**, 393 (2018)

Santos, N.C., Israelian, G., Mayor, M.: Astron. Astrophys. **415**, 1153 (2004a)
Santos, N.C., Bouchy, F., Mayor, M., et al.: Astron. Astrophys. **426**, 19 (2004b)
Scargle, J.D.: Astrophy. J. **263**, 835 (1982)
Schlaufman, K.C.: Astrophy. J. **719**, 602 (2010)
Schlaufman, K.C.: Astrophy. J. **790**, 91 (2014)
Schlichting, H.E., Sari, R., Yalinewich, A.: Icarus, **247**, 81 (2015)
Seager, S., Mallén-Ornelas, G.: Astrophy. J. **585**, 1038 (2003)
Seager, S., Kuchner, M., Hier-Majumder, C.A., Militzer, B.: Astrophy. J. **669**, 1279 (2007)
Shallue, C.J., Vanderburg, A.: Astron. J. **155**, 9 (2018)
Silburt, A., Gaidos, E., Wu, Y.: Astrophy. J. **799**, 180 (2015)
Sing, D.: Observational techniques with transiting exoplanetary atmospheres. In: Bozza, V., Mancini, L. Sozzetti, A. (eds.) Astrophysics of Exoplanetary Atmospheres: 2nd Advanced School on Exoplanetary Science, Astrophysics and Space Science Library, vol. 450, p. 3. Springer International Publishing Switzerland (2018)
Sinukoff, E., Howard, A.W., Petigura, E.A., et al.: Astron. J. **153**, 271 (2017)
Snellen, I.A.G., Stuik, R., Navarro, R., et al.: Proc. SPIE **8444**, 84440I (2012)
Sousa, S.G., Santos, N.C., Mayor, M., et al.: Astron. Astrophys. **487**, 373 (2008)
Sousa, S.G., Adibekyan, V., Santos, N.C., et al.: Mon. Not. R. Astron. Soc. **485**, 3981 (2019)
Southworth, J.: Mon. Not. R. Astron. Soc. **417**, 2166 (2011)
Southworth, J.: Astron. Astrophys. **557**, A119 (2013)
Sozzetti, A., Torres, G., Latham, D.W., et al.: Astrophy. J. **697**, 544 (2009)
Sozzetti, A., Bernagozzi, A., Bertolini, E., et al.: EPJWC **47**, 03006 (2013)
Sozzetti, A., Damasso, M., Bonomo, A.S., et al.: Astron. Astrophys. **648**, A75 (2021)
Spalding, C., Batygin, K.: Astrophy. J. **811**, 82 (2015)
Spalding, C., Batygin, K.: Astrophy. J. **830**, 5 (2016)
Spalding, C., Millholland, S.: Astron. J. **160**, 105 (2020)
Stauffer, J., Cody, A.M., McGinnis, P., et al.: Astron. J. **149**, 130 (2015)
Stefansson, G., Mahadevan, S., Maney, M., et al.: Astron. J. **160**, 192 (2020)
Steffen, J.H., Farr, W.M.: Astrophy. J. **774**, 12L (2013)
Steffen, J.H., Ragozzine, D., Fabrycky, D.C., et al.: Proc. Natl. Acad. Sci. **109**, 7982S (2012)
Szabó, G., M., Kiss, L.L.: Astrophy. J. **727**, L44 (2011)
Takaishi, D., Tsukamoto, Y., Suto, Y.: Mon. Not. R. Astron. Soc. **492**, 5641 (2020)
Tanaka, H., Takeuchi, T., Ward, W.R.: Astrophy. J. **565**, 1257 (2002)
Terquem, C., Papaloizou, J.C.B.: Mon. Not. R. Astron. Soc. **482**, 530 (2019)
Thompson, S.E., Coughlin, J.L., Hoffman, K., et al.: Astrophy. J. Suppl. Ser. **235**, 38 (2018)
Tinetti, G., Drossart, P., Eccleston, P., et al.: Exp. Astr. **46**, 135 (2018)
Torres, G., Fischer, D.A., Sozzetti, A., et al.: Astrophy. J. **757**, 161 (2012)
Traub, W.A.: Astrophy. J. **745**, 20 (2012)
Tregloan-Reed, J., Southworth, J., Tappert, C.: Mon. Not. R. Astron. Soc. **428**, 3671 (2013)
Tregloan-Reed, J., Southworth, J., Burgdorf, M., et al.: Mon. Not. R. Astron. Soc. **450**, 1760 (2015)
Tremaine, S., Dong, S.: Astron. J. **143**, 94 (2012)
Triaud, A.H.M.J.: Handbook of Exoplanets. Springer International Publishing AG, id. 2 (2018)
Tsiaras, A., Waldmann, I.P., Tinetti, G., et al.: Nat. Astron. **450**, 1760 (2019)
Tuomi, M.: aap, **543**, A52 (2012)
Tuomi, M., Jones, H.R.A., Barnes, J.R.: Mon. Not. R. Astron. Soc. **441**, 1545 (2014)
Turbet, M., Ehrenreich, D., Lovis, C., et al.: Astron. Astrophys. **628**, A12 (2019)
Turbet, M., Bolmont, E., Ehrenreich, D., et al.: Astron. Astrophys. **638**, 41 (2020)
Turrini, D., Zinzi, A., Belinchon, J.A.: Astron. Astrophys. **636**, A53 (2020)
Udalski, A., et al.: Acta Astron. **52**, 1 (2002)
Udry, S., Mayor, M., Santos, N.C.: Astron. Astrophys. **407**, 369 (2003)
Udry, S., Dumusque, X., Lovis, C., et al.: Astron. Astrophys. **622**, A37 (2019)
Valencia, D., Sasselov, D.D., O'Connell, R.J.: Astrophy. J. **665**, 1413 (2007)
Valencia, D., Ikoma, M., Guillot, T., Nettelmann, N.: Astron. Astrophys. **516**, 20 (2010)

Valsecchi, F., Rasio, F.A.: Astrophy. J. **786**, 102 (2014)
Valsecchi, F., Rappaport, S., Rasio, F.A., et al.: Astrophy. J. **813**, 101 (2015)
Van Eylen, V., Albrecht, S.: Astrophy. J. **808**, 126 (2015)
Van Eylen, V., Agentoft, C., Lundkvist, M.S., et al.: Mon. Not. R. Astron. Soc. **479**, 4786 (2018)
Van Eylen, V., Albrecht, S., Huang, Xu., et al.: Astron. J. **157**, 61 (2019)
Van Grootel, V., Gillon, M., Valencia, D., et al.: Astrophy. J. **786**, 2 (2014)
Vandenburg, A., et al.: Astrophy. J. Suppl. Ser. **222**, 14 (2016)
Vanderburg, A., Becker, J.C., Buchhave, L.A., et al.: Astrophy. J. **154**, 237 (2017)
Venturini, J., Guilera, O.M., Haldemann, J., et al.: Astron. Astrophys. **643**, L1 (2020)
Volk, K., Malhotra, R.: Astron. J. **160**, 98 (2020)
Wang, J., Fischer, D.A.: Astron. J. **149**, 14 (2015)
Ward, W.R.: Astrophy. J. **482**, L211 (1997)
Weiss, L.M., Marcy, G.W.: Astrophy. J. **783**, L6 (2014)
Weiss, L.M., Rogers, L.A., Isaacson, H.T., et al.: Astrophy. J. **819**, 83 (2016)
Weiss, L.M., Isaacson, H.T., Marcy, G.W., et al.: Astron. J. **156**, 254 (2018a)
Weiss, L.M., Marcy, G.W., Petigura, E.A., et al.: Astron. J. **155**, 48 (2018b)
Weiss, L.M., Petigura, E.A.: Astrophy. J. **893**, L1 (2020)
Welsh, W.F., Orosz, J.A.: Handbook of Exoplanets. Springer International Publishing AG, id. 34 (2018)
Wheatley, P.J., Pollacco, D.L., Queloz, D., et al.: EPJ Web Conf. **47**, 13002 (2013)
Wheatley, P.J.: European Planetary Science Congress, 10, EPSC2015-908 (2015)
Wilson, R.F., Teske, J., Majewski, S.R., et al.: Astron. J. **155**, 68 (2018)
Winn, J.: Exoplanets. In: Seager S. (ed.) University of Arizona Press (2010)
Winn, J.N., Fabrycky, D.C.: Ann. Rev. Astron. Astrophys. **53**, 409 (2015)
Winn, J.N., Holman, M.J., Henry, G.W., et al.: Astron. J. **133**, 1828 (2007)
Winn, J.N., Johnson, J.A., Albrecht, S., et al.: Astrophy. J. **703**, L99 (2009)
Winn, J.N., Fabrycky, D., Albrecht, S., Johnson, J.A.: Astrophy. J. **718**, L145 (2010)
Winn, J.N., Matthews, J.M., Dawson, R.I., et al.: Astrophy. J. Lett. **737**, L18 (2011)
Winn, J.N., Petigura, E.A., Morton, T.D., et al.: Astron. J. **154**, 270 (2017a)
Winn, J.N., Sanchis-Ojeda, R., Rogers, L., et al.: Astrophy. J. **154**, 60 (2017b)
Winn, J.N., Sanchis-Ojeda, R., Rappaport, S.: New Astron. Rev. **83**, 37 (2018)
Wittenmyer, R.A., Tinney, C.G., Butler, R.P., et al.: Astrophy. J. **738**, 81 (2011)
Wittenmyer, R.A., Wang, S., Horner, J., et al.: Mon. Not. R. Astron. Soc. **492**, 377 (2020)
Wolfgang, A., Rogers, L.A., Ford, E.B.: Astrophy. J. **825**, 19 (2016)
Wolszczan, A., Frail, D.A.: Nature **355**, 145 (1992)
Wright, J.T., Upadhyay, S., Marcy, G.W., et al.: Astrophy. J. **693**, 1084 (2009)
Wright, J.T., Fakhouri, O., Marcy, G.W., et al.: Proc. Astron. Soc. Pac. **123**, 412 (2011)
Wright, J.T., Marcy, G.W., Howard, A.W., et al.: Astrophy. J. **753**, 160 (2012)
Wright, J.: Handbook of Exoplanets. Springer International Publishing AG, id. 4 (2018)
Wu, Y., Murray, N.: Astrophy. J. **589**, 605 (2003)
Wu, Y.: Astrophy. J. **874**, 91 (2019)
Wu, Y., Lithwick, Y.: Astrophy. J. **772**, 74 (2013)
Xie, J.-W., Dong, S., Zhu, Z., et al.: Proc. Natl. Acad. Sci. **113**, 11431 (2016)
Xu, W., Lai, D.: Mon. Not. R. Astron. Soc. **468**, 3223 (2017)
Xuan, J.W., Wyatt, M.C.: Mon. Not. R. Astron. Soc. **497**, 2096 (2020)
Yang, J.-Y., Xie, J.-W., Zhou, J.-L.: Astron. J. **159**, 164 (2020)
Youdin, A.N.: Astrophy. J. **742**, 38 (2011)
Zechmeister, M., Kürster, M.: Astron. Astrophys. **49**, 577 (2009)
Zeng, L., Sasselov, D.D.: Proc. Astron. Soc. Pac. **125**, 227 (2013)
Zeng, L., Sasselov, D.D., Jacobsen, S.B.: Astrophy. J. **819**, 127 (2016)
Zeng, L., Jacobsen, S.B., Sasselov, D.D., et al.: Proc. Natl. Acad. Sci. **116**, 9723 (2019)
Zhou, G., Latham, D.W., Bieryla, A., et al.: Mon. Not. R. Astron. Soc. **460**, 3376 (2016)
Zhou, G., Bakos, G.Á., Bayliss, D., et al.: Astron. J. **157**, 31 (2019a)

Zhou, G., Huang, C.X., Bakos, G.Á.: Astron. J. **158**, 141 (2019b)
Zhou, G., Winn, J.N., Newton, E.R., et al.: Astrophy. J. Lett. **892**, L21 (2020)
Zhu, W., Wang, J., Huang, C.: Astrophy. J. **832**, 196 (2016)
Zhu, W., Wu, Y.: Astron. J. **156**, 92 (2018)
Zhu, W., Petrovich, C., Wu, Y., et al.: Astrophy. J. **860**, 101 (2018)
Zhu, W.: Astrophy. J. **873**, 8 (2019)
Zhu, W.: Astron. J. **159**, 188 (2020)
Zink, J.K., Hansen, B.M.S.: Mon. Not. R. Astron. Soc. **487**, 246 (2019)
Zink, J.K., Christiansen, J.L., Hansen, B.M.S.: Mon. Not. R. Astron. Soc. **483**, 4479 (2019)
Zinzi, A., Turrini, D.: Astron. Astrophys. **605**, 4 (2017)

Part IV
Wide-Orbit Exoplanets

Chapter 4
The Demographics of Wide-Separation Planets

B. Scott Gaudi

Abstract I begin this review by first defining what is meant by exoplanet demographics, and then motivating why we would like as broad a picture of exoplanet demographics as possible. I then outline the methodology and pitfalls to measuring exoplanet demographics in practice. I next review the methods of detecting exoplanets, focusing on the ability of these methods to detect wide separation planets. For the purposes of this review, I define wide separation as separations beyond the "snow line" of the protoplanetary disk, which is at $\simeq$ 3au for a sun-like star. I note that this definition is somewhat arbitrary, and the practical boundary depends on the host star mass, planet mass and radius, and detection method. I review the approximate scaling relations for the signal-to-noise ratio for the detectability of exoplanets as a function of the relevant physical parameters, including the host star properties. I provide a broad overview of what has already been learned from the transit, radial velocity, direct imaging, and microlensing methods. I outline the challenges to synthesizing the demographics using different methods and discuss some preliminary first steps in this direction. Finally, I describe future prospects for providing a nearly complete statistical census of exoplanets.

4.1 Introduction: The Demographics of Exoplanets

The demographics of exoplanets can be defined as the distribution of exoplanets as a function of a set of physical properties of the planets, their host stars, or the environment of the planetary systems. It can be distinguished from exoplanet characterization by the depth of information that is measured about the exoplanet. Demographic surveys generally only focus on the most basic properties of the planets, host stars, or their environment. These properties are generally measured directly as parameters intrinsic to the detection technique being used, or are inferred from these param-

B. Scott Gaudi (✉)
Department of Astronomy, The Ohio State University, Columbus, OH, USA
e-mail: gaudi.1@osu.edu

Jet Propulsion Laboratory, California Institute of Technology, Pasadena, CA, USA

© Springer Nature Switzerland AG 2022
K. Biazzo et al. (eds.), *Demographics of Exoplanetary Systems*,
Astrophysics and Space Science Library 466,
https://doi.org/10.1007/978-3-030-88124-5_4

eters, sometimes requiring auxiliary information. In contrast, characterizing exoplanets generally requires more sophisticated techniques to determine the detailed properties of the exoplanets. Of course, the distinction between exoplanet demographics and characterization is somewhat arbitrary, but nevertheless it provides a useful framework for outlining the basic goals of studies of exoplanets.

The primary motivation for measuring exoplanet demographics is to test planet formation theories. A complete, ab-initio theory of planet formation must be able to describe the physical processes by which micron-sized dust grains grow by $\sim$13–14 orders of magnitude in size and $\sim$38–41 orders of magnitude in mass to their final radii and masses between the radius and mass of the Earth and the radius and mass of Jupiter. As is described in Chap. 1 and references therein, as protoplanets grow by these many orders of magnitude in size and mass, the physical mechanisms that govern their growth and migration vary significantly. In principle, the signature of many of these physical mechanisms should be imprinted in the distribution of properties of mature planetary systems, including the properties of the planets, the nature and diversity of system architectures, and their dependencies on the host star properties and environment. Thus, a *robust* and *unbiased* measurement of the demographics of exoplanets over a broad range of planet and host star properties provides one of the most fundamental empirical tests of planet formation theories.

This review discusses methods for characterizing the population of wide-separation planets, and constraints on the demographics of such planets from various surveys using these methods. For the purposes of this review, I will define wide-separation as semi-major axes and periods greater than the "snow line" in the protoplanetary disk

$$a_{\rm sl} \simeq 3 \text{ au} \left(\frac{M_*}{M_\odot}\right)^{\sim 1}, \quad P_{\rm sl} \simeq 5 \text{ year} \left(\frac{M_*}{M_\odot}\right)^{\sim 5/4}, \qquad (4.1)$$

where M_* is the mass of the host star. Equation (4.1) are motivated by the estimated distance of the snow line in the solar protoplanetary disk (e.g., Morbidelli et al. 2016), and the scaling of the snow line with host star mass estimated by Kennedy and Kenyon (2008), although I note that this scaling is uncertain, and the location of the snow line itself is a function of the age of the protoplanetary disk. This definition is useful to the extent that planet formation likely proceeds differently beyond the snow line due to the larger surface density of solid material, but I note that this is also uncertain. For a more comprehensive discussion of the snowline in protoplanetary disks, its time evolution, and its possible effect on planet formation, see the discussion in Chap. 1. Despite these uncertainties, I will adopt Eq. (4.1) as approximate boundaries between close and wide separation planets. As discussed in Sect. 4.4, the practical boundary between close- and wide-separation planets depends on the detection method and specifics of the survey, as well as on the planet mass and radius. Thus this boundary should be not applied too rigidly.

There is a large body of literature on topics related to exoplanet demographics, which cannot be comprehensively covered in this relatively short review chapter. For a more comprehensive (if somewhat outdated) introduction to exoplanet demographics, I refer the reader to Winn and Fabrycky (2015). A more recent review

4 The Demographics of Wide-Separation Planets

is forthcoming in Zhu and Dong (2021). Chapter 3 of this book provides another excellent source of recently reviewed literature on the subject.

The plan for this chapter is as follows. Section 4.2 summarizes the mathematical formalism for constraining exoplanet demographics. Section 4.3 discusses a few of the possible pitfalls when constraining exoplanet demographics in practice. Section 4.4 briefly summarizes the primary methods of detecting planets, focusing on the scaling of the sensitivity of each method with the planet and host star properties. Section 4.5 highlights some of the main results from surveys for wide-separation planets, including results from radial velocity (Sect. 4.5.1), transit (Sect. 4.5.2), direct imaging (Sect. 4.5.3), and microlensing (Sect. 4.5.4) surveys. Section 4.5.5 reviews the relatively few attempts to synthesize results from multiple methods, and Sect. 4.6 discusses a few notable comparisons between the predictions of the demographics of planets from ab-initio planet formation theories and results from exoplanet surveys. Section 4.7 discusses future prospects for determining wide-orbit exoplanet demographics. Finally, I briefly conclude in Sect. 4.8.

4.2 Mathematical Formalism

Although this review focuses on methods for measuring the demographics of wide-separation planets and highlights some observational results to this end, it is worthwhile to present the general mathematical formalism for measuring the demographics of planets (of all separations). Following Clanton and Gaudi (2014a), in very general terms, I can mathematically define the goal of demographic surveys of exoplanets to be a measurement of the distribution function $d^n N_{\rm pl}/d\boldsymbol{\alpha}$, where $\boldsymbol{\alpha}$ is a vector containing the set of n physical parameters upon which the planet frequency intrinsically depends. These can include (but are not limited to): parameters of the planets (mass $M_{\rm p}$, radius $R_{\rm p}$), their orbits (period P, semi-major axis a, eccentricity e), properties of the host stars (mass M_*, radius R_*, luminosity L_*, effective temperature $T_{\rm eff}$, metallicity [Fe/H], multiplicity), and many others. The total number of planets $N_{\rm pl}$ in the domain covered by $\boldsymbol{\alpha}$ is

$$N_{\rm pl} = \int_{\alpha_1} d\alpha_1 \int_{\alpha_2} d\alpha_2 \cdots \int_{\alpha_n} d\alpha_n \cdot \frac{d^n N_{\rm pl}}{d\boldsymbol{\alpha}} \qquad (4.2)$$

Therefore, in principle, one could simply count the number of planets as a function of the parameters $\boldsymbol{\alpha}$ and then differentiate with respect to those parameters to derive $d^n N_{\rm pl}/d\boldsymbol{\alpha}$. This is essentially equivalent to binning in the parameters and counting the number of planets per bin. This distribution function represents the most fundamental quantity that describes the demographics of a population of planets. With such a distribution function in hand, one can then compare to the predictions for this distribution that are the outputs of ab-initio planet formation theories (e.g., Ida et al. 2013; Mordasini et al. 2015) and determine how well they match the observations. Furthermore, one can then vary the input physics in these models, or parameters of semi-analytic parameterizations of the physics, to provide the best match to the observed distribution of physical parameters. Thus these models can be refined, improving our understanding of planet formation and migration.

Unfortunately, it is generally not possible to measure $d^n N_{pl}/d\alpha$ directly, for several reasons. First, all surveys are limited in the range of parameters to which they are sensitive. Some of these are intrinsic to the detection method itself, while others are due to the survey design. Second, all surveys suffer from inefficiencies and detection biases. Thus the measured distribution of planet properties is not equal to the true distribution. The effects of these first two issues can be accounted for by carefully determining the survey completeness. Finally, each exoplanet detection method is sensitive to a different set of parameters of the planetary system. Some of these parameters belong to the set of the physical parameters of interest α, others are a function of these parameters, others are parameters that are not directly constrained by the data under consideration and must be accounted for using external information or derived from external measurements, and others are essentially nuisance parameters that must be marginalized over.

As a concrete example, consider radial velocity (RV) surveys for exoplanets. As we saw in Chap. 2, Sect. 2.1, the RV of a star arising from the reflex motion due to a planetary companion can be described by 6 parameters: the RV semi-amplitude K, the orbital period P, eccentricity e, argument of periastron ω, and time of periastron T_p (or the time of some fiducial point in the orbit for circular orbits, such as the time of periastron), and the barycentric velocity γ. Of these parameters, generally, only P and e are fundamental physical parameters that (potentially) contain information about the formation and evolution of the system. The parameter ω is a geometrical parameter that describes the orientation of the orbit with respect to the plane of the sky from the perspective of the observer,[1] and T_P is an (essentially) arbitrary conventional definition for the zero point of the radial velocity time series. The parameter γ does not contain any information about the planet or its orbit, although (if absolutely calibrated) does contain information about the Galactic orbit of the system and thus the Galactic stellar population to which the host star belongs. Finally, K depends on P, e, the mass of the planet M_p, the mass of the host star M_*, and the inclination i of the orbital plane with respect the plane of the sky, with $i = 90°$ is edge-on). The inclination is another geometrical parameter, and thus is not of intrinsic interest. However, M_p and M_* are fundamental physical parameters of interest. The latter cannot be estimated from RV measurements of the star alone, and thus must be determined from external information. With an estimate of M_* in hand, the remaining physical quantity of interest, M_p, still cannot be determined directly, since the inclination i is not constrained by RV measurements. Assuming $M_p \ll M_*$, one can then infer the minimum mass of the companion $M_p \sin i$. In order to infer M_p, one must use the known a priori distribution of i (namely that $\cos i$ is uniformly distributed), *and* an assumed prior on the distribution of true-planet masses dN/dM_p,

---

[1] I note that in systems with multiple planets, any apsidal alignment can be inferred by the individual values of ω. The existence of an apsidal alignment (or alignments) can provide constraints on the formation and/or evolution of that system. See, e.g., Chiang et al. (2001).

to infer the posterior distribution of M_p for any given detection.[2] However, dN/dM_p is typically a distribution one would like to infer, and thus one must deconvolve a distribution of minimum masses of a sample of detections to infer the posterior distribution dN/dM_p (see, e.g., Zucker and Mazeh 2001).

In order to deal with the difficulties in inferring the true distribution function $d^n N_{pl}/d\alpha$ of some set of physical parameters α, one must account for not only the survey completeness but also for the fact that the detection method (or methods) being used to survey for planets are generally not directly sensitive to (all of the) parameters of interest α. To deal with the latter point, several approaches can be taken. First, one can attempt to transform the observable parameters into the physical parameters of interest. This often requires introducing external information (such as the properties of the star), or adopting priors for, and then marginalizing over, parameters that are not directly constrained. Alternatively, one can simply choose to constrain the distribution functions of the observable parameters that are most closely related to the physical parameters of interest. It is also possible to adopt both approaches.

An important point is that the completeness of each individual target is obviously a function of the observable parameters, not the transformed physical parameters. Therefore the completeness of each target must first be determined in terms of the observable parameters, and then transformed to the physical (or more physical) parameters of interest (if desired).

In order to determine the completeness of a given survey of a set of targets N_{tar}, one must first specify the criteria adopted to define a detection. An essential (but often overlooked) point is that the criteria used to determine the completeness of the entire sample of N_{tar} targets must *strictly* be the same criteria used to detect the planets in the survey data to begin with. Failure to adhere to this requirement can lead to (and indeed, has led to) erroneous inferences about the distribution of planet properties. This is particularly important in regions of parameter space where the number of detected planets is a strong function of the specific detection criteria.

Typically one computes the completeness or efficiency (hereafter referred to efficiency for definiteness) of a given target as a function of the primary observables whose distributions affect the detectability of the planets. I define this set of observables as β. I note that this set can be subdivided into two subsets. I define β_I to be the subset of the β observable parameters of interest whose distributions one would like to constrain (or observables which will subsequently be transformed to other, more physical parameters, whose distributions are to be constrained), and β_N to be the remainder of the set of observable 'nuisance' parameters (e.g., the parameters that affect the detectability but are either considered nuisance parameters or parameters whose distributions are not specifically of interest). The efficiency of a given target j is given by

---

[2] I note that the posterior distribution of M_p given a measurement $M_p \sin i$ is often estimated by simply assuming that $\cos i$ is uniformly distributed. This leads to the familiar result that the median of the true mass is only $(\sin[\arccos(0.5)])^{-1} \simeq 1.15$ larger than the minimum mass. Unfortunately, this is only correct if the distribution of true planet masses is such that $dN/d\log M_p$ is a constant (Ho and Turner 2011; Stevens and Gaudi 2013). For other distributions, the true mass can be larger or smaller than this naive estimate.

$$\epsilon_j(\boldsymbol{\beta}_I) = \int_{\beta_N} d\beta_{N,1} \int_{\beta_N} d\beta_{N,2} \cdots \int_{\beta_{N,n}} d\beta_{N,n} \frac{d^n N_{\mathrm{pl}}}{d\boldsymbol{\beta}_N} \mathcal{D}(\boldsymbol{\beta}_N). \qquad (4.3)$$

Here $d^n N_{\mathrm{pl}}/d\boldsymbol{\beta}_N$ is the assumed prior distribution of the observable 'nuisance' parameters $\boldsymbol{\beta}_N$, and $\mathcal{D}(x)$ is the set of detection criteria used to select planet candidates. It is worth noting that the prior distribution for the nuisance parameters $\boldsymbol{\beta}_N$ is generally trivial, and does not contain any valuable physical information. In the simplest case of a χ^2 threshold as the only detection criterion, then $\mathcal{D} = \mathcal{H}[\Delta\chi^2(\boldsymbol{\beta}_N) - \Delta\chi^2_{\min}]$, where $\mathcal{H}(x)$ is the Heaviside step function and $\Delta\chi^2$ is the difference between the χ^2 of a fit to the data with a planet and the null hypothesis of no planet, and $\Delta\chi^2_{\min}$ is the minimum criterion for detection. In reality, most surveys employ several detection criteria. The most robust method of determining the efficiency ϵ_j for each target is to inject planet signals into the data according to the distribution function $d^n N_{\mathrm{pl}}/d\boldsymbol{\beta}$, and then attempt to recover these signals using the same set of detection criteria $\mathcal{D}$ (e.g., the same pipeline) used to construct the original sample of detected planets.

In order to determine the overall survey efficiency $\Phi(\boldsymbol{\beta}_I)$, one must compute the efficiency for every target, whether or not the target contains a signal that passes the set of detection criteria $\mathcal{D}$. The total survey efficiency is then

$$\Phi(\boldsymbol{\beta}_I) = \sum_{j=1}^{N_{\mathrm{tar}}} \epsilon_j(\boldsymbol{\beta}_I). \qquad (4.4)$$

Given the total survey efficiency Φ, it is possible to marginalize over the parameters $\boldsymbol{\beta}_I$ to estimate the total number of expected planet detections $N_{\mathrm{pl,exp}}$ in the survey, given a prior assumption for the distribution function $d^n N_{\mathrm{pl}}/d\boldsymbol{\beta}_I$:

$$N_{\mathrm{pl,exp}} = \int_{\beta_{I,1}} d\beta_{I,1} \int_{\beta_{I,2}} d\beta_{I,2} \cdots \int_{\beta_{I,n}} d\beta_{I,n} \frac{d^n N_{\mathrm{pl}}}{d\boldsymbol{\beta}_I} \Phi(\boldsymbol{\beta}_I). \qquad (4.5)$$

Of course, it is precisely the distribution function $d^n N_{\mathrm{pl}}/d\boldsymbol{\beta}_I$ that we wish to infer. Thus, in the usual Bayesian formalism, we must adopt a prior distribution of the parameters of interest, and we then determine if this prior distribution is consistent (in the likelihood sense) with the posterior distribution of these parameters. If not, then we adjust the prior distribution and repeat. This is the way one 'learns' in the Bayesian formalism.

It is also possible to convert the individual target efficiencies $\epsilon_j(\boldsymbol{\beta}_I)$ as a function of the observable parameters of interest $\boldsymbol{\beta}_I$ to efficiencies as a function of physical parameters (or more physical parameters) $\boldsymbol{\alpha}$ by simply transforming from $\boldsymbol{\beta}_I$ to $\boldsymbol{\alpha}$. In this case, one would replace $\boldsymbol{\beta}_I \rightarrow \boldsymbol{\alpha}$ in Eq. (4.5) after transforming to $\boldsymbol{\alpha}$. In general, such transformations can be relatively straightforward or fairly complicated, depending on the details of what is known about the properties of the target sample, and the variables that are being transformed. It is also important to note that such

4 The Demographics of Wide-Separation Planets

transformations can introduce additional sources of uncertainty; for example, if one wants to convert from planet/host star mass ratio to planet mass, one must explicitly account for the uncertainty in the host star mass. Finally, it is important to include the Jacobian of the transformation of the adopted prior distribution of the observables $d^n N_{\rm pl}/d^n \boldsymbol{\beta}_I$ to the prior distribution of the physical parameters $d^n N_{\rm pl}/d^n \boldsymbol{\alpha}$, as a given prior distribution on an observable parameter $\beta_{I,i}$ does not guarantee the desired prior distribution on the physical parameters $\boldsymbol{\alpha}$.

With the individual values of ϵ_j and total survey Φ efficiencies in hand, there are two basic approaches that are typically taken to infer the posterior distribution $d^n N_{\rm pl}/d\boldsymbol{\beta}_I$ (or the distribution of transformed variables $d^n N_{\rm pl}/d\boldsymbol{\alpha}$).

Binning: The simplest and most obvious method of inferring the distribution function is simply to define bins in the parameters of interest $\boldsymbol{\beta}_I$ (or $\boldsymbol{\alpha}$). For definiteness, I assume one is working in the space of the observable parameters of interest $\boldsymbol{\beta}_I$. One then counts the number of detected planets in each bin k of $\boldsymbol{\beta}_I$, $N_{{\rm det},k}$, and then divides by the number of planets expected to be detected in each bin using Eqs. (4.4) and (4.5), where, in this case, the integrals are over the span of each bin. The estimated frequency of planets (more specifically, the number of planets per star) in each bin k is then

$$\left.\frac{d^n N_{\rm pl}}{d\boldsymbol{\beta}_I}\right|_{{\rm post},k} \equiv \frac{N_{{\rm det},k}}{N_{{\rm pl,exp},k}}, \tag{4.6}$$

where the symbol $|_{{\rm post},k}$ serves to indicate that the value of $d^n N_{\rm pl}/d\boldsymbol{\beta}_I$ inferred from Eq. (4.6) is the *posterior* distribution inferred for bin k. This gives the frequency of planets in that bin, weighted by the assumed prior distribution function of the bin parameters. In this case, the uncertainty in the inferred planet frequency of each bin is given by Poisson statistics based on the number of detections in each bin, weighted by the prior distribution. Although binning is generally not recommended (for reasons discussed in Sect. 4.3), it can provide a useful method to visualize the results of the survey. I note that, despite some claims to the contrary, binning is neither a 'non-parametric' nor a 'prior-free' method of inferring the distribution of planet properties. In reality, all inferences about the distribution of planet properties from survey data are parametric and assume priors (whether those parameters or priors are explicitly stated or not). This is clear from the discussion above, as the value of $N_{{\rm pl,exp},k}$ is evaluated using Eq. (4.5), which itself depends on the assumed *prior* distribution of $d^n N_{\rm pl}/d\boldsymbol{\beta}_I$.

Although this method is conceptually quite straightforward, it is worthwhile to point out that a somewhat different approach has been taken to implement this method in practice, particularly for early results gleaned from the *Kepler* data. This approach, dubbed the inverse detection efficiency method (IDEM) by Foreman-Mackey (2014), requires only evaluating the planet detection sensitivity for those stars around which planets have been discovered. Specifically, the planet frequency (again, more specifically the number of planets per star) in a given bin k is given by

$$\left.\frac{d^n N_{\text{pl}}}{d\boldsymbol{\beta}_I}\right|_{\text{post},k} \equiv \frac{1}{N_{\text{tar}}} \sum_{j=1}^{N_{\text{det},k}} \frac{1}{\epsilon_j}. \tag{4.7}$$

I again note that, while it may appear that this method is 'non-parametric' or 'prior-free', this is not the case, as one must define the boundaries of each bin.

As discussed in detail in Foreman-Mackey (2014) and Hsu et al. (2018), this approach is not optimal, and specifically is likely to yield biased inferences for $N_{\text{pl,exp}}$. This issue is particularly acute for transit surveys, for which a small number of detections must be multiplied by large correction factors to estimate the intrinsic frequency of planets. The fact that this approach is not optimal stems from several reasons, the most important being that the stars with the largest intrinsic sensitivity are those for which it is most likely that planets will be detected. I note that several analyses of ground-based transit surveys published prior to any results from *Kepler* (e.g., Mochejska et al. 2005; Burke et al. 2006; Hartman et al. 2009) utilized the (more optimal) first method described above (i.e., the method where one estimates and utilizes the sensitivity of all the target stars), rather than the IDEM method.

Maximum Likelihood: A second method to estimate the distribution function $d^n N_{\text{pl}}/d\boldsymbol{\beta}_I$ (or the distribution of transformed variables $d^n N_{\text{pl}}/d\boldsymbol{\alpha}$) is to use the maximum likelihood method. The mathematics of this method can be derived by starting with the method of binning described above, and then decreasing the bin size such that there is only one detection in each bin (see, e.g., Appendix A of Cumming et al. 2008 or Sect. 3.1 of Youdin 2011). Here one seeks to constrain the parameters of a distribution function F, such that

$$\frac{d^n N_{\text{pl}}}{d\boldsymbol{\beta}_I} = F(\boldsymbol{\beta}_I; \boldsymbol{\Theta}), \tag{4.8}$$

where $\boldsymbol{\Theta}$ are the variables that parameterize the distribution function F.[3] A common form for F when two variables are being considered is a double power-law. In this case, the vector $\boldsymbol{\Theta}$ would consist of three parameters: a normalization, and one exponent for each of the two parameters. Note this form implicitly assumes a range for each of the two parameters, the minimum and maximum of which can be considered as additional, hidden parameters, the choice of which can affect the maximum likelihood inferences for the other, explicit parameters.

Given this parameterized form for the distribution function $d^n N_{\text{pl}}/d\boldsymbol{\beta}_I$, then one can determine the likelihood of the observations given the parameters $\boldsymbol{\Theta}$ of the distribution function F (see the references below):

$$\mathcal{L}(\boldsymbol{\Theta}) = \exp^{-N_{\text{pl,exp}}} \prod_{j=1}^{N_{\text{det}}} F(\boldsymbol{\beta}_{I,j}; \boldsymbol{\Theta}) \Phi(\boldsymbol{\beta}_{I,j}), \tag{4.9}$$

---

[3] Again, one might instead wish to instead transform $\boldsymbol{\beta}_I \to \boldsymbol{\alpha}$ and then constrain $d^n N_{\text{pl}}/d\boldsymbol{\alpha}$. For brevity, I will no longer specifically call out this alternate method.

where N_{det} is the number of planets detected in the survey (more specifically, the number of detections that pass all of the detection criteria $\mathcal{D}$). Equation (4.9) can then be used to estimate the likelihood that the function $F(\boldsymbol{\beta}_I; \boldsymbol{\Theta})$ with a given set of parameters $\boldsymbol{\Theta}$ describe the data. Using the usual methods of exploring likelihood space (e.g., Markov Chain Monte Carlo), one can determine both the maximum likelihood, as well the confidence intervals of the parameters. These methods are generally well known and have been explored in many publications, and so I will not reiterate them here.

I note that the above mathematical description of the methodology to estimate the distribution of exoplanet properties (e.g., to determine the "demographics of exoplanets") was exceptionally abstract. This approach was taken intentionally in order to make the discussion as general as possible. Nevertheless, it may be difficult to apply this formalism to specific exoplanet surveys. I, therefore, invite the reader to consult the following publications that apply the formalism discussed above (including both binning and maximum likelihood) to specific exoplanet detection methods and surveys. For readers interested in radial velocity surveys, I suggest starting with Tabachnik and Tremaine (2002), Cumming et al. (2008), Howard et al. (2010), for those interested in transit surveys, I suggest starting with Burke et al. (2006); Youdin (2011), Dong and Zhu (2013), Dressing and Charbonneau (2013), Howard et al. (2012), Burke et al. (2015), for those interested in microlensing surveys, I suggest starting with Gaudi et al. (2002), Gould et al. (2010), Suzuki et al. (2016), and for those interested in direct imaging surveys, I suggest starting with Bowler (2016), Nielsen et al. (2019).

Finally, I note that much more sophisticated statistical methodologies for determining exoplanet demographics have been developed (e.g., hierarchical Bayesian modeling, approximate Bayesian computation), particularly in application to *Kepler* (see, e.g., Foreman-Mackey 2014; Hsu et al. 2018, 2019). Additional discussions of results from all the above methodologies are provided in Chap. 3.

4.3 Pitfalls to Measuring Exoplanet Demographics in Practice

In this section, I outline a few of the difficulties in measuring exoplanet demographics in practice. Many of these should be fairly obvious given the discussion in the previous section, but some are much more subtle. I do not claim this to be a comprehensive list of where one might go wrong, but merely a partial list of common pitfalls that one should avoid.

- **Adopting different detection criteria to estimate the survey efficiency than were used to identify the planet candidates.** This is likely one of the most prevalent sources of systematic error when determining exoplanet demographics. An excellent example is the *Kepler* survey (Borucki et al. 200). The earliest estimates for the intrinsic distribution function of planet parameters assumed simple, ad hoc assumptions for the detection criteria. This approach was required due to the fact that the algorithms used by the *Kepler* team to detect the announced planet candidates were not initially publicly available. As a result, the adopted detection criteria were often significantly different than the true detection criteria, leading to inferences about the planet distribution that were sometimes egregiously incorrect. This issue is particularly important near the "edges" of the survey sensitivity, where the number of detected planets can be a strong function of the detection criteria (e.g., the signal-to-noise ratio). Later analyses circumvented this issue by developing independent pipelines to identify planet candidates. These were then used to determine the detection efficiency by injecting transit signals into the data and recovering them using the same pipelines (e.g., Dressing and Charbonneau 2013; Petigura et al. 2013). Inferences about the exoplanet distribution made in this way are generally much more robust. See Chap. 3, Sect. 3.2 for additional details.

- **Use of 'by-eye' candidate selection**. Quite often, candidates from exoplanets surveys are subjected to human vetting. While this can be an excellent method of eliminating false positives, it is obviously difficult to reproduce in an automated way. While it is possible to inject a large number of signals into the data sets and then have the 'artificial' candidates vetted by humans (e.g., Gould et al. 2006a), this is both labor intensive and can impose biases (e.g., if the humans doing the vetting know that they are searching for injected signals). It is possible to remove the step of human vetting, as has been done using, e.g., the Robovetter tool developed by the *Kepler* team (Thompson et al. 2018).
- **Ignoring reliability**. An implicit assumption that was made in Sect. 4.2 was that all signals that passed the detection criteria $\mathcal{D}$ were due to real planets. This ignores false positives, which can be astrophysical, instrumental, or simple statistical false alarms. The reliability of a sample of candidate planets is simply the fraction of candidates that are true planets. Thus, in order to determine the true distribution of planet parameters from a survey, one must not only estimate the efficiency or completeness of the survey (the fraction of true planets that are detected), but one must also assess the reliability (the fraction of detections that are due to true planets).
- **Overly optimistic detection criteria**. One of the more straightforward and frequently-used detection criteria is a simple signal-to-noise ratio or $\Delta \chi^2$ cut, or some related statistic that quantifies how much better the data is modeled with a planet signal than the null hypothesis of no planet. In principle, one should set the threshold value of such detection criteria such that there are few or no statistical false alarms. One might naively assume that the data have uncertainties that are Gaussian distributed and uncorrelated, in which case the threshold value can generally be estimated analytically (or semi-analytically) by simply computing the probability (say 0.1%) of a given value of the statistic arising by random fluctuations. In practice, setting the appropriate threshold value is significantly more complicated. First, one requires detailed knowledge of the noise properties of the data, which are often not Gaussian distributed and are also often correlated on various timescales. Second, one requires an estimate of the number of independent

trials used to search for planetary signals to determine the threshold probability. In practice, both can be difficult to estimate: the noise properties of the data may be poorly behaved and difficult to characterize, and the number of independent trials often cannot be estimated analytically and must be estimated via, e.g., Monte Carlo simulations. A reasonably robust method of estimating the appropriate threshold is to inject simulated planetary signals into the data, and compare the distribution of the desired statistic (e.g., $\Delta\chi^2$) in the data where no planets were injected to the distribution in the data where planets were injected.[4] The optimal threshold can therefore be chosen such that the completeness is maximized while minimizing the number of false positives.

- **The dangers of binning.** It is often tempting to adopt the procedure outlined above where one collects the detected planets in parameter bins of a given size, and then estimates the intrinsic frequency of planets in that bin by dividing the number of detected planets in that bin by the number of expected detections if every star has a planet in that bin (see, e.g., Chap. 3, Sect. 3.2 and Fig. 4.14). Indeed, this is a useful tool that can enable one to visualize the intrinsic distribution function of planet properties, which can then inform the parameterized models to be fitted. However, it should be noted that binning is not a well-defined procedure. In particular, one can ask: What is the optimal bin size? If one makes the bin size too large (in order to, e.g., decrease the Poisson uncertainty), then one risks smoothing over real features in the underlying distribution of planet properties. On the other hand, if the bin size is too small, the Poisson uncertainties blow up. Indeed, there is *no optimal bin size* in this sense, and thus it is generally advised to fit parametric models using the maximum likelihood approach briefly summarized above and derived previously by many authors.

- **Not accounting for uncertainties in the parameters of the detected planets** Note that in Eq. (4.9), the distribution function $F(\boldsymbol{\beta}_I, \boldsymbol{\Theta})$ and the total survey sensitivity are evaluated at the point estimates of the values of the parameters $\boldsymbol{\beta}_{I,k}$ for each of the $k = [1, N_{\mathrm{det}}]$ planet detections. This implicitly assumes that the distributions of these parameters for each detected planet are unimodal with zero uncertainty, neither of which is generically true. Therefore, one must marginalize the likelihood over the uncertainties in the parameters $\boldsymbol{\beta}_I$ (properly accounting for covariances between these parameters), and account for multimodal degenerate solutions, as applicable. See Foreman-Mackey (2014); Hsu et al. (2018) for discussions of the deleterious effects of ignoring the uncertainties on the planet parameters.

- **Poor knowledge of the properties of the target sample.** When converting between the observable parameters of detected exoplanet systems to the physical properties of the planets, one must often assume or infer properties of the host star. When inferring exoplanet demographics as a function of the physical properties of the planets, one must typically assume or infer properties of the entire target sample. Any systematic errors in these inferences will propagate directly

---

[4] In general this method works only if the majority of the targets do not have a planetary signal that is significant compared to the intrinsic noise distribution in the data.

into systematic errors in the inferred exoplanet distributions. Even if systematic uncertainties are not present, the statistical uncertainties in the host star properties should be taken into account by marginalizing over these uncertainties. As a concrete example, transit surveys are directly sensitive to the transit depth δ, which is just the square of the ratio of the radius of the planet to the radius of the star, $\delta = (R_p/R_*)^2$. Thus, to infer the demographics of planets as a function of (R_p, P), one must estimate the radii of *all* the target stars (not simply the ones that have detected transit signals). Prior to Gaia (2016), it was generally not possible to robustly estimate radii for most stars; rather these had to be estimated using other observed properties of the star (e.g., $T_{\rm eff}$, $\log g$, [Fe/H]), combined with theoretical evolutionary tracks. This led to systematic differences in the inferred radii of subsamples of the *Kepler* targets, which in turn led to discrepancies in the inferred radii of the detected planets, and the inferred planet radius distribution (Mann et al. 2012; Pinsonneault et al. 2012; Dressing and Charbonneau 2013; Huber et al. 2014; Gaidos et al. 2016).[5]

- **Properly distinguishing between the fraction of stars with planets and the average number of planets per star**. A commonly overlooked subtlety in determining exoplanet demographics from surveys is the distinction between estimates of the average number of planets per star (NPPS) and estimates of the fraction of stars with planets (FSWP). Depending on the distribution of exoplanet multiplicities, these two quantities can be significantly different (Youdin 2011; Brakensiek and Ragozzine 2016). Generally, studies that consider all the planets detected in a survey (including multi-planet systems) can be thought of as measuring the NPPS (Youdin 2011; Brakensiek and Ragozzine 2016; Hsu et al. 2018).

- **Including post-detection data or detections from other surveys**. When inferred planet occurrence rates from a survey, it is critical that the survey be 'blind'. In other words, the data that were used to detect the planets and characterize the survey completeness must not have been influenced by the presence (or absence) of significant planetary signals. For true surveys with predetermined and fixed observation strategies (such as *Kepler*), this is generally not an issue. However, for targeted surveys, it is common to acquire more data on targets that show tentative evidence of a signal in order to bolster the significance of the signal (or determine that is not real). It is also common to take additional data once a robust detection is made, in order to better characterize the parameters of the system. Using this data may result in biased estimates of the occurrence rate of planets. Similarly, it is often the case that targets will stop being observed or will be observed at a lower cadence after a certain time if there no evidence of a signal. Again, such a strategy can lead to biased estimates of the occurrence rates. For similar reasons, one must not use detections from other surveys or data from other surveys to confirm marginal detections, as this may also lead to biased estimates of the occurrence rates.

---

[5] Fortunately, with the availability of near-UV to near-IR absolute broadband photometry, combined with Gaia parallaxes and stellar atmosphere models, it is possible to measure the radii of most bright stars nearly purely empirically to relatively high precision (Stassun et al. 2017; Stevens et al. 2017).

- **The dangers of extrapolation.** No single survey or detection method, and indeed even the totality of the surveys that have been conducted with all the detection methods at our disposal, have yielded a "complete" statistical census of exoplanets (see Sect. 4.4 and the discussion therein). It is therefore tempting to extrapolate from regions of parameter space where exoplanet demographics have been relatively well-measured to more poorly-studied regions of parameter space. Aside from theoretical arguments why such extrapolations may not be well-motivated, the dangers of such extrapolations have already been empirically demonstrated. For example, initial extrapolations of power-law fits to the occurrence rate of giant planets with semi-major axes of $\lesssim 3$ au as estimated from RV surveys (Cumming et al. 2008) out to very large separations implied that ground-based direct imaging surveys for young, giant planets should yield a large number of detections. This prediction has not been borne out, and it is now known that very wide-separation giant planet are relatively rare (see Sect. 4.5.3). Perhaps more strikingly, Dulz et al. (2020) found that, by extrapolating the double power-law fit to the distribution of planets detected by *Kepler* found by Kopparapu et al. (2018) with periods of $\lesssim 600$ days to longer periods, many (and depending on the adopted uncertainties in the fit by Kopparapu et al. (2018), even the majority) of the systems were dynamically unstable.

4.4 Methods of Detecting Exoplanets: Inherent Sensitivities and Biases

There are four primary methods that have been used to detect exoplanets: radial velocities, transits, direct imaging, and microlensing. Another well-known method of detecting exoplanets is astrometry. However, to date, there have been no confirmed detections of planetary companions using astrometry, although there are some candidates that have yet to be confirmed. Nevertheless, astrometry is an extremely promising method for detecting planets, particularly wide-orbit planets. In particular, as we discuss in Sect. 4.7, Gaia (2016) is expected to detect tens of thousands of giant planets on relatively wide orbits (Casertano et al. 2008; Perryman et al. 2014). Therefore, I will also discuss astrometry in this section. There are, of course, many other methods of detecting exoplanets (see Fig. 4.1). However, none of these methods has yielded large samples of planets to date, and so I will not be discussing these in this review.

Figures 4.2 and 4.3 show the distribution of confirmed exoplanets as a function of mass (or minimum mass in the case of RV detections) and period (Fig. 4.2), or radius and period (Fig. 4.3). There are several noteworthy features:

- Most of the planets with (minimum) mass measurements and orbital periods greater than ~ 100 days were discovered via RV, and thus do not have radius measurements. Conversely, the majority of planets with mass measurements and periods of less than ~ 100 days were discovered by ground-based transit surveys (e.g., "hot

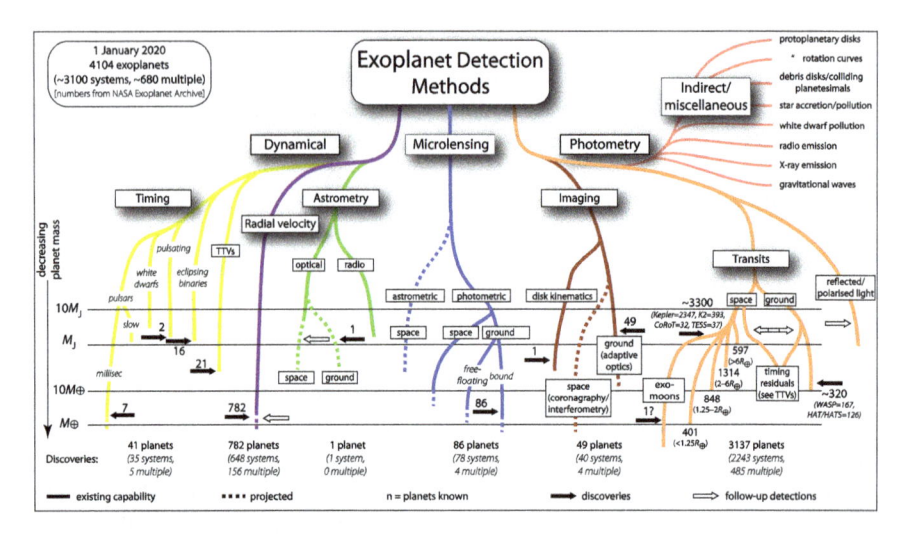

Fig. 4.1 A graphical summary of exoplanet detection methods, including the number of discoveries by each method as of January 1, 2020. Courtesy of Perryman (2018) and reproduced with permission

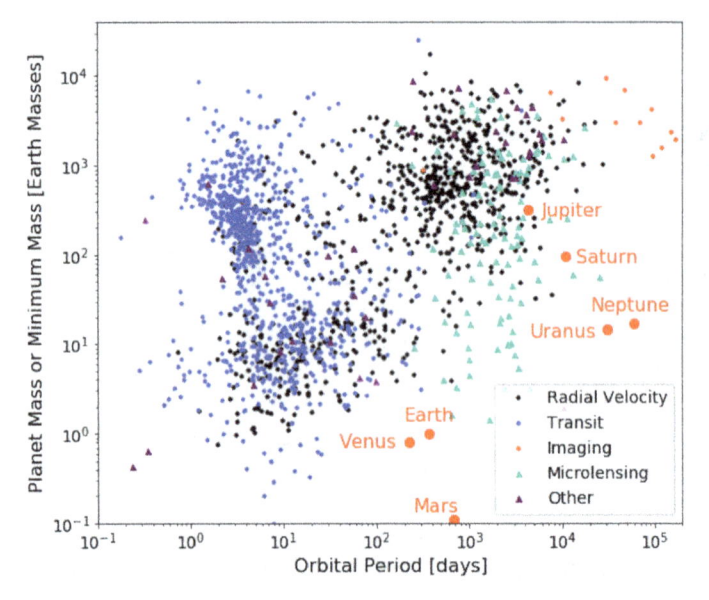

Fig. 4.2 The distributions of the ~4300 confirmed low mass companions to stars as a function of their mass (or minimum mass in the case of RV detections) and period. For planets detected by microlensing and direct imaging, the projected semi-major axis has been converted to period using Newton's version of Kepler's Third Law. The color-coding denotes the method by which planets were detected. I note that ~25 planets detected by direct imaging are not shown in this plot because they have periods that are greater than $\gtrsim 10^6$ days. This figure is based on data from the NASA Exoplanet Archive: https://exoplanetarchive.ipac.caltech.edu/. Courtesy of Jesse Christiansen with assistance from Radek Poleski, reproduced with permission

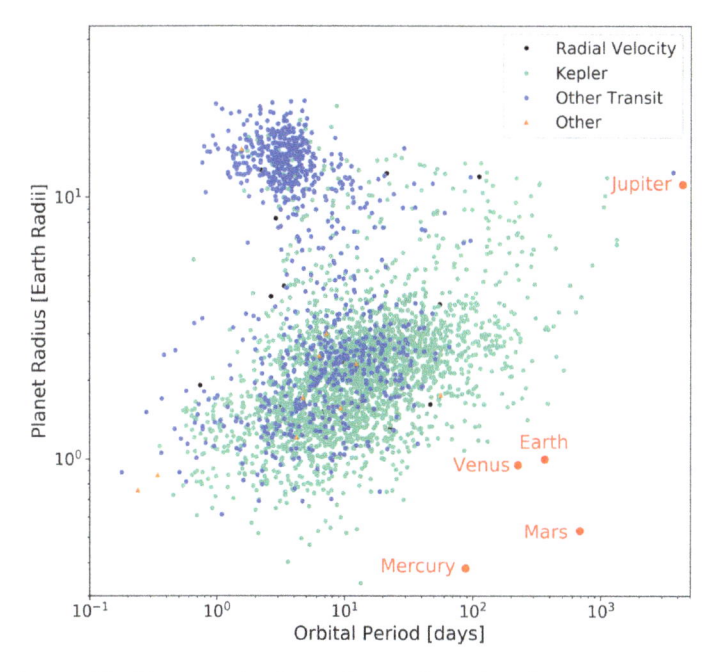

4 The Demographics of Wide-Separation Planets 251

Fig. 4.3 Same as Fig. 4.2, except showing the known planets in radius and period space. As with Fig. 4.2, the color coding denotes the method by which planets were detected. This figure is based on data from the NASA Exoplanet Archive: https://exoplanetarchive.ipac.caltech.edu/. Courtesy of Jesse Christiansen, reproduced with permission

Jupiters"). This is because, although *Kepler* discovered far more transiting planets than ground-based transit surveys, most of the host stars were too faint for RV follow-up, and only a relatively small subset exhibited transit timing variations that allow for a measurement of the planet mass.
- The number of hot Jupiters detected via RV is much smaller than the number of hot Jupiters detected by ground-based transit surveys, and much smaller than the number of Jovian planets (mostly discovered with RV) with periods $\gtrsim 100$ days. This has several implications. From RV surveys, it is known that hot Jupiters are relatively rare compared to the population of Jupiters with longer periods (Wright et al. 2012). On the other hand, when hot Jupiters detected by transits are also considered, the fact that the two populations appear comparable in number is due to the strong selection bias of transit surveys toward short-period planets (e.g., Gaudi et al. 2005; Gaidos and Mann 2013).
- The vast majority of planets with measured radii do not have mass measurements. As mentioned above, this is due to the fact that the *Kepler* host stars are generally too faint for RV follow-up, and a minority of systems exhibit transiting timing variations. The majority of planets that have both mass and radius measurements were discovered by ground-based transit surveys.

- The fact that the spread in the radii of hot Jupiters is considerably smaller than the spread in mass is a consequence of the fact that the radii of objects with masses that span roughly the mass of Saturn up to the hydrogen-burning limit are all approximately constant, with radii of $\sim R_{\rm Jup}$.
- Finally, and most relevant to this review, there is paucity of planets in the lower right corner of both figures (small masses/radii and long periods). This is purely a selection effect, as I discuss below. It is also worth noting that this is the area that is spanned by the planets in our solar system. In particular, essentially no detection method is currently sensitive to analogs of any of the planets in our solar system except for Jupiter (and, in the case of microlensing, Saturn and possibly analogs of the ice giants).

4.4.1 A Note About Stars

Before discussing the sensitivities and biases of each detection method, I will first make a few comments about stars. Although initial RV surveys focused primarily on solar-type (FGK) stars, to date a wide range of stellar host types have been surveyed for exoplanetary companions. These hosts span a broad range of properties that can be relevant for exoplanet detection, including radius, mass, effective temperature, projected rotation velocity, luminosity, activity, and local number density, to name a few. Thus, as is well appreciated, in order to understand the impact of the sensitivities and biases of a given detection method on the population of planets that can be detected, it is essential to understand how these depend on the properties of the host stars ("Know thy star, know thy planet"). Consequently, it is also important to have a reasonably in-depth understanding of stellar properties and how they vary with stellar type, as well as have a detailed accounting of the distribution of stellar properties of any given exoplanet survey.

For the purposes of this review, I focus on relatively unevolved solar-type FGKM main sequence stars. For such stars, the mass-luminosity and mass-radius relation can very roughly be approximated by Torres et al. (2010)

$$L_* = L_\odot \left(\frac{M_*}{M_\odot}\right)^4 \tag{4.10}$$

$$R_* = R_\odot \left(\frac{M_*}{M_\odot}\right) \tag{4.11}$$

$$T_{\rm eff} = T_{\rm eff,\odot} \left(\frac{M_*}{M_\odot}\right)^{1/2}. \tag{4.12}$$

The mass-radius relation holds from roughly the hydrogen-burning limit to $\sim 2\,M_\odot$ for stars near the zero-age main sequence, whereas the luminosity-mass relation holds from roughly the fully convective limit of $\sim 0.35\,M_\odot$ up to $\sim 10\,M_\odot$. Below the fully convective limit, the mass-luminosity relation is shallower than Eq. (4.10), and similarly, in this regime, the effective temperature-mass relationship deviates from the scaling relation above. I will use these approximate relations to express the sensitivities of the various detection methods as a function of the host star mass.

4.4.2 A General Framework for Characterizing Detectability

In the following few sections, I will discuss the scaling of the sensitivity of each of the five methods discussed in this section (RV, transits, direct imaging, microlensing, and astrometry) with planet and host star parameters, highlighting the intrinsic sensitivities and biases of each method. I will not attempt to provide an in-depth discussion of these detection methods, as this material has been covered in numerous other books and reviews (Seager 2010; Perryman 2018; Wright and Gaudi 2013; Fischer et al. 2014; Bozza et al. 2016).

As outlined in Wright and Gaudi (2013), although the criteria to detect a planet depends on the details of the planet signal, the data properties (e.g., cadence, uncertainties), and the precise quantitative definition of a detection, one can often estimate the scaling of the signal-to-noise ratio with the planet and host star properties by decomposing the signal into two contributions. These are the overall magnitude of the signal, and the detailed form of the signal itself. The magnitude of the signal typically depends on the parameters of the system (planet and host), and largely dictates the detectability of the planet. The detailed form of the signal typically depends on geometrical or nuisance parameters, but generally has a small impact on the magnitude of the signal itself. These two contributions can often be relatively cleanly separated, such that one only needs to consider the magnitude of the signal to gain intuition about the scaling of the signal-to-noise ratio of a given method with the stellar and planetary parameters, and thus detectability with these parameters. In the language of Sect. 4.2, the magnitude of the signal generally depends on the set of the (more) physical parameters of interest $\boldsymbol{\beta}_I$, whereas the form of the signal depends on set of parameters $\boldsymbol{\beta}_N$, although I stress that this separability does not strictly apply over all the detection methods.

Assuming this separability, the approximate signal-to-noise ratio of a planet signal can be written in terms of the magnitude or amplitude of the signal A, the number of observations N_{obs}, and the typical measurement uncertainty σ, such that

$$(S/N) \simeq A(\boldsymbol{\beta}_I) \frac{N_{obs}^{1/2}}{\sigma} g(\boldsymbol{\beta}_N), \qquad (4.13)$$

where $g(\boldsymbol{\beta}_N)$ is a function of $\boldsymbol{\beta}_N$ whose value depends on the details of the signal, but is typically of order unity. Thus, to roughly determine the scaling of the (S/N) with the physical parameters of the planet and star (for an arbitrary survey), one must simply consider the scaling of the amplitude A on the physical parameters.

4.4.3 Radial Velocity

The set of parameters that can be measured with radial velocities are $\boldsymbol{\beta} = \{K, P, e, \omega, T_p, \gamma\}$ (see Chap. 3, Sect. 2.1). Assuming uniform and dense sampling of the RV curve over a time span that is long compared to the period P, the signal amplitude is $A = K(P, M_*, M_p, e, i)$, and the total signal-to-noise ratio scales as

$$(S/N)_{RV} \propto A \propto M_p P^{-1/3} M_*^{-2/3} \propto M_p a^{-1/2} M_*^{-1/2}. \tag{4.14}$$

with a relatively weak dependence on eccentricity for $e \lesssim 0.5$.

Thus radial velocity surveys are more sensitive to more massive planets, with the minimum detectable mass[6] at fixed (S/N) scales as $P^{1/3}$ for planet periods of $P < T$, where T is the duration of the survey.

For $P > T$, it becomes increasingly difficult to characterize each of the parameters $\boldsymbol{\beta}$ individually. For $P \gg T$, the signal of the planet is an approximately constant acceleration $\mathcal{A}_*$, with a magnitude that is $\mathcal{A}_* = (2\pi K/P) f(\phi, \omega, e)$, where $f(x)$ is a function that depends on ω, e, and the phase of the orbit ϕ. Thus, a measurement of the acceleration can constrain the combination $M_p/P^{4/3}$. By combining a measurement of $\mathcal{A}_*$ with the direct detection of the companion causing the acceleration, one can derive a lower limit on M_p (Torres 1999; Crepp et al. 2012).

4.4.4 Transits

The set of parameters than can be measured with transits are $\boldsymbol{\beta} = \{P, \delta, T, \tau, T_0, F_0\}$, where T is the full-width half-maximum duration of the transit, τ is the ingress/egress duration,[7] T_0 is a fiducial reference time, and F_0 is the out-of-transit baseline flux (see Chap. 3, Sect. 2.2). If the radius of the host star can be estimated, then the radius of the planet can also be inferred via $R_p = \delta^{1/2} R_*$. If the radial velocity of the host star can also be measured, it is then possible to determine the orbital eccentricity and planet mass M_p, and thus the density of the planet ρ_p. Of course, with additional follow-up observations, it is also possible to study the atmospheres for some transiting planets (e.g., Seager 2010).

---

[6] Formally, the minimum detectable $M_p \sin i$.

[7] I note that a common alternative parameterization is to use T_{14}, the time between first and fourth contact, and T_{23}, the time between second and third contact (also referred to as T_{full} and T_{flat}). I strongly advise against adopting this parameterization, for several reasons. First, the algebra required to transform from this parameterization to the physical parameters is significantly more complicated (compare Seager and Mallén-Ornelas 2003 and Carter et al. 2008). Second, T_{14} and T_{23} are generally much more highly correlated than T and τ, making the analytical interpretation of fits using the former parameterization much more difficult than using the latter parameters. Finally, the timescale estimated from the Boxcar Least Squares algorithm (Kovács et al. 2002) is much more well approximated by T than T_{14} or T_{23}.

Assuming uniform sampling over a time span that is long compared to the transit period P, we have that the signal-to-noise ratio is $(S/N)_{TR} = N_{tr}^{1/2}(\delta/\sigma)$, where $N_{tr} = (N_{tot}/\pi)(R_*/a)$ is the number of data points in transit and N_{tot} is the total number of data points. Therefore, $A = \delta = f(R_p, R_*, M_*, P)$, or $A = f(R_p, R_*, M_*, a)$. The signal-to-noise ratio of the transit when folded about the correct planet period scales as (Wright and Gaudi 2013)

$$(S/N)_{TR} \propto A \propto R_p^2 P^{-1/3} M_*^{-5/3} \propto M_p a^{-1/2} M_*^{-3/2}. \tag{4.15}$$

Furthermore, the transit probability scales as

$$P_{tr} \propto \frac{R_*}{a} \propto P^{-2/3} M_*^{2/3} \propto a^{-1} M_*. \tag{4.16}$$

Of course, the planet must transit and the signal-to-noise ratio requirement must be met to detect the planet. Finally, transit surveys require at least two transits to estimate the period of the planet, and often require at least three to aid in the elimination of false positives. Thus the final requirement to detect a planet via transits is $P \leq T/3$.

Thus transit surveys are more sensitive to large planets, with the minimum detectable radius at fixed signal-to-noise ratio scaling as $P^{1/6}$ up until $P = T/3$. Planets with periods longer than this are essentially undetectable with traditional transit selection cuts (but see Sect. 4.5.2). Furthermore, the transit probability decreases as $P^{-2/3}$. Because of these two effects, transit surveys are very "front loaded" for reasonable planet distributions that do not rise sharply with increasing period, meaning that, e.g., doubling the duration of the survey will generally not double the yield of planets.

4.4.5 Microlensing

Unlike most of the other detection methods discussed in this chapter, the detectability of planets via microlensing does not lend itself as well to the simple analytic description described above. In particular, it is not possible to write down a simple scaling of the signal-to-noise ratio with the planet and/or host star properties. I will therefore simply review the essentials of microlensing and the parameter space of planets and stars to which it is most sensitive. For more detail, I refer the reader to the following reviews and references therein (Gaudi 2012; Gould 2016).

Briefly, a microlensing event occurs whenever a foreground compact object (the lens, which could be, e.g., a planet, brown dwarf, star, or stellar remnant) passes very close to an unrelated background source star. In general, for a detectable microlensing event to occur, the lens must pass within an angle of roughly the angular Einstein ring radius θ_E of the lens,

$$\theta_E \equiv (\kappa M \pi_{rel})^{1/2}, \tag{4.17}$$

where $\pi_{\rm rel} = \pi_l - \pi_s = {\rm au}/D_{\rm rel}$ is the relative lens (π_l)-source(π_s) parallax, $D_{\rm rel}^{-1} \equiv D_l^{-1} - D_s^{-1}$, D_l and D_s are the distances to the lens and source, respectively, and $\kappa \equiv 4G/(c^2 {\rm au}) = 8.14$ mas $M_\odot^{-1}$ is constant. For a typical stellar mass of $M_* = 0.5 M_\odot$, a source in the Galactic center with a distance of $D_s = 8$ kpc, and a lens half way to the Galactic center with $D_l = 4$ kpc, $\theta_{\rm E} \simeq 700$ µas. The minimum lens-source alignment must be exquisite for a detectable microlensing event to occur, and given the typical number density of lenses along the line of sight and typical lens-source relative proper motions $\mu_{\rm rel}$, microlensing events are exceedingly rare. Thus most microlensing surveys focus on crowded fields toward the Galactic center, where there are many ongoing microlensing events per square degree at any given time.

When a microlensing event occurs, the lens creates two images of the source, whose separations are of order $2\theta_{\rm E}$ during the event, and are thus generally unresolved (c.f. Dong et al. 2019). However, the background source flux is significantly magnified if the lens passes within a few $\theta_{\rm E}$ of the source, resulting in a transient brightening of the source: a microlensing event. The characteristic timescale of a microlensing event is the Einstein ring crossing time,

$$t_{\rm E} \equiv \frac{\theta_{\rm E}}{\mu_{\rm rel}}, \tag{4.18}$$

where $\mu_{\rm rel}$ is the relative lens-source proper motion. The typical timescale of observed microlensing events toward the Galactic bulge is $t_{\rm E} \sim 20$ days (Mroz et al. 2019a), but can range from a few days to hundreds of days.

A bound planetary companion to the lens can be detected during a microlensing event if it happens to have a projected separation and orientation that is close to the paths that the two images create by the host lens trace on the sky. The planet will then further perturb the light rays from the source, causing a short duration deviation to the otherwise smooth, symmetric microlensing event due to the more massive host (Mao and Paczynski 2011; Gould and Loeb 1992; Bennett and Rhie 1996). These deviations are also of order hours to days for terrestrial to gas giant masses. Since the planets must be located close to the paths of one of the two images to create a significant perturbation, and the images are always close to the Einstein ring during the primary microlensing events, the sensitivity of microlensing is maximized for planets with angular separations of $\sim \theta_{\rm E}$. At the distance of the lens, $\theta_{\rm E}$ corresponds to a linear Einstein ring radius of

$$r_{\rm E} \equiv \theta_{\rm E} D_l \tag{4.19}$$

$$= 2.85 {\rm au} \left(\frac{M}{0.5 M_\odot}\right)^{1/2} \left(\frac{D_s}{8 \text{ kpc}}\right)^{1/2} \left[\frac{x(1-x)}{0.25}\right]^{1/2}, \tag{4.20}$$

where $x \equiv D_l/D_s$. Thus the sensitivity of microlensing surveys for bound exoplanets peaks for planets with semi-major axes of ~ 3 au $(M_*/0.5 M_\odot)^{1/2}$, which corresponds to orbital separations relative to the snow line (as defined in Eq. 4.1) of

$\sim 2\,(M_*/0.5\,M_\odot)^{-1/2}$. Planets with a significantly smaller semi-major axes become difficult to detect due to the fact that they perturb faint images. Planets with significantly larger semi-major axis can be detected if the lens-source trajectory is aligned with the host star-planet projected binary axis. However, the probability of having the requisite alignment decreases with increasing semi-major axis. Eventually, the probability of detecting the magnification due to both the host and planet becomes exceedingly small. Thus very wide separation planets are generally only detected as isolated, short-timescale microlensing events, which produce light curves that are essentially indistinguishable from events due to free-floating planets (Han et al. 2005).[8] Thus microlensing is, in principle, sensitive to planets with separations out to infinity, e.g., including free-floating planets (Mroz et al. 2018, 2019b, 2020a; Kim et al. 2020; Ryu et al. 2020). The timescales of microlensing events caused by free-floating or widely-bound planets with terrestrial to gas giant masses also range from hours to days (since $t_E \propto \theta_E \propto M_p^{1/2}$).

Microlensing events are rare and unpredictable. Furthermore, planetary perturbations to these events are brief, unpredictable and generally uncommon. Typical detection probabilities given the existence of a planet located with a factor of ~ 2 of θ_E are a few percent to tens of percent for terrestrial to gas giant planets (Gould and Loeb 1992; Bennett and Rhie 1996). As with free-floating planets, the duration of planetary perturbations caused by bound planets scales as $\sim M_p^{1/2}$, and thus range from hours to days. Therefore, microlensing surveys for exoplanets must observe many square degrees of the Galactic bulge on timescales of tens of minutes to both detect the primary microlensing events and monitor them with the cadence needed to detect the shortest planetary perturbations.

The magnitudes of planetary perturbations are essentially independent of the mass of the planet, for planets with angular Einstein ring radii that are larger than the angular size of the source, i.e., when $\theta_E \gtrsim \theta_*$. For a source with $R_* \sim R_\odot$ at a distance of $D_s \sim 8$ kpc, $\theta_* \sim 1$ μas. The angular Einstein ring radius for an Earth-mass lens at a distance of 4 kpc and source distance of 8 kpc is $\theta_E \sim 1$ mas. Thus, for planets with mass $\lesssim M_\oplus$, finite source effects will begin to dominate, as the planet is only magnifying a fraction of the source at any given time.

In summary, while the magnitude of the planetary perturbations to microlensing events, as well as the peak magnification of microlensing events due to widely-separated and free-floating planets, is essentially independent of planet mass for planets with mass $\gtrsim M_\oplus$, these signals become rarer and briefer with decreasing planet mass. Nevertheless, planets with $M_p \gtrsim M_\oplus$ can be detected from current ground-based surveys (Suzuki et al. 2018), and planets with $M_p \gtrsim 0.01 M_\oplus$ (or roughly the mass of the moon) can be detected with a space-based microlensing survey (Bennett and Rhie 2002; Penny et al. 2019).

The parameters that can be measured from a microlensing event with a well-sampled planetary perturbation are $\boldsymbol{\beta} = \{t_0, t_E, u_0, F_s, F_b, \alpha_{\mu L}, q, s\}$. Here t_0 is the time of the peak of primary microlensing event, u_0 is the minimum lens-source

---

[8] I note that, in some cases, it is possible to disambiguate free-floating planets from bound planets by detecting (or excluding) light from the host.

angular separation in units of θ_E (which occurs at a time t_0), F_s is the flux of the source, F_b is the flux of any light blended with the source but not magnified, which can include light from the lens, companions to the lens and/or source, and unrelated stars, $\alpha_{\mu L}$ is the angle of the source trajectory relative to the projected planet-star axis,[9] $q \equiv M_p/M_*$ is the planet-star mass ratio, and s is the instantaneous projected separation in units of θ_E. Note that F_s and F_b can be (and usually are) measured in several bandpasses. The primary physical parameters of interest are F_s, F_b, t_E, q, and s. For most planetary perturbations, the effect of the finite size of the source on the detailed shape of the perturbation can also be detected, which allows one to also infer $\rho = \theta_*/\theta_E$. Since θ_* can be estimated from the color and flux of the source, θ_E can generally be inferred, leaving a one-parameter degeneracy between the lens mass and distance (assuming D_s can be estimated, as is usually the case). This degeneracy can be broken in a number of ways (see Gaudi et al. 2002 for a detailed discussion), but the most common method is to measure the flux of the lens (Bennett et al. 2007). This requires isolating light from the lens in the blend flux F_b, and thus typically requires high angular resolution from the ground using adaptive optics, or from space, in order to resolve light from unrelated stars from the lens and source flux. For space-based microlensing surveys, it is expected that the lens flux will be detectable for nearly all luminous lenses (Bennett et al. 2007), and thus it will be possible to estimate M_*, M_p, D_l, and the instantaneous projected separation between the star and planet in physical units, for the majority of planet detections. Since microlensing events can be caused by stars (or planets) all along the line of sight toward the Galactic bulge, it will be possible to determine the frequency of bound and free-floating planets as a function of Galactocentric distance, and, in particular, determine if the planet population is different in the Galactic disk and bulge (Penny et al. 2016).

4.4.6 Direct Imaging

It is generally more complicated to characterize direct imaging observations. In part, this is because there are three general cases of emission from the planet that must be considered. The first, and the most relevant to current ground-based direct imaging surveys, is the case where the luminosity of the planet is dominated by the residual heat from formation and is thus decoupled from the luminosity of, and distance from, its host star. In this case, it is possible to measure the angular separation of the planet from its host star θ. The amplitude of the signal is simply the flux of the planet in a specific band $A = F_{p,\lambda}$, and thus $\boldsymbol{\beta} = \{F_{p,\lambda}, \theta\}$. With a distance to the star, it is possible to determine the monochromatic luminosity of the planet as well as the instantaneous physical separation projected on the sky. Using detailed cooling models of exoplanets as well as an estimate of the age of the system, it is possible

---

[9] I note that in the microlensing literature, this variable is typically simply referred to as α. Since I have already defined α above, I adopt the form $\alpha_{\mu L}$ to avoid confusion.

to use the former to obtain a (model-dependent) estimate of the mass of the planet. With multiple epochs spanning a significant fraction of the orbit of the planet, it is possible to measure its Keplerian orbital elements (up to a two-fold degeneracy in the longitude of the ascending node).

The second and third cases correspond to when the energy output from the residual heat from formation is negligible compared to the energy input due to the irradiation from the host star (e.g., when the planet is in thermal equilibrium with the host star irradiation). In this case, the spectrum of the planet has two components: reflected starlight and the thermal emission from starlight that is absorbed and then re-radiated as thermal emission. In the case of detection by reflected starlight, the basic parameters that can be measured are again the instantaneous angular separation from the host star and the amplitude of the signal which is simply given by $A = F_{p,\lambda}$. Thus $\boldsymbol{\beta} = \{F_{p,\lambda}, \theta\}$. As before, the orbital elements can be inferred via a distance to the system and multiple epochs of astrometric observations of the planet. In reflected light, the flux of the planet at any given epoch depends on the radius of the planet, its albedo, and its phase function. The phase function can be estimated from the measurement of the planet flux over multiple epochs of its orbit, leaving a degeneracy between the planet's radius and albedo. The albedo can be roughly estimated from a spectrum of the planet, with a precision that depends on the particular properties of the system. For a detection in thermal emission, the directly-measured parameters are (assuming that a spectrum can be obtained) $\boldsymbol{\beta} = \{\theta, F_{p,\lambda}, T_{\text{eq}}\}$, where T_{eq} is the equilibrium temperature of the planet. As with detection in reflected light, the orbital elements can be determined with multiple epochs and a distance to the system. In this case, the amplitude of the signal is also given by $A = F_{p,\lambda}$. By combining the detection of a planet in both reflected light and thermal emission, it is possible to determine R_p, T_{eq} and the albedo.

In addition to these three kinds of emission, there are also multiple sources of noise in direct imaging surveys. These include, but are not limited to: Poisson noise from the planet itself, noise from imperfectly removed light from the host, noise from the local zodiacal light, noise from the zodiacal light in the target system (the "exozodi"), and any other sources of background noise (e.g., read noise and dark current). Which of these dominate (if anyone dominate) depends on many factors, including, e.g., the planet/star flux ratio, the limiting contrast floor, the amount of exozodiacal dust in the target system, and the angular resolution of the telescope.

Rather than repeat the discussion of the parameters that can be measured in each case, or how the various physical parameters interplay to affect the detectability of directly-imaged planets, I will simply refer the reader to the discussion in Wright and Gaudi (2013). However, in contrast with Wright and Gaudi (2013), I will not assume any specific source of noise, and thus will not assume any specific relation between the noise and the properties of the host star. Therefore, the scalings of the signal-to-noise ratio below reflect only the contributions due to the planet flux signal, and **do not** include any contributions to the noise from the planet or host star, and thus do not include any scalings of the properties of the planet and/or host star with their potential contribution to the noise.

With these caveats in mind, we have that amplitude of the signal is given by the planet flux $A = F_{p,\lambda}$. Considering the three different cases discussed above, we have that, for planets that are not necessarily in thermal equilibrium with their host star, $F_{p,\lambda} \propto R_p^2 T_p$, where T_p is the temperature of the planet and I have assumed observations in the Rayleigh-Jeans tail (where the flux of a blackbody is linearly proportional to its temperature). For planets in thermal equilibrium with their host stars, we have that the planet flux is $F_{p,\lambda} \propto L_* R_p^2 a^{-2}$ in reflected light, and $F_{p,\lambda} \propto R_p^2 T_p \propto R_p^2 L_*^{1/4} a^{-1/2}$ in the Rayleigh-Jeans tail. Thus the scalings are

$$(S/N)_{dir} \propto A \propto R_p^2 a^{-2} M_*^4 \propto R_p^2 P^{-4/3} M^{10/3} \quad \text{(Reflected Light, Equilibrium)} \tag{4.21}$$

$$(S/N)_{dir} \propto A \propto R_p^2 a^{-1/2} M_* \propto R_p^2 P^{-1/3} M_*^{5/6} \quad \text{(Thermal, Equilibrium, RJ)} \tag{4.22}$$

$$(S/N)_{dir} \propto A \propto R_p^2 T_p \quad \text{(Thermal, RJ),} \tag{4.23}$$

where the last two expressions are only valid for the Raleigh-Jeans tail.

In addition, the planet must have a maximum angular separation that is outside the inner working angle θ_{IWA} of the direct imaging survey, which is essentially the minimum angular separation from the star that the planet can be detected. This leads to the requirement that

$$a \gtrsim \theta_{IWA} d, \tag{4.24}$$

where d is the distance to the star. The inner working angle is generally tied to the wavelength of light λ of the direct imaging survey, and the effective diameter D of the telescope, or for interferometers, the distance between the individual apertures. Specifically

$$\theta_{IWA} \sim N \frac{\lambda}{D}, \tag{4.25}$$

where N is a dimensionless number that is typically between 2 and 4 that primarily depends on the detailed properties of the starlight suppression system (coronagraph, starshade, or interferometer).

In general, the above scalings imply that ground-based direct imaging surveys are generally most sensitive to massive, young giant planets at projected separations of $\gtrsim 10$ au from their host star. Direct imaging surveys for planets in thermal equilibrium have a more complicated selection function. In terms of semi-major axis, they are typically most sensitive to planets on semi-major axis that are just outside of the inner working angle. At fixed semi-major axis and distance, they are more generally sensitive to planets orbiting more massive stars. However, more massive stars are generally rare and thus more distant (and thus are less likely to meet the inner working angle requirement). Furthermore, the contrast between the planet and star is larger for more massive stars (at fixed planet radius and semi-major axis). Thus if residual stellar flux is the dominant source of noise, lower mass stars are preferred. The net

4.4.7 Astrometry

Assuming a large number of astrometric measurements that cover a time span that is significantly longer than the period of the planet, the parameters than can be measured from the astrometric perturbation of star due to the orbiting planet are $\beta = \{\theta_{\text{ast}}, P, e, \omega, T_p, i, \Omega\}$, where Ω is the longitude of the ascending node,[10] d is the distance to star, and the amplitude of the astrometric signal due to the planet is $A = \theta_{\text{ast}}$, where

$$\theta_{\text{ast}} \equiv \frac{a}{d} \frac{M_p}{M_*}. \quad (4.26)$$

Note that Eq. (4.26) assumes $M_p \ll M_*$.

In practice, for planetary-mass companions, the magnitude of the stellar proper motion μ_* and parallax π is significantly larger than θ_{ast}, and thus if the astrometric perturbation from the planet can be detected, so can μ_* and d. The mass of the star can be estimated using the usual methods, and thus M_p and a can be inferred, along with the Keplerian orbital elements and orientation of the orbit on the sky (up to the two-fold degeneracy Ω).

Again, assuming a large number of astrometric measurements that cover a time span that is significantly longer than the period of the planet, the signal-to-noise ratio scales as

$$(S/N)_{\text{AST}} \propto A \propto M_p a M_*^{-1} d^{-1} \propto M_p P^{2/3} M_*^{-2/3} d^{-1}. \quad (4.27)$$

Thus astrometry is more sensitive to more massive planets, as well as planets on longer period orbits. However, unlike the RV method, the sensitivity of astrometry declines precipitously for planets periods P greater than the duration of the survey T. This is because such planets only produce an approximately linear astrometric deviation of the host star (particularly for $P \gg T$), which is then absorbed when fitting for the (much larger) stellar proper motion (Casertano et al. 2008; Gould 2008).

Thus the detection and characterization of exoplanets via astrometry have an additional criterion, namely that $P \lesssim T$. The net result is that the sensitivity function of astrometry in $M_p - P$ space has a 'wedge-shaped' appearance, with the minimum detectable planet mass a given (S/N) decreasing as $P^{-2/3}$ until $P \sim T$, and then increasing precipitously for $P > T$.

---

[10] Note that, with only astrometric observations, there is a two-fold ambiguity in Ω.

4.4.8 Summary of the Sensitivities of Exoplanet Detection Methods

Considering the five primary methods of detecting exoplanets: radial velocities, transits, microlensing, direct imaging, and astrometry, it is clear that all methods are (not surprisingly) more sensitive to more massive or larger planets.

Radial velocity and transit surveys are generally more sensitive to shorter-period planets. The sensitivity of the radial velocity method (in the sense of the minimum detectable planet mass) declines as $P^{1/3}$, and maintains this sensitivity scaling up to the survey duration T. Thus, long-period or wide-separations planets can be detected only in radial velocity surveys with sufficiently long baselines. For planets with $P > T$, planets produce accelerations or "trends", which can constrain a combination of the planet mass and period. The sensitivity of transit surveys (in the sense of the minimum detectable planet mass) decline as $P^{1/6}$ up until roughly $T/3$, assuming three transits are required for a robust detection. Planets with periods longer than this are undetectable under this criterion. Thus radial velocity and transit surveys are generally less well-suited to constraining the demographics of long-period or wide-separation exoplanets.

Astrometric surveys are more sensitive to longer-period planets, with their sensitivity (in the sense of the minimum detectable planet radius) increasing as $P^{2/3}$, up to $P \sim T$. For planets with $P > T$, the sensitivity of astrometric surveys drops precipitously, e.g., it is very difficult to detect and characterize planets with $P > T$.

The sensitivity functions for microlensing and direct imaging surveys are generally more complicated. Microlensing surveys are most sensitive to planets with semi-major axes that are within roughly a factor of two of the Einstein ring radius, which is $r_E \sim 3$ au$(M_*/M_\odot)^{1/2}$ for typical lens and source distances. Microlensing is relatively insensitive to planets with semi-major significantly smaller than r_E, but maintains some sensitivity to planets with separation significantly larger than r_E, although the detection probability drops as $\sim a^{-1}$. Microlensing is the only method capable of detecting old free-floating planets with masses significantly less than $\sim M_{\rm Jup}$. The sensitivity functions of direct imaging surveys are similarly complicated. Current ground-based surveys, which typically detect young planets via thermal radiation from their residual heat from formation are typically sensitive to planets that are relatively widely separated from (the glare of) their host stars. Future space-based direct imaging surveys designed to detect mature planets in thermal equilibrium with their host stars in reflected light are generally only sensitive to planets with semi-major axes greater than ~ 1 au (due to the requirement that the planet angular semi-major axis is outside the inner working angle, which is typically a few λ/D), and their sensitivity declines as $R_p \propto a$. Space-based direct imaging mission surveys designed to detect mature planets in thermal equilibrium with their host stars via their thermal emission are also generally only sensitive to planets with semi-major axis $\gtrsim 1$ au, and their sensitivity declines as $R_p \propto a^{1/4}$.

Thus, the strongest constraints on the demographics of long-period or wide-separation planets are expected to come from long-running radial velocity surveys, microlensing surveys, astrometric surveys, and direct imaging surveys. Relatively short-duration radial velocity surveys and transit surveys, in general, do not constrain the demographics of long-period planets.

Finally, I will make a few comments about the sensitivity of the various methods to the mass of the host star. All of the methods discussed here are more sensitive to planets at fixed mass M_p or radius R_p and period P that orbit lower-mass stars, with the exception of direct imaging surveys for planets in thermal equilibrium. For radial velocity surveys, the scaling of the (S/N) at fixed period is $\propto M_*^{-2/3}$. For transits, the scaling of the (S/N) at fixed period is $\propto M_*^{-5/3}$. For microlensing, the scaling is approximately $\propto M_*^{-1/2}$. For direct imaging surveys, the scaling ranges from being independent of the host star mass (for young planets detected in thermal emission) to $\propto M_*^4$ (for planets in thermal equilibrium detected in reflected light). For astrometric surveys, the (S/N) at fixed period scales as $\propto M_*^{-2/3}$.

4.5 The Demographics of Wide-Separation Planets

I now turn to summarize some of the most important extant constraints on the demographics of long-period or wide separation planets. I will first summarize results from each of the individual detection techniques that have placed at least some constraints on wide-separation planets, and then discuss efforts to synthesize results from multiple surveys using the same technique, and multiple surveys using different techniques. There have been very few attempts to synthesize results from multiple surveys (regardless of whether or not they use the same detection technique) in general, for many reasons, some of which are discussed in Sect. 4.3. However, because each technique is more or less sensitive to a particular region of the planet (M_p or R_p and P or a) and host star parameter space, such a synthesis is needed to assemble as complete a picture of exoplanet demographics as possible. Such a broad synthesis of exoplanet demographics, when performed correctly, provides the empirical ground truth to which all theories of planet formation and evolution must match. Thus synthesizing results from multiple surveys remains a fruitful avenue of future research.

I note that, while I will attempt to highlight the most relevant results from the literature, it is simply not possible to provide a comprehensive and complete summary of all exoplanet demographics surveys in the limited amount of space available here. In particular, I will generally not discuss results regarding the demographics of short-period planets, as this topic is already covered in Chap. 3. Specifically, I will not be discussing the vast literature on inferences about the demographics of relatively short-period planets from *Kepler*.

4.5.1 Results from Radial Velocity Surveys

Radial velocity surveys for exoplanets have been ongoing since the late 1980s (Campbell et al. 1988; Latham et al. 1989; Mayor et al. 1992; Marcy and Butler 1992), with the first widely-accepted discovery of an exoplanet using the radial velocity technique announced in 1995 with the discovery of 51 Pegasi b (Mayor and Queloz 1995). A few of these radial velocity surveys have been ongoing since the early 1990s and thus have accrued a baseline of roughly 30 years (corresponding to a semi-major axis of ∼10 au for a solar-type star, or roughly the semi-major axis of Saturn). The minimum achievable precision of these surveys have generally decreased with time, but the most relevant precision for detecting long-period planets is roughly the worst precision, which for the surveys mentioned above was in the range of a few m/s. For a Jupiter-sun analog, $K \sim 13$ m/s, $P \sim 12$ year whereas for a Saturn-sun analog, $K \sim 3$ m/s, $P \sim 30$ year. Thus these long-term surveys are readily sensitive to Jupiter analogs orbiting Sun-like stars, but Saturn analogs orbiting Sun-like stars are just at the edge of their sensitivity (Bryan et al. 2016). Thus, extant radial velocity surveys can only constrain the population of giant planets ($M_p \gtrsim 0.15$ M_{Jup} beyond the snow line of Sun-like stars.

Surveys for planets orbiting M dwarfs generally have shorter baselines, but cover the complete orbits of planets with semi-major axes out to 3 au (Johnson et al. 2010a; Bonfils et al. 2013). Giant planets with orbits longer than the baseline of these surveys can be detected via their trends, or, for sufficiently massive planets and sufficiently precise radial velocity observations, the planet properties can be characterized using partial orbits (Bonfils et al. 2013; Montet et al. 2014). However, as with surveys for planets orbiting solar-like stars, these surveys can generally only constrain the demographics of relatively massive planets beyond the snow line of M dwarfs.

Using the sample of planets discovered in the HARPS and CORALIE radial velocity surveys of planets orbiting primarily Sun-like stars from Mayor et al. (2011), and accounting for the survey completeness, Fernandes et al. (2019) infer that the frequency of giant planets with masses in the range of 0.1–20 M_{Jup} rises with increasing period (in agreement with previous results, e.g., Cumming et al. 2008) up to roughly 2–3 au (e.g., roughly the snow line for Sun-like stars), at which point the frequency of such planets begins to decrease with increasing period (see Fig. 4.4).[11] Wittenmyer et al. (2016) analyzed 17 years of data from the Anglo-Australian Planet Search radial velocity survey of Sun-like stars. They estimated the frequency of "Jupiter analogs", which they define as planets with semi-major axes between $a = 3$–7 au, mass of $M_p > 0.3$ M_{Jup}, and eccentricity $e < 0.3$, to be $6.2^{+2.8}_{-1.6}\%$. This is consistent with the result from Fernandes et al. (2019), implying that true Jupiter analogs are relatively rare. Furthermore, extrapolating the declining frequency of giant planets beyond the snow line as inferred by Fernandes et al. (2019) results in a frequency of giant plan-

---

[11] As noted in Chap. 3, Sect. 3.1 and Fig. 4.12, Wittenmyer et al. (2020) find instead that long-period giant planet frequency beyond the snowline does not appear to decline, while Fulton et al. (2021) provide evidence that the turnover in occurrence rate of gas giants might start closer to 10 au, and it might be mass-dependent.

ets accessible to direct imaging surveys that is consistent with detection rates from those surveys (with some caveats), which is not the case if one simply extrapolates the increasing frequency of giant planets inferred by Cumming et al. (2008) from the analysis of the shorter-baseline California-Carnegie planet survey out to separations where direct imaging surveys would be sensitive to (young analogs) of them.

Using a database of 123 known exoplanetary systems (primarily hot and warm Jupiters) monitored by Keck for nearly 20 years, combined with NIRC2 K-band adaptive optics (AO) imaging, Bryan et al. (2016) estimated the frequency of giant planets with masses between $M_p = 1$–20 M_{Jup} and $a = 5$–20 au to be $\sim 52 \pm 5\%$. This frequency is quite high, and in particular higher than the extrapolation of the results from Cumming et al. (2008). This suggests that the existence of hot/warm Jupiters and cold Jupiters may be correlated. In other words, the conditional probability of a particular star hosting a cold Jupiter, given the existence of a hot/warm Jupiter, is not random.

The constraints on the population of wide-separation planets orbiting low-mass stars are generally weaker. Surveys for giant planets orbiting low-mass M stars ($M_* \lesssim 0.5\ M_\odot$) have generally found a paucity of planets on relatively short orbital periods interior to the snow line (Johnson et al. 2010a). This is seemingly a victory for the core-accretion model of planet formation, which generally predicts that giant planets should be rare around low-mass stars, due to their lower-mass disks and longer dynamical timescales at the distances where giant planets are thought to have formed (Laughlin et al. 2004; Johnson et al. 2010b). However, these constraints were generally only applicable for planets with semi-major axes less than a few au, implying that giant planets could still form efficiently around low-mass stars, but perhaps do not migrate to the relatively close-in orbits where they can be detected via the relatively short time baseline radial velocity surveys for exoplanets orbiting low-mass stars. Indeed, a closer inspection of the results from the HARPS (Bonfils et al. 2013) and California-Carnegie (Montet et al. 2014) surveys for planets orbiting M dwarfs hint at a significant population of giant planets at orbits with periods roughly equal to the survey duration. See Fig. 4.4.

4.5.2 Results from Transit Surveys

As discussed in Sect. 4.4.4, transit surveys are generally not sensitive to planets with periods longer than $T/3$ of the survey duration T. This is because such surveys impose an arbitrary (but reasonable) criterion that at least three transits must be detected. A robust detection of at least two transits is generally required to infer the period of the planet, whereas the detection of a third transit largely eliminates most false positives.

However, as first pointed out by Yee and Gaudi (2008), based on the arguments from Seager and Mallén-Ornelas (2003), the robust detection of a single transit can be used to estimate the period of the planet, given an estimate of the density of the star ρ_* and assuming zero eccentricity. Since single transits from long-period

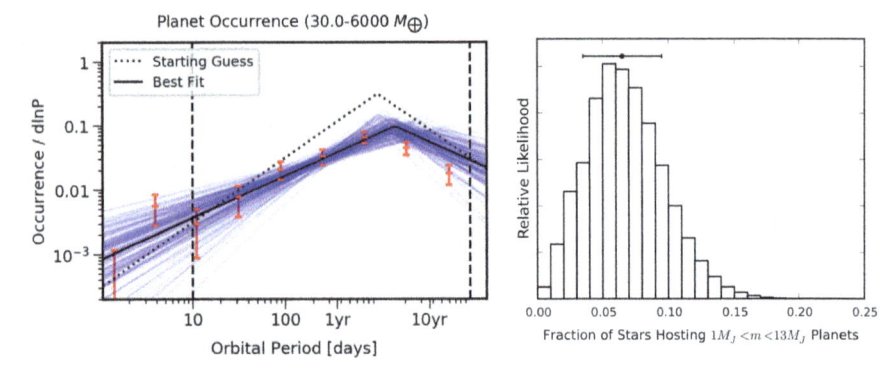

Fig. 4.4 *Left-hand panel*: Occurrence rate per ln P as a function of orbital period P based on the analysis of the HARPS/CORALIE survey for exoplanets (Mayor et al. 2011). The red points show the binned occurrence rate, whereas the black broken power law shows the maximum likelihood fit to the unbinned data. The blue broken power laws show samples from the 1σ range of fits. The dashed line shows the initial starting value of the fit. Finally, the vertical lines show the range of periods considered in the fit. From Fernandes et al. (2019). ©AAS. Reproduced with permission. *Right-hand panel*: the distribution of the giant planet occurrence rates inferred from the California-Carnegie radial velocity survey of 111 M dwarfs. The median and 68% confidence interval of the distribution implies that 6.5% ± 3.0% of M dwarfs host a planet with a mass between 1 $M_{\rm Jup}$ − 13 $M_{\rm Jup}$, with separations <20 au. From Montet et al. (2014). ©AAS. Reproduced with permission

planets typically have long durations, they also typically have quite large signal-to-noise ratios, which are sufficient to allow an estimate of the planet period, under the assumptions above (Yee and Gaudi 2008). Indeed, Yee and Gaudi (2008) used this fact to argue that the follow-up of single transit events could be used to extend the period sensitivity of transit surveys, and in particular that of *Kepler*. As Yee and Gaudi (2008) argued, not only does this require an estimate of ρ_*, but it also generally requires nearly immediate radial velocity observations of the target star to measure the acceleration of the star due to the planetary companion.

The suggestion of Yee and Gaudi (2008) was largely ignored during the primary *Kepler* mission. Likely this was because the yield of such single-transit events was expected to be small. Nevertheless, two groups (Foreman-Mackey 2016; Herman et al. 2019) endeavored to estimate the planet occurrence rate for planets with periods beyond the nominal range of the *Kepler* primary mission using the methods outlined in Yee and Gaudi (2008). In particular, Herman et al. (2019) used updated information about the planet host stars from the Gaia DR2 release (Gaia Collaboration 2018) to improve the purity of the long-period single-transit sample (see Fig. 4.5). They inferred a frequency of cold giant planets (radii between 0.3 and 1 $R_{\rm Jup}$ and periods of ∼2–10 of $0.70^{+0.40}_{-0.20}$. This rate is consistent with the results of Cumming et al. (2008), albeit with larger uncertainties. They also infer a radius distribution of planets beyond the snow line of

$$\frac{dN}{d\log R_{\rm p}} = R_{\rm p}^{-1.6^{+1.0}_{-0.9}}, \tag{4.28}$$

4 The Demographics of Wide-Separation Planets

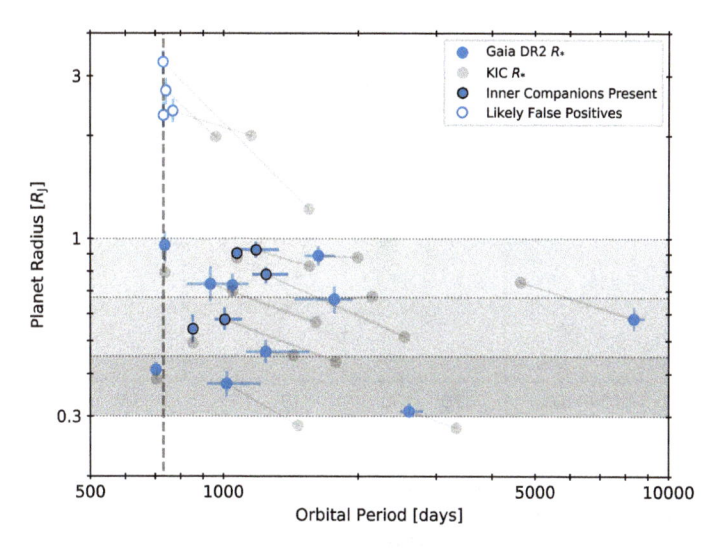

Fig. 4.5 The results of the single-transit candidate search from Herman et al. (2019). The parameters of the candidates using revised stellar parameters are plotted in blue, while gray points indicate the parameters inferred using the older *Kepler* input catalog values. The vertical dashed line denotes the maximum possible period to exhibit at least three transits during the *Kepler* primary mission. From Herman et al. (2019). ©AAS. Reproduced with permission

consistent with the conclusion from microlensing surveys that Neptunes are more common than Jovian planets beyond the snow line (Gould et al. 2006b). Finally, they note that 5 of the 13 long-period planet candidates they identify have confirmed inner transiting planets, indicating that there is a strong correlation between the presence of cold planets with warm/hot inner planets and that the mutual inclinations between the inner and outer planets must be small. The results of Herman et al. (2019) demonstrate the importance of a holistic picture of exoplanet demographics.

4.5.3 Results from Direct Imaging Surveys

A relatively recent, cogent, and exceptionally comprehensive review of direct imaging surveys has been provided by Bowler (2016). Rather than attempt to reproduce or replicate the contents of that exceptional review, I will simply highlight the most important conclusions from direct imaging surveys as summarized therein.

As shown in Fig. 4.6, in the current state of the art direct imaging surveys for exoplanets (Hinkley et al. 2011; Macintosh et al. 2014; Beuzit et al. 2019; Skemer et al. 2014) are primarily sensitive to planets with masses of $\gtrsim M_{\mathrm{Jup}}$ and orbits of $\gtrsim 10$ au.

These surveys have searched for planetary companions orbiting roughly ∼400, relatively young (<300 Myear) stars, with spectral types ranging from low-mass M

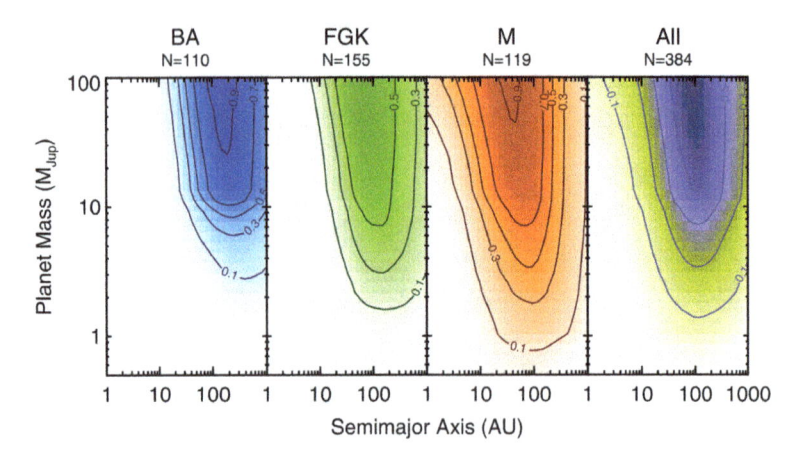

Fig. 4.6 Completeness contours from an ensemble analysis of ~380 unique stars with published high-contrast imaging observations. The contours denote 10, 30, 50, 70, and 90% completeness limits. Note that low-mass stars provide stronger constraints for planets at a fixed mass and semi-major axis. From Bowler (2016) © Publications and the Astronomical Society of the Pacific. Reproduced with permissions

stars to B stars. Roughly 8 planetary ($\lesssim 13\ M_{\rm Jup}$) candidates with semi-major axes of $\lesssim 100$ au have been found orbiting main-sequence stars (Lagrange et al. 2010; Macintosh et al. 2015; Rameau et al. 2013; Marois et al. 2008, 2010; Kraus and Ireland 2012).

The majority of these candidates are very dissimilar to the giant planets in our solar system: they are typically more massive and have much larger separations. This has led to the speculation that these planets formed via a different mechanism than the core accretion mechanism (Pollack et al. 1996) that is typically invoked to explain the formation of shorter-period giant planets detected by other methods. Indeed, the massive, long-period planets detected by direct imaging have been suggested to be evidence of planet formation via gravitational collapse (Boss 1997; Dodson-Robinson et al. 2009). However, it has been argued that there are serious theoretical difficulties with this formation mechanism (Rafikov 2005). If gravitational instability can create long-period giant planets, the theoretical prediction is that there should be a larger population of brown dwarfs (Kratter et al. 2010). Indeed, this has been observed (Nielsen et al. 2019). There is some evidence that the orbits of directly-imaged planets are distinct from those found via the radial velocity method, perhaps suggesting that there are indeed two mechanisms for giant planet formation (Bowler et al. 2020).

Even if there are two channels of forming giant planets, the population of planets formed by these two different channels may not occupy distinct regions of parameter space. For example, the directly-imaged planet candidate 51 Eridani b (Macintosh et al. 2015), which has an estimated mass of $\sim 2\ M_{\rm Jup}$ and a semi-major axis of ~ 13 au, has properties that are quite similar to the giant planets detected via radial velocity surveys, as well as the giant planets in our solar system.

4 The Demographics of Wide-Separation Planets

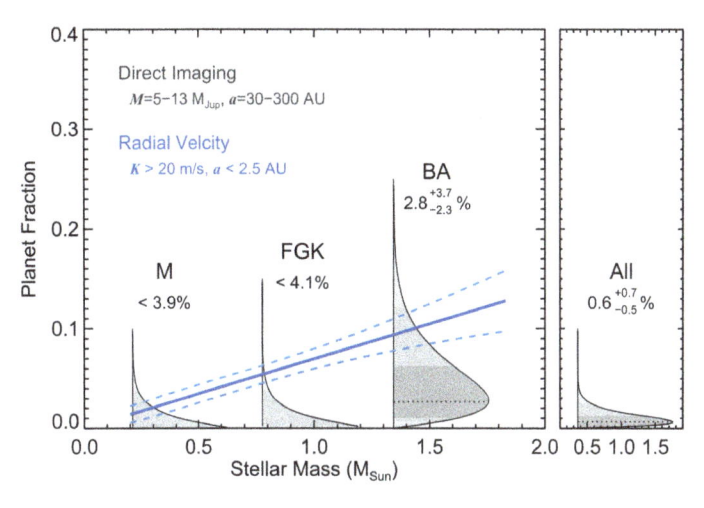

Fig. 4.7 Distributions of the occurrence rate of giant planets from an ensemble analysis of direct imaging surveys in the literature, as compared to the results from radial velocity surveys for giant plants at small separations ($\lesssim 2.5$ au) (Johnson et al. 2010b). There is no statistically significant evidence for a correlation between the giant planet occurrence and host star spectral type for planets at large separations probed by direct imaging surveys. Determining whether or not this is consistent with the significant trend inferred by Johnson et al. (2010b) will require a larger sample of directly-imaged planets. From Bowler (2016). © Publications of the Astronomical Society of the Pacific. Reproduced with permission

Finally, it is worth noting that there is no statistically significant evidence for an increase in the frequency of long-period giant planets with host star spectral type (a proxy for host star mass) from direct imaging surveys, as shown in Fig. 4.7. However, given the relatively small number of giant-planet candidates detected by direct imaging surveys, this result is not statistically discrepant with the result from radial velocity surveys that the frequency of shorter-period giant plants increases with increasing host mass (Johnson et al. 2010b).

4.5.4 Results from Microlensing Surveys

Microlensing surveys for exoplanets cannot choose their target stars—rather, the sample of hosts around which microlensing surveys can constrain the properties of planetary systems is dictated by the number density of compact objects (brown dwarfs, stars, and remnants) weighted by the event rate, which scales as $\theta_E \propto M_*^{1/2}$. Figure. 4.8 shows the predicted distribution of host masses probed by microlensing (Henderson et al. 2014; Gould 2000). Ignoring brown dwarfs and remnants, the median mass of main-sequence (MS) stars probed by microlensing surveys is $\sim 0.4 \, M_\odot$. Thus, although a little over half the MS host stars will be M dwarfs, microlensing surveys still probe the frequency of planets beyond the snow line orbiting G and K type stars, a fact that is not widely appreciated.

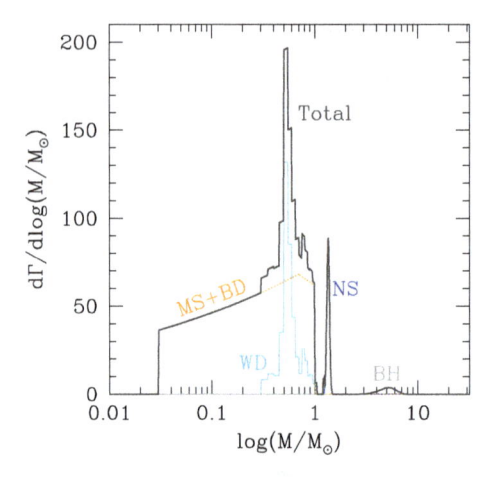

Fig. 4.8 The predicted distribution of host masses for microlensing surveys for exoplanets. Brown dwarfs BD), main-sequence (MS) stars, and remnants [including white dwarfs (WD), neutron stars (NS), and black holes (BH)] are included. Considering only MS stars, the median host star mass surveyed by microlensing is ∼0.4 $M_\odot$. Note that MS stars with masses above the bulge turn-off have been suppressed; in reality, the distribution of host star masses will include a small contribution from more massive MS stars in the Galactic disk. From Gaudi et al. (2002), adapted from Gould (2000). © ARAA. Reproduced with permission

The first constraints on the frequency of planets from microlensing surveys were by Gaudi et al. (2002), based on five years of data from the Probing Lensing Anomalies with a world-wide NETwork (PLANET) collaboration (Albrow et al. 1998). Although they did not detect any planets, they were able to place a robust upper limit on the frequency of massive companions. They concluded that <33% of hosts have ∼M_{Jup} companions with separations between 1.5 and 4 au, and less than < 45% of hosts have ∼3M_{Jup} with separations between 1 and 7 au. As the majority of these hosts were M dwarfs, this result provided the first significant limits on planetary companions to M dwarfs.

The first conclusive discovery of an exoplanet by microlensing was in 2004 (Bond et al. 2004). Additional detections followed soon after (Udalski et al. 2005; Beaulieu et al. 2006; Gould et al. 2006b). Of the first four planets detected via microlensing, two had Jovian mass-ratios, whereas the other two had super-Earth/Neptune mass ratios. Given the decreasing sensitivity of microlensing for smaller mass ratios (see 4.4.5), this immediately implied that cold low-mass (super-Earth to Neptune) mass planets were much more common than cold Jovian planets (Gould et al. 2006b).

To date, over 100 planets have been detected by microlensing.[12] Individual surveys have detected a sufficiently large sample of planets that they have been able to place robust constraints on the population of cold planets orbiting main-sequence stars (Sumi et al. 2010; Gould et al. 2010; Cassan et al. 2012; Shvartzvald et al. 2016),

---

[12] See https://exoplanetarchive.ipac.caltech.edu/.

4 The Demographics of Wide-Separation Planets

see Fig. 4.9 for a graphical representation of the constraints derived by Gould et al. (2010) and Cassan et al. (2012).

The most recent and thorough analysis with the largest sample of planets is that by Suzuki et al. (2016), who analyzed six years of data from the second generation Microlensing Observations in Astrophysics (MOA-II) collaboration (Sumi 2013). The analysis included 23 planets detected from 1474 alerted microlensing events. They find that the distribution of mass ratios q and projected separations s is well-described by a broken power law, with the form:

$$\frac{dN}{d\log q \, d\log s} = 0.61^{+0.21}_{-0.16} \left[\left(\frac{q}{q_{br}}\right)^{-0.93 \pm 0.13} \mathcal{H}(q-q_{br}) + \left(\frac{q}{q_{br}}\right)^{0.6^{+0.5}_{-0.4}} \mathcal{H}(q_{br}-q) \right] s^{0.49^{+0.47}_{-0.49}}, \quad (4.29)$$

where, as before, $\mathcal{H}(x)$ is the Heaviside step function. Thus, Suzuki et al. (2016) find that the mass ratio function of cold exoplanets with mass ratios above that of $q_{br} \sim 1.7 \times 10^{-4}$ (roughly the mass ratio of Neptune to the Sun) is steeper than that mass function for shorter-period giant planets found by Cumming et al. (2008). See Fig. 4.9. In addition, the distribution of orbits for planets beyond the snow line is consistent with a log-uniform distribution, and planets with Neptune/sun mass ratios are likely the most common planets beyond the snow line. Suzuki et al. (2016) also synthesized their MOA-II constraints with those of Cassan et al. (2012), which also included constraints from Sumi et al. (2010) and Gould et al. (2010). The combined results from these four surveys (Sumi et al. 2010; Gould et al. 2010; Cassan et al. 2012; Suzuki et al. 2016) are shown in Fig. 4.13.

Using a sample of seven planets detected by microlensing with well-measured mass ratios of $q < 10^{-4}$ and unique solutions, Udalski et al. (2018) confirmed the result from Suzuki et al. (2016) that the mass ratio function for planets with mass ratio below q_{br} declines with decreasing mass ratio. They inferred a power-law index in this regime of $1.05^{+0.78}_{-0.68}$, as compared to the value of $0.6^{+0.5}_{-0.4}$ found by Suzuki et al. (2016). By combining their results with those of Suzuki et al. (2016) and Udalski et al. (2018), they refine the power-law index in this regime to $0.73^{+0.42}_{-0.34}$.

However, using a sample of 15 planets with well-measured mass ratios of $q < 3 \times 10^{-4}$, Jung et al. (2019) arrived at a somewhat different conclusion than Suzuki et al. (2016) and Udalski et al. (2018). Again assuming the same double power-law form as Eq. (4.29), they find a smaller value for the break in the mass function of $q_{br} = 0.55 \times 10^{-4}$, e.g., a factor of ~ 3 times smaller than found by Suzuki et al. (2016). Furthermore, they inferred a very large difference in the slope of the mass function for mass ratios above and below q_{br}, with a best fit of 5.5 (>3.3 at 1σ). Assuming the power-law index of -0.93 for $q > q_{br}$ found by Suzuki et al. (2016), which they did not constrain, this implies a very steep power-law index for $q < q_{br}$ of 4.6 (with an upper limit of 2.4 at 1σ), as compared to the value of $0.6^{+0.5}_{-0.4}$ and $0.73^{+0.42}_{-0.34}$ found by Suzuki et al. (2016) and Udalski et al. (2018), respectively. Jung et al. (2019) also note an apparent 'pile-up' of planets with mass ratio similar to that of Neptune to the sun. Specifically, four of their 15 planets have mass ratios between 0.55×10^{-4} and 5.9×10^{-4}, a span of only $\Delta \log q = 0.030$, as compared to the full range spanned by their sample of $\Delta \log q = 0.875$. However, they were

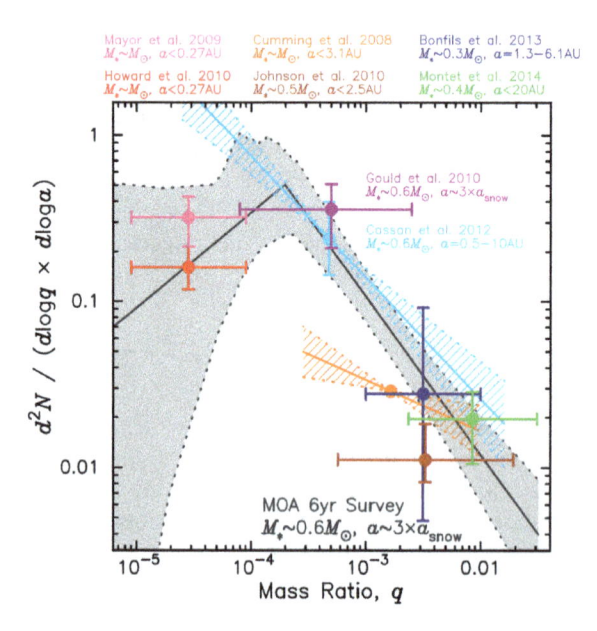

Fig. 4.9 The distribution of planet/star mass ratios inferred from Suzuki et al. (2016). The black line shows best-fit broken power-law mass-ratio function (see Eq. 4.29), whereas the gray shaded region shows the uncertainties about this best-fit model. This result is compared to compendium of demographic constraints from other radial velocity and microlensing surveys (Mayor et al. 2011; Howard et al. 2010; Cumming et al. 2008; Johnson et al. 2010a; Bonfils et al. 2013; Montet et al. 2014). Note that the typical primary host mass and semimajor axis range vary amongst the various results. From Suzuki et al. (2016), adapted and updated from Gould (2000) and Gaudi et al. (2002). ©AAS. Reproduced with permission

unable to determine conclusively whether this 'pile-up' is real or due to a statistical fluctuation.

As discussed in Sect. 4.4.5, microlensing is uniquely sensitive to widely-bound and free-floating planets. These are detectable as isolated, very short timescale (hours-to-days) microlensing events. The first constraints on the frequency of free-floating or widely bound planets from microlensing were by Sumi et al. (2011). Based on an excess of short timescale (∼1 day) events, the argued for the existence of a population of free-floating or widely-separation planets with masses of $\sim M_{\rm Jup}$, with a frequency of roughly twice that of stars in the Milky Way. By comparing to the frequency of giant planets found by direct imaging surveys, Clanton and Gaudi (2017) were able to demonstrate that $\gtrsim 70\%$ of these events must be due to free-floating planets. There are significant difficulties with producing such a large population of free-floating planets, which I will not expound upon here (but see Veras and Raymond 2012; Ma et al. 2016). This is because a subsequent analysis of the Optical Gravitational Lensing Experiment (OGLE, Udalski et al. 2015) data conclusively demonstrated that the purported excess of short-timescale microlensing events by MOA-II was spurious (Mroz et al. 2017).

4 The Demographics of Wide-Separation Planets

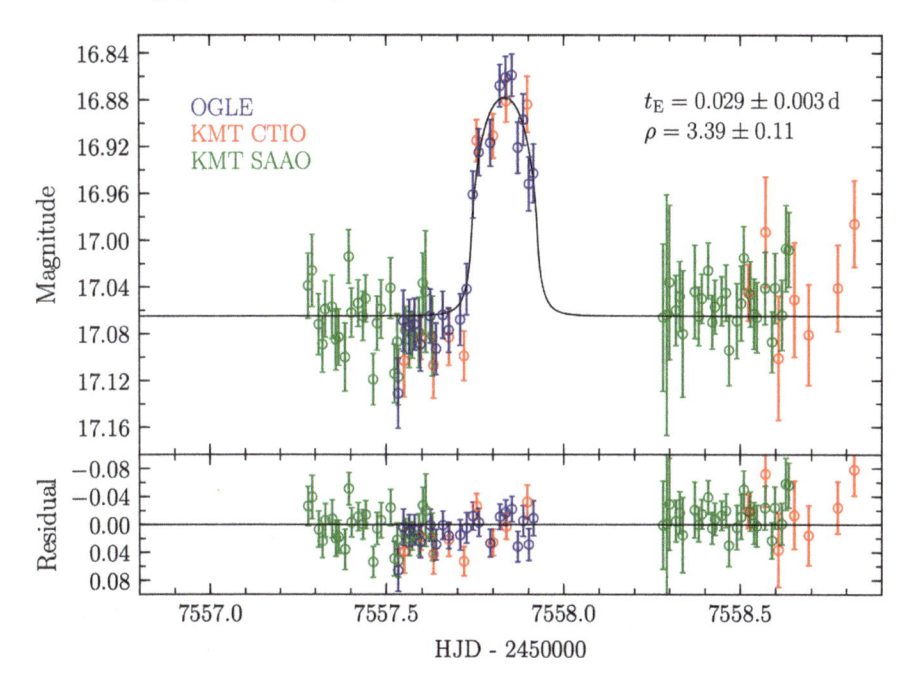

Fig. 4.10 Data for microlensing event OGLE-2016-BLG-1928, which has the shortest Einstein timescale of any microlensing event detected to date. It is likely the lowest-mass free-floating or widely-bound ($\gtrsim 10$ au) planet detected to date, with an estimated mass between Mars and the Earth. From Mroz et al. (2020a). © AAS. Reproduced with permission

Curiously, Mroz et al. (2017) do find an excess of very short timescale ($\lesssim 0.5$ days) candidate microlensing events, which may be an indication of population of free-floating or widely-bound planets with planets of mass of ~ 1–$10\ M_\oplus$. Some theories of planet formation predict the ejection of a significant number of such low-mass planets during the chaotic phase of planet formation.

Since the Mroz et al. (2017) result, a total of seven robust free-floating planet or wide-orbit planets have been discovered via microlensing (Mroz et al. 2018, 2019b, 2020a; Kim et al. 2020; Ryu et al. 2020; Mroz et al. 2020b) primarily using data from the OGLE and Korea Microlensing Telescope Network (KMTNet) collaborations (Henderson et al. 2014; Kim et al. 2016).

Figure 4.10 shows the data for the shortest timescale microlensing event detected to date, with an Einstein timescale of only $t_E \sim 40$ min (Mroz et al. 2020a). The lens likely has a mass in the Mars- to Earth-mass regime, with lower masses being favored. If the planet is bound to a host star, it must have a projected separation of $\gtrsim 10$ au.

4.5.5 Synthesizing Wide-Separation Exoplanet Demographics

Ultimately, because of the intrinsic sensitivities and biases of all exoplanet detection methods (as discussed in Sect. 4.4), no single method or survey can provide broad constraints on exoplanet demographics that are needed to properly constrain and refine planet formation theories. Thus, multiple surveys using multiple detection methods must be "stitched together" to provide the needed empirical constraints.

We are fortunate that the various detection methods at our disposal are largely complementary, and can, in principle, constrain exoplanet demographics over nearly the full range of planet and host-star properties needed to fully test planet formation theories. However, combining the results from various surveys and methods is not trivial. Often, exoplanet researchers have the relevant expertise in only one or perhaps two detection methods. Surveys often do not report the details needed to combine their results with other surveys, such as providing the appropriate information about their target sample, or providing the individual detection sensitivities for each of their targets (including those for which no planetary candidates were detected).

For these reasons and others, the obstacles to synthesizing the demographics of exoplanets are significant, which is likely the reason why very little progress in this area has been made. Nevertheless, exoplanet surveys now have significant overlap in terms of the parameter space of the planet and host star properties, and thus there is an opportunity to make significant progress in constructing a broad statistical census of exoplanet demographics, using results that are already in hand.

In this section, I will highlight a few notable attempts to synthesize the results of wide-orbit demographics from multiple surveys using multiple methods. However, I emphasize that much more work needs to be done in this area.

Two of the first rigorous attempts to compare the frequency of giant planets as constrained by radial velocity and microlensing surveys were performed by Montet et al. (2014) and Clanton and Gaudi (2014a, b). Both groups focused on low-mass stellar hosts and used the results from trends found in radial velocity surveys of relatively low-mass hosts (e.g., evidence for companions with periods longer than the duration of the survey) to constrain the frequency of long-period giant planets. In particular, Montet et al. (2014) used adaptive-optics imaging to constrain the mass of companions causing such trends to be in the planetary regime. They then used their constraints to determine if the population of long-period giant planets was consistent with that found by microlensing. In contrast, Clanton and Gaudi (2014a, b) took a different approach and mapped the distribution of planets orbiting low-mass stars as inferred from microlensing to that expected from RV surveys of low-mass stars. Both groups concluded that the demographics of long-period giant planets as determined by radial velocity and microlensing surveys were consistent. In particular, both groups found that there exists a significant population of Jovian planets at relatively long periods. Specifically, Clanton and Gaudi (2014b) found that the frequency of Jupiters and super-Jupiters ($1 < M_p \sin i/M_{\mathrm{Jup}} < 13$) with periods $1 < P/\mathrm{days} < 10^4$ is $0.029^{+0.013}_{-0.015}$. This is a median factor of 4.3 smaller than the

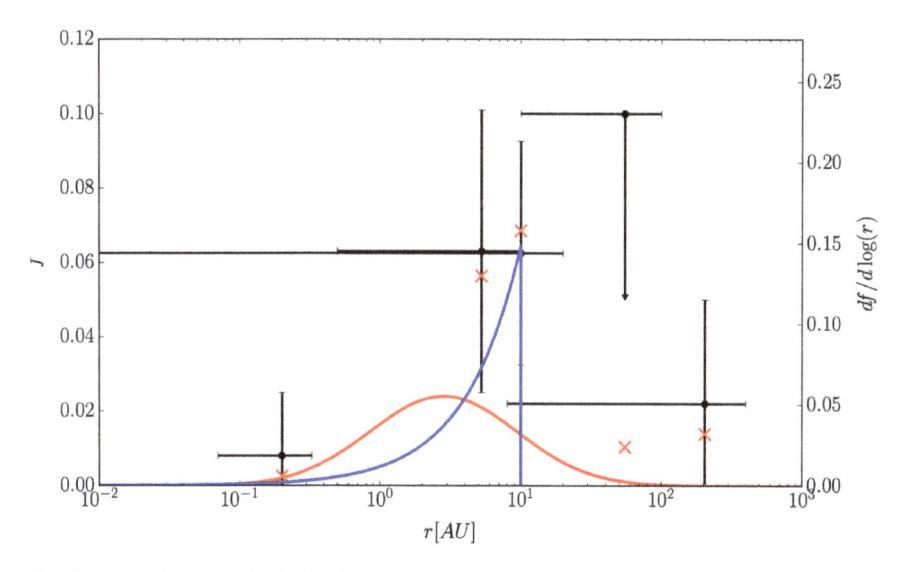

4 The Demographics of Wide-Separation Planets 275

Fig. 4.11 Semimajor axis distributions of roughly Jovian-mass companions to M dwarfs. The vertical axis shows point estimates of the semi-major axis distribution from several surveys (red "X"s and uncertainties and/or upper limits shown as black error bars), as well as parametric fits to these distributions from Clanton and Gaudi (2016) (blue curve) and Meyer et al. (2018) (red curve). From Meyer et al. (2018). Reproduced with permission ©ESO

inferred frequency of such planets around FGK stars of 0.11 ± 0.02 (Cumming et al. 2008). Thus, although low-mass stars do indeed host giant planets, they are less common than giant planets orbiting sun-like stars and tend to be at larger separations (compared to the snow line; Kennedy and Kenyon 2008).

Several authors have combined results from radial velocity, microlensing, and direct imaging surveys to constrain the population of giant planets with large semimajor axes orbiting low-mass stars. In particular, Clanton and Gaudi (2016) synthesized constraints on the population of long-period planets from five different exoplanet surveys using three independent detection methods: microlensing, radial velocity, and direct imaging. Adopting a power-law form for properties of long-period (>2 au) planets, they found

$$\frac{d^2 N}{d \log M_p \, d \log a} = 0.21^{+0.20}_{-0.15} \left(\frac{M_p}{M_{Sat}}\right)^{-0.86^{+0.21}_{-0.19}} \left(\frac{a}{2.5 \text{ au}}\right)^{1.1^{+1.9}_{-1.4}}, \quad (4.30)$$

with an outer cutoff of $a_{out} = 10^{+26}_{-4.7}$ au. This result was for "hot-start" models, but the results for "cold-start" models are very similar. This is because the typical host stars are quite old, and as such the luminosity at fixed mass of planets assuming "hot start" and "cold start" models have largely converged.

A similar analysis was performed by Meyer et al. (2018), although they fit a (likely) more well-motivated log-normal model for the semi-major axis distribution of giant planets orbiting M dwarfs. Their results are shown in Fig. 4.11. Their conclusions are broadly consistent with those of Clanton and Gaudi (2017): generally speaking, the frequency of giant planets orbiting M dwarfs increases with increasing semi-major axis up to a few au and then declines for semi-major axes beyond ∼10 au. An interesting but unanswered question is how the distribution of giant planets found by Clanton and Gaudi (2017) and Meyer et al. (2018) differs from that found by radial velocity surveys of giant planets for solar-type (FGK) stars, and whether or not they are consistent with the expectations of ab-initio planet formation theories.

I note that the analyses of both Clanton and Gaudi (2017) and Meyer et al. (2018) did not include the most recent and comprehensive microlensing constraints on the demographics of planets from Suzuki et al. (2016). Therefore, there is a clear opportunity for improving and updating the syntheses provided in Clanton and Gaudi (2017) and Meyer et al. (2018).

There have been several other studies that have attempted to synthesize the demographics of exoplanets determined by various methods.

- Howard et al. (2012) compared early demographic constraints from *Kepler* for planets with periods of $\lesssim$50 days with the constraints from the Keck/HIRES Eta-Earth RV survey for planets with periods in the same range as Howard et al. (2010). By adopting a deterministic density-radius relation and restricting the analysis to planets with masses $\gtrsim$3 $M_\oplus$ and radii $\gtrsim$2 $R_\oplus$ (where the Eta-Earth and [then available] *Kepler* results were mostly complete), they were able to map the radius distribution inferred from *Kepler* to the $M_p \sin i$ distribution from the Eta-Earth survey. They found good agreement, particularly when they assumed that the density of planets increased with decreasing radii. See the discussion in Chap. 3, Sect. 3.1 and Sect. 3.2.

- Both Gould et al. (2010) and Suzuki et al. (2016) compared constraints on the frequency of planets inferred from microlensing surveys to several estimates of the frequency of shorter-period planets found by several RV surveys of both solar-type FGK stars as well as M stars (see, e.g., Fig. 4.9).

- As discussed in more detail in Sect. 4.5.3, Bowler (2016) compared the frequency distribution giant planets as a function host star spectral type found by RV surveys (Johnson et al. 2010b) with the distribution found by direct imaging surveys (see Fig. 4.7).

- Using the DR25 *Kepler* catalog of planets between ∼1–6 $R_\oplus$ and $P < 100$ days, and converting planet radii to planet masses using the Chen and Kipping (2017) planet mass-radius relation, Pascucci et al. (2018) found a break in the mass ratio function of planets at $q \sim 0.3 \times 10^{-4}$, independent of host star mass. This break is at a mass ratio that is ∼3–10 times lower than the break in the mass-ratio function for longer-period planets found by microlensing as estimated by Suzuki et al. (2016) and Udalski et al. (2018) (Fig. 4.9), but is similar (a factor of ∼2 smaller) than that inferred by Jung et al. (2019). Assuming the latter result holds, this implies that Neptune mass-ratio planets are the most common planets both interior and exterior to the snow line, at least down to the mass ratios that have been probed by transits and microlensing of $\gtrsim 0.5 \times 10^{-6}$, or slightly larger than the Earth/sun mass ratio.

- Using results from both RV surveys and *Kepler*, Zhu and Wu (2018) studied the relationship between the population of small-separation ($a \lesssim 1$ au) super-Earths (M_p roughly between that of the Earth and Neptune) and 'cold' Jupiters ($a > 1$ au and $M_p > 0.3\, M_{\rm Jup}$) orbiting sun-like stars. They found that the conditional probability of a system with a super-Earth hosting a cold Jupiter was $\sim 30\%$, three times higher than the frequency of cold Jupiters orbiting typical sun-like field stars (e.g., Cumming et al. 2008). Given the prevalence of super-Earths found from *Kepler*, this implies that nearly every star that hosts a cold Jupiter also hosts an inner super-Earth. Since our solar system has a cold Jupiter but does not host an inner super-Earth, the corollary to this result is that solar systems with architectures like ours are rare, $\sim 1\%$.

- Using a meta-analysis to combine the results of many demographics studies of transiting planets detected by *Kepler*, the NASA Exoplanet Exploration Program Analysis Group (ExoPAG) Study Analysis Group (SAG) 13 determined a consensus estimate of the occurrence rates for planets with relatively short periods of $P = 10$–640 days, and radii from roughly that of the Earth to that of Jupiter. This estimate, and the process by which it was determined, is described in detail in Kopparapu et al. (2018). The double power-law fit to the SAG 13 occurrence rate was extrapolated to longer periods by Dulz et al. (2020), who also synthesized these results with the frequency of relatively long-period gas giants planets as determined by Cumming et al. (2008), Bryan et al. (2016), and Fernandes et al. (2019). Taking care to eliminate systems that were dynamically unstable, Dulz et al. (2020) provide a comprehensive synthesis of the demographics of planets with masses of 0.1–$10^3\, M_\oplus$ and semimajor axes of 0.1–30 au, albeit with significant reliance on extrapolation.

These and other results begin the process of "stitching together" the demographics constraint of multiple surveys using multiple detection methods. However, the studies itemized above, as well as the majority of other similar studies (additional pointers to the relevant literature on the subject are provided in See Chap. 3 Sect. 3.2), have primarily focused on comparing the demographics of relatively short-period planets detected by different surveys, or by comparing the demographics of close and wide-orbit companions. Thus, while very important, a comprehensive review of such studies is beyond the scope of this chapter.

4.6 Comparisons Between Theory and Observations

Remarkably, despite the large number of exoplanets that have been confirmed to date (~ 4300), there have been relatively few studies that rigorously compare the predictions of ab-initio exoplanet population synthesis models to empirical exoplanet demographic data. The most important reason for this is that the majority of

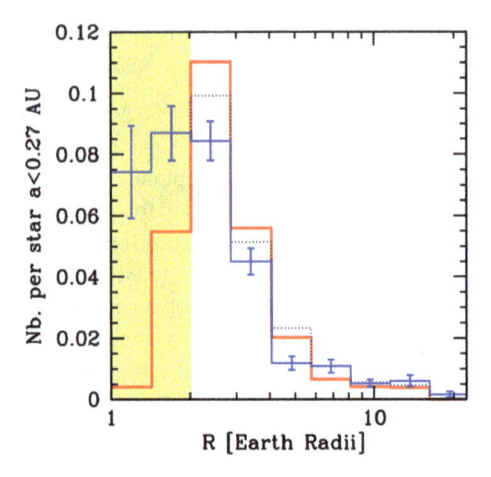

Fig. 4.12 Number of planets per star in a given radius bin for semi-major axes of $a < 0.27$ au. The thick red histogram shows the predictions from the ab-initio population synthesis models of Mordasini et al. (2012). The blue histogram with error bars shows the empirical distribution inferred by Howard et al. (2010) from an early release of *Kepler* data. The black dotted histogram is a preliminary analysis from the initial *Kepler* data. The *Kepler* data from the yellow shaded region ($R_p < 2R_\oplus$) was substantially incomplete at the time of this study. From Mordasini et al. (2012). Reproduced with permission ©ESO

the observational results on exoplanet demographics do not supply the data needed to compare these theories with the empirical predictions. Such data include (but are not limited to) the details of the target samples, the detection efficiencies (or completeness) of all of the target sample (not just those that have detected planets), and the quantification of the false positive rate (or reliability).

One example of an attempt to compare the prediction of ab-initio models of planet formation with empirical constraints is provided by Mordasini et al. (2012). They compare an estimate of the radius distribution of planets from some of the first results from *Kepler* (Howard et al. 2012) with the predictions from their population synthesis models for planets with semi-major axis of $a \lesssim 0.3$ au. They find reasonable agreement between the predictions and the empirical results (see Fig. 4.12). I note that this result applies to relatively short-period planets, and is thus formally out of the scope of this review. Nevertheless, it does provide an important example of the quantitative comparison of empirical results of the demographics of exoplanets with ab-initio theories.

A direct comparison between the ab-initio predictions of multiple independent planet formation theories (see Ida et al. 2013 and Mordasini et al. 2015 and references therein) and the demographics of cold exoplanets as constrained by microlensing was explored in Suzuki et al. (2016). As shown in Fig. 4.13, they find that the generic prediction of the core accretion (or nucleated instability) model of giant formation predicts a paucity of planets with masses roughly between that of Neptune and Jupiter (e.g., Pollack et al. 1996). This prediction is not confirmed by the results from microlensing surveys. This result appears to be fairly robust against some of

4 The Demographics of Wide-Separation Planets

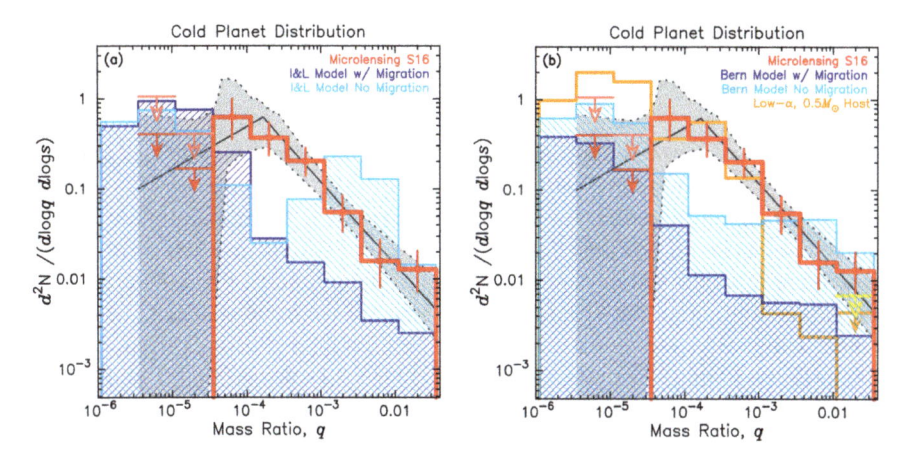

Fig. 4.13 The planet/host star mass ratio distribution as measured by microlensing surveys for exoplanets (Gould 2000; Cassan et al. 2012; Suzuki et al. 2016), compared to the predicted mass ratio distribution function from ab-initio models of planet formation (see Ida et al. 2013 and Mordasini et al. 2015 and references therein). The red histogram shows the measured mass-ratio distribution from microlensing, along with the best-fit broken power-law model and 1σ uncertainty indicated by the solid black line and gray shaded regions. The dark and light blue histograms show the predicted mass-ratio functions from the population synthesis models with migration, and the alternative migration-free models. *Left-hand panel*: comparison to models from Ida and Lin (e.g., (Ida et al. 2013) and references therein). *Right-hand panel*: comparison to models from the Bern group (e.g., Mordasini et al. 2015 and references therein). The gold histogram shows the results for a lower-viscosity disk model and 0.5 $M_\odot$ host stars. From Suzuki et al. (2018) © AAS. Reproduced with permission

the model assumptions, including the treatment of migration and the viscosity of the protostellar disk. Some possible resolutions to this discrepancy are discussed in Suzuki et al. (2018).

One potential complication is that the condition for core-nucleated instability (e.g., runaway gas growth to become a giant planet) may be more sensitive to planet mass than planet mass ratio.[13] If this is the case, then the mass gap predicted by the core-accretion theory may be smoothed out in the microlensing mass-ratio distribution, given the relatively broad range of host masses probed in microlensing surveys (see Fig. 4.8). This can be tested by measuring the masses of the host stars of planets detected by microlensing (e.g., Bennett et al. 2007). Indeed, Bhattacharya et al. (2018) measure the host and planet star masses of the microlensing planet OGLE-2012-BLG-0950Lb, finding a host star mass of $M_* = 0.58 \pm 0.04$ $M_\odot$ and a planet mass of $M_p = 39 \pm 8$ $M_\oplus$, placing the planet in the middle of the mass gap predicted by generic models of giant planet formation via core accretion.

---

[13] Formally, the condition for runaway growth is that the mass of the gaseous envelop becomes larger than that of the core, leading to a Jeans-like instability and rapid gas accretion (e.g., Mizuno 1980; Stevenson 1982; Pollack et al. 1996). However, the simplest models of protoplanetary disks generally predict core masses of ~ 10 $M_\oplus$, and thus the total critical mass for runaway accretion of ~ 20 $M_\oplus$, e.g., somewhat larger than the masses of the ice giants.

4.7 Future Prospects for Completing the Census of Exoplanets

While substantial progress has been made in determining the demographics of wide-separation planets, it is nevertheless the case that it is this regime of $M_p - P$ and $R_p - P$ parameter space that remains the most incomplete, particularly for planets with masses and radii less than that of Neptune (see Figs. 4.2 and 4.3). Fortunately, there are several planned or candidate surveys on the horizon that will largely fill in this region of parameter space, enabling a nearly complete statistical census of planets with masses/radii greater than that of the Earth, and periods from < 1 day to essentially infinity, including free-floating planets. In this section, I will briefly summarize these future prospects.

- **Radial Velocity surveys**. The longest-running RV surveys have been monitoring a sample of bright FGK stars for roughly 30 years, with precisions of a few to $\sim$10 m/s. These surveys are now sensitive to Jupiter analogs with $P \sim 12$ years and $M_p \gtrsim M_{\rm Jup}$ (e.g., Wittenmyer et al. 2016). For stars with the longest baselines and RV precisions of a few m/s, Saturn analogs ($P \sim 30$ years and $M_p \gtrsim M_{\rm Sat}$) are barely detectable, with $K \sim 3$ m/s. RV surveys are unlikely to be sensitive to analogs of the ice giants in our solar system (Kane 2011) in the foreseeable future, and are similarly unlikely to be sensitive to Neptune-mass planets on orbits beyond $\sim$3 au, which would have RV semi-amplitudes of $\sim$1 m/s and periods of $P \sim 5$ years. Beyond the difficulties with maintaining such RV precisions over such long time spans, it is likely to be the case that the majority of the (already oversubscribed) precision radial velocity resources will be focused on following-up planets detected in current and future transit surveys such as the Transiting Exoplanet Survey Satellite (TESS, Ricker et al. 2015) and the PLAnetary Transits and Oscillation of stars mission (PLATO, Rauer et al. 2014), as well as search for Earth analogs (e.g., Burt et al. 2020). Therefore, I reluctantly conclude that radial velocity surveys are unlikely to contribute to significantly expanding our knowledge of the population of wide-separation planets.
- **The Transiting Exoplanet Survey Satellite**. TESS is a NASA Medium-Class Explorers (MIDEX) mission (Ricker et al. 2015), whose goal is to survey nearly the entire sky ($\sim$83%) to find systems with of transiting planets orbiting bright host stars, which are the most amenable to detailed characterization of the planet and host star. In particular, many of the planets detected by TESS will be ideal targets for atmospheric characterization with the James Webb Space Telescope (JWST). See, e.g., Beichman et al. (2014). Because TESS is designed to survey the brightest stars in the sky for transiting planets, its dwell time for the majority ($\sim$74%) of its survey area is only 27 days for the prime mission (see Fig. 4.14). For stars in the two continuous viewing zones at the ecliptic poles, TESS will obtain nearly continuous observations for $\sim$ 350 days. Thus, even considering extended missions, TESS's region of sensitivity in the $M_p - P$ planet will be entirely within that of *Kepler*. Simply put, although TESS will (of course) provide important demographic constraints, and will provide them with higher fidelity than

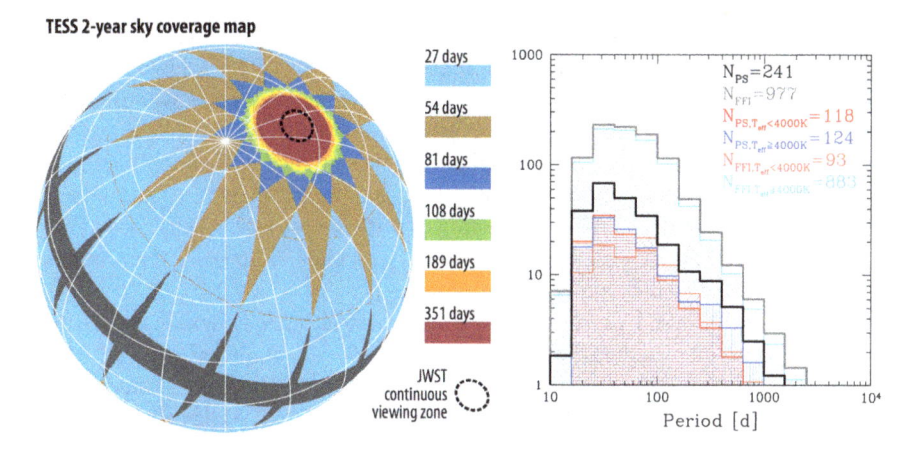

Fig. 4.14 *Left-hand panel*: the original planned Transiting Exoplanet Survey Satellite (TESS) 2-year mission sky coverage, as a function of equatorial coordinates (solid white lines). TESS originally planned to monitor ∼83% of the sky, excluding regions within ∼6° of the ecliptic equator. ∼74% of its survey area was planned to be monitored for 27 days during the prime mission, whereas a region around the ecliptic poles (including the JWST continuous viewing zone) would be monitored for up to ∼350 days. Stray light issues forced the TESS team to deviate somewhat from this survey strategy in the northern ecliptic hemisphere. Courtesy of G. Ricker, reproduced by permission. *Right-hand panel*: an estimate of the number and period distribution of single-transit events expected from the TESS primary mission. Single transit events from the 2-minute cadence postage stamp targets are shown in the black histogram, whereas those from stars in the 30-minute cadence full-frame images (FFIs) are shown in grey. Single transit events found for stars with $T_{\rm eff} > 4000$ K are shown in blue, whereas those for stars with stars $T_{\rm eff} < 4000$ K are shown in red. The darker shades are for the postage-stamp targets, while the lighter shades are for the FFI stars. A total of 241 single-transit events due to planets are predicted in the postage stamp data and at least 977 single-transit events are expected in the FFIs. Reproduced from Villanueva et al. (2019). ©AAS. Reproduced with permission

Kepler (primarily because of the brightness of the target stars), in general TESS will not significantly expand our knowledge of exoplanet demographics beyond what has been learned by *Kepler*. Therefore, it will generally not contribute to the demographics of wide-orbit planets.

However, one area where TESS may contribute to the demographics of wide-orbit planets is via single transit events. As discussed in Sect. 4.5.2, single transit events can be used to constrain the demographics of planets with periods beyond the survey baseline. Because TESS will be looking at a much larger number of stars for a shorter period of time than the primary *Kepler* survey, its yield of single-transit events is expected to be significantly higher than that of *Kepler*. The yield of single transit events in the 2-year TESS primary mission has been predicted by Villanueva et al. (2019). Their results are shown in Fig. 4.14. They find that over 1000 single-transit events due to planets are expected from the TESS primary mission, with 241 of these coming from the 2 min cadence targeted postage-stamp data and a lower limit of 977 coming from stars in the full-frame images (FFIs).

- **Gaia**. The primary goal of the European Space Agency's (ESA) Gaia mission is to provide exquisite astrometric measurements of $\sim 2 \times 10^9$ stars down to a magnitude of $V \sim 20.7$ (Gaia Collaboration 2016). These measurements will produce a sample of accurate and precise stellar proper motions and distances that is orders of magnitude larger than is currently available. A 'by-product' of this unprecedented database will be the detection of many thousands of giant planets at intermediate periods via their astrometric perturbations on their host stars. There have been many studies of the expected yield of planets with Gaia; here I will focus on two papers (but see also Sozzetti et al. 2014).

  Using double-blind experiments, Casertano et al. (2008) demonstrated that planets can be detected if they induce astrometric signatures of roughly $3\sigma_{ast}$, where σ_{ast} is the single-measurement precision, provided that the period of the planet is less than the survey duration (5 years for the primary Gaia mission). They also noted that the threshold for reliable inference about the orbital parameters was significantly higher. They determined that at twice the detection limit of $3\sigma_{ast}$, the uncertainties in orbital parameters and planet masses were typical of order 15–20%. They also demonstrated that for planets with periods longer than the survey duration, the planet detectability dropped off more slowly with increasing planet period than the precision with which the parameters of the system, and in particular the planet mass, can be measured (see Fig. 4.15 and the discussion in Sect. 4.4.7). Thus, although the astrometric signal of planets with periods longer than the survey may be detectable, these detections will be significantly compromised as they will have large uncertainties in the orbital elements and, in particular, the planet mass.

  A more recent estimate of the yield of planets from Gaia was performed by Perryman (2018). In particular, these authors used updated planet occurrence rate estimates, as well as updated estimates of the expected Gaia single-measurement astrometric precision. They consider the detectability of known planetary companions with Gaia (see Fig. 4.15), as well as projections for as-yet undetected planets. They find that Gaia should discover $\sim 10^4$ planets with masses in the range of ~ 1–15 M_{Jup}, the majority of which will have semi-major axes in the range of ~ 2–5 au. They also estimate that Gaia should find ~ 25–50 intermediate-period $P \sim 2$–3 year transiting systems.

  Thus, although Gaia will likely deliver a very large number of planets, including a large number of planetary systems, it will not be sensitive to a currently unexplored region of $M_p - a$ parameter space. Nevertheless, Gaia will be sensitive to the mutual inclinations of planets in multi-planet systems, an important property of planetary systems that has been poorly explored to date. Furthermore, because Gaia should detect a very large number of planets, it will be sensitive to the "tails" of the distribution of planetary properties, e.g., the "oddball" exoplanet systems. Such systems often provide unique insights into the physics of planet formation and evolution. Finally, the fact that it will uncover temperate transiting giant planets will enable the estimate of the mass-radius relationship for such planets, which will provide important priors on the properties of the giant planets detected by future direct imaging surveys.

- **The Nancy Grace Roman Space Telescope**. The Nancy Grace Roman Space Telescope, or *Roman* (née WFIRST), was the highest-priority large space mission from the National Academies of Science 2010 Astronomy Decadal Survey (National Research Council 2010). As outlined by the Astro2010 Decadal Survey, one of *Roman*'s primary goals is to "open up a new frontier of exoplanet studies by monitoring a large sample of stars in the central bulge of the Milky Way for changes in brightness due to microlensing by intervening solar systems. This census, combined with that made by the *Kepler* mission, will determine how common Earth-like planets are over a wide range of orbital parameters." This application of *Roman*, based originally on the concept and simulations by Bennett and Rhie (2002), promises to probe a broad region of planet mass/semi-major axis parameter space that is inaccessible by any other exoplanet detection methods or surveys, including ground-based microlensing surveys. Initial estimates of the yield of such a space-based microlensing survey can be found in Bennett and Rhie (2002), Green et al. (2011, 2012) and Spergel et al. (2013, 2015). The most up-to-date estimate of the yield of bound planets by microlensing is provided by Penny et al. (2019). In summary, *Roman* is expected to detect roughly ~1500 bound and free-floating planets with masses $\gtrsim M_\oplus$ with separations >1 au. At the peak of its sensitivity, *Roman* will be sensitive to bound planets with masses as low as ~ 0.02 $M_\oplus$, or roughly twice the mass of the moon and roughly the mass of Ganymede. *Roman* will also be sensitive to free-floating planets, as explored by Johnson et al. (2020). They predict that *Roman* could detect ~250 free-floating planets with masses down to that of Mars, including ~60 with masses less than that of the Earth.

  By combining the demographic constraints from *Kepler* with those from *Roman*, it will be possible to obtain a nearly complete statistical census of exoplanets with mass $\gtrsim M_\oplus$ with arbitrary separations. See Fig. 4.16. It is also worth noting that *Roman* will also have some sensitivity to potentially habitable planets, and thus may provide constraints on $\eta_\oplus$. Finally, it will also be sensitive to massive satellites to wide-separation bound planets (Bennett and Rhie 2002), thus complementing the sensitivity of transit surveys to massive satellites orbiting shorter-period planets (e.g., Kipping 2009a, b; Kipping et al. 2009).

- **The PLAnetary Transits and Oscillations of stars mission**. PLATO (Rauer et al. 2014) is an ESA M-class mission that is designed to find transiting planets with periods of $\lesssim 2$ years orbiting relatively bright stars. The PLATO mission can be thought of as an intermediate mission between *Kepler* and TESS. It will survey stars with a longer baseline than TESS, and will survey brighter stars than those surveyed by *Kepler*. Because the hosts of the transiting planets discovered by PLATO will be brighter than *Kepler*, they will be more amenable to follow-up. Given the current mission architecture and survey design, while PLATO is predicted to expand upon *Kepler*'s sensitivity range in the $R_p - P$ plane, it will nevertheless not contribute significantly to our understanding of the demographics of wide-orbit planets, excepting for the possibility of following-up single transit events (e.g., Yee and Gaudi 2008; Herman et al. 2019; Villanueva et al. 2019).

- **Future Direct Imaging Surveys**. A comprehensive review of the potential of future ground and space-based direct imaging surveys to contribute to our knowl-

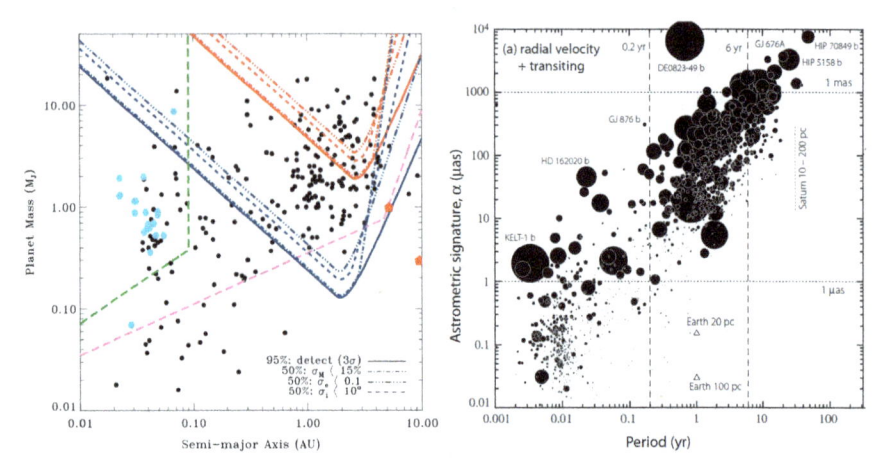

Fig. 4.15 *Left-hand panel*: Gaia sensitivity as a function of planet mass and semi-major axis. The red curves show the estimated Gaia sensitivity for a $\sim M_\odot$ primary at 200 pc. The meaning of the different line styles is indicated in the figure. The blue curves are the same, except for a $\sim 0.5\,M_\odot$ primary at 25 pc. The pink line shows the sensitivity of an RV survey assuming a precision of 3 m/s, a $M_\odot$ primary, and a 10-year survey. The green curve shows the sensitivity of a transit survey assuming $\gtrsim 1000$ data points (approximately uniformly sampled in phase), each with a photometric precision of ~ 5 mmag, and a $M_\odot$ and $R_\odot$ primary. Black dots indicate the inventory of exoplanets as of September 2007. Transiting systems are shown as light-blue filled pentagons. Jupiter and Saturn are also shown as red pentagons. From Casertano et al. (2008). Reproduced with permission ©ESO. *Right-hand panel*: astrometric signature versus period for the planets listed in the exoplanet.eu archive as of September 1, 2014. The sizes of the circles are proportional to the planet mass. The vertical lines roughly bracket the range in periods in which Gaia will be most sensitive. The horizontal line delineates an astrometric signature of 1 μas. From Perryman (2018). Courtesy of Michael Perryman. Reproduced with permission

edge of the demographics of exoplanets is well beyond the scope of this chapter. Rather, I refer the reader to the review in this chapter of National Academies of Science Exoplanet Science Strategy Report (National Research Council 2018), and in particular Fig. 4.3 of that report. I will, however, make a few general comments. First, current ground-based direct imaging surveys (e.g., Macintosh et al. 2014; Beuzit et al. 2019) are expected to continue to improve upon their current sensitivity in terms of inner working angle and contrast, but are unlikely to do so by an order of magnitude. Second, direct-imaging surveys using JWST will provide modest improvements over ground-based surveys (Beichman and Greene 2018), but are nevertheless not expected to reach the contrast ratios needed to detect mature planets in reflected light or thermal emission. Third, the next generation of ground-based Giant Segmented Mirror Telescopes (GSMTs) offer opportunities for dramatic improvements in the capabilities of direct-imaging surveys relative to current ground-based facilities. These opportunities include the potential to directly detect mature "warm Jupiters" in reflected light, as well as to detect temperate planets in thermal emission (e.g., Wang et al. 2019). However, these improvements in capabilities also require dramatic breakthroughs in several key

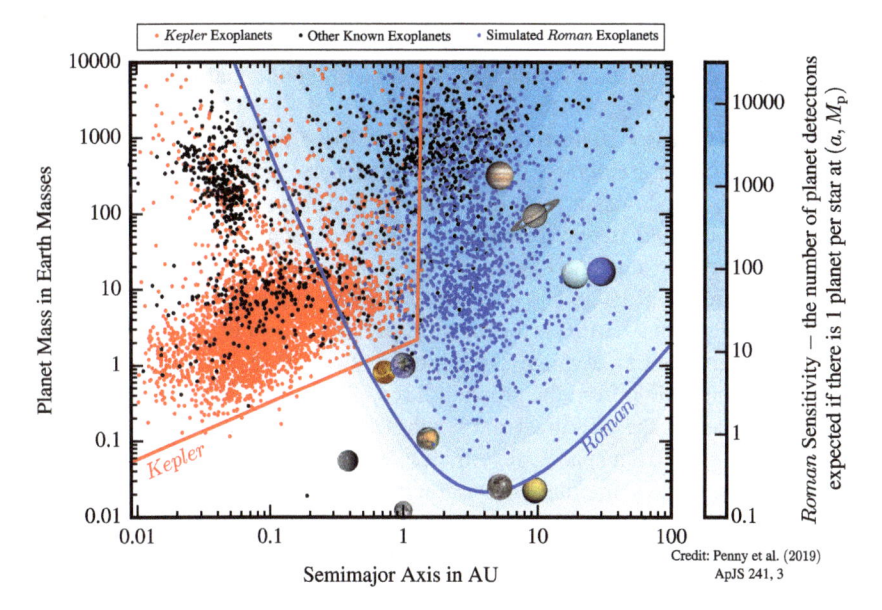

4 The Demographics of Wide-Separation Planets 285

Fig. 4.16 (Left) The region of sensitivity in the $M_p - a$ plane of the *Kepler* prime mission (red solid line) versus the predicted region of sensitivity of the *Roman* (née WFIRST) Galactic Exoplanet Survey (RGES survey, blue solid line). The red dots show *Kepler* candidate and confirmed planets, while the black dots show all other known planets extracted from the NASA exoplanet archive as of 2/28/2018. The blue dots show a simulated realization of the planets detected by the RGES survey. Solar system bodies are shown by their images, including the satellites Ganymede, Titan, and the Moon at the semi-major axis of their hosts. Images of the solar system planets courtesy of NASA. Adapted from Penny et al. (2019). Courtesy of M. Penny. ©AAS. Reproduced with permission

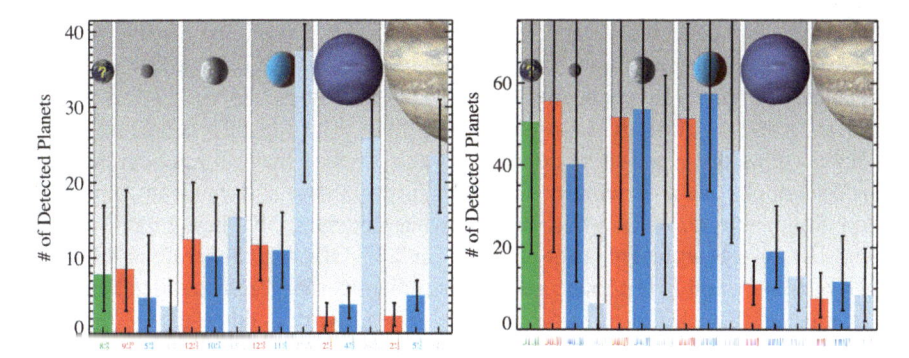

Fig. 4.17 Predicted exoplanet detection yields with uncertainties for the HabEx and LUVOIR mission concepts. Planet class types from left to right are: exoEarth candidates (green bar), rocky planets, super-Earths, sub-Neptunes, Neptunes, and Jupiters. Red, blue, and ice blue bars indicate hot, warm, and cold planets, respectively. The predicted yields are indicated under each bar. (Left) Predictions for HabEx. (Right) Predictions for LUVOIR A. Both figures courtesy of Christopher Stark. Reproduced with permission

technology areas. Finally, I will note that there exists an enormous opportunity for a future space-based direct imaging mission in the thermal mid-infrared (e.g. Quanz et al. 2019).

- **Future Space-Based Reflected Light Direct Imaging Surveys.** Direct imaging surveys for mature planets orbiting sun-like stars in reflected light generally require space-based missions. This is because, for planets with radii and separations similar to those in our solar system, the planet/star contrast ratio is $\lesssim 10^{-8}$ and the planet/star angular separation for a system at ~ 10 pc is $\lesssim 0.5''$. For a Jupiter/sun analog at 10 pc, the contrast is $\sim 10^{-9}$ with an angular separation of $\sim 0.5''$, whereas for a Earth/sun analog at 10 pc, the contrast is $\sim 10^{-10}$ at an angular separation of $\sim 0.1''$. Achieving these contrasts at these angular separations is essentially impossible from the ground (Guyon 2005).

Indeed, achieving these contrast ratios at these angular separation is exceptionally challenging, even from space. Nevertheless, there are two promising techniques for doing so, namely internal coronographs (see, e.g., Guyon et al. 2006, and references therein) and external occulters, or starshades (Spitzer 1962; Cash 2006). Both techniques have been extensively studied; a comprehensive review of these methods is well beyond the scope of this chapter. Suffice it to say that, after decades of laboratory demonstrations, there does not appear to be any insurmountable obstacles to achieving the required contrasts at the required angular separations using either technique.

In preparation for the National Academies of Science 2020 Astronomy Decadal Survey (Astro2020), NASA constituted studies of four large space mission concepts, currently named the Habitable Exoplanet Observatory (HabEx, Gaudi et al. 2020), the Large UV-Optical-InfraRed Telescope (LUVOIR, The LUVOIR Team 2019), Lynx (The Lynx Team 2018), and Origins (Meixner et al. 2019). Two of these mission concepts, namely HabEx and LUVOIR, would be capable of providing important constraints on the demographics of both close and wide-orbit planets orbiting relatively nearby ($\sim$10–20 pc) FGK stars.

HabEx and LUVOIR differ primarily in ambition. The fiducial architecture of the HabEx telescope is a 4-m monolithic, off-axis primary that utilizes both internal coronagraphy and an external starshade to directly image and characterize planets orbiting nearby stars. LUVOIR studied two architectures, but here I focus on the most ambitious architecture for context. The LUVOIR A architecture baselines a 15-m diameter on-axis primary mirror that utilizes coronagraphy to directly image exoplanets. Because of its larger aperture, LUVOIR A would be able to detect a much larger sample of planets than the baseline architecture of HabEx. On the other hand, because HabEx utilizes a (primarily) achromatic starshade, it would be able to better characterize the (smaller number) of exoplanets it would detect. The estimated yield of both mission concepts is shown in Fig. 4.17.

Regardless of the specific architecture, both HabEx and LUVOIR would be able to determine the demographics of planetary systems orbiting nearby stars over a wide range of planet orbits and sizes *for individual systems*. This in contrast to the statistical compendium of different systems that will be enabled by combining the demographic results from, e.g., *Kepler* and *Roman*. In particular, both HabEx and

LUVOIR would be able to address such fundamental questions as: "What is the conditional probability that, given the detection of a potentially habitable planet, there exists an outer gas giant planet?" The answers to such questions are likely to be essential for understanding the context of habitability.

4.8 Conclusion

The demographics of exoplanets, i.e., the distribution of exoplanets as a function of the physical parameters, may encode the physical processes of planet formation and evolution, and thus provides the empirical ground truth that all ab-initio planet formation theories must reproduce. There exist numerous challenges to determining the demographics of exoplanets over as broad a region of parameter space as possible. In particular, the exoplanet detection methods at our disposal are sensitive to planets in different regions of (planet and host star) parameter space, as well as being sensitive to planets orbiting different host star parameters. While this presents a challenge, in that we must construct robust statistical methodologies to combine the results of different surveys and different detection methods, it also provides an opportunity to more completely survey the demographics of exoplanets over the relevant regions of the planet and host star parameter space. There exist many exciting future opportunities to expand our understanding of the demographics of exoplanets.

The future of exoplanet demographics is bright. I expect that, within the next few decades, we will have a nearly complete statistical census of exoplanets with masses/radii greater than roughly than that of the earth, with essentially arbitrary separations (including free-floating planets). As well as providing the empirical ground truth for theories of the formation and evolution of exoplanetary systems, this census will provide the essential context for our detailed characterization of exoplanet properties, and ultimately of our understanding of the conditions of exoplanet habitability.

Acknowledgements I would like to thank the organizers of the 3rd Advanced School on Exoplanetary Science: Demographics of Exoplanetary Systems for inviting me to attend a very enjoyable school held at an extraordinarily beautiful venue, and for giving me the opportunity to present on the topic of "Wide-Separation Exoplanets". I am very much indebted to the editors of the proceedings (L. Mancini, K. Biazzo, V. Bozza, and A. Sozzetti) for their exceptional patience as I wrote this chapter. I would like to thank the many people who have shaped my thinking about exoplanet demographics over the past 20+ years, including (but not limited to) Thomas Beatty, Chas Beichman, David Bennett, Gary Blackwood, Brendan Bowler, Chris Burke, Jennifer Burt, Dave Charbonneau, Jesse Christiansen, Christian Clanton, Andrew Cumming, Martin Dominik, Subo Dong, Courtney Dressing, Debra Fischer, Eric Ford, Andrew Gould, Calen Henderson, Andrew Howard, Marshall Johnson, Bruce Macintosh, Eric Mamajek, Michael Meyer, Matthew Penny, Michael Perryman, Peter Plavchan, Radek Poleski, Aki Roberge, Penny Sackett, Sara Seager, Yossi Shvartzvald, Karl Stapelfeldt, Christopher Stark, Keivan Stassun, Takahiro Sumi, Andrzej Udalski, Steven Villanueva, Jr., Ji Wang, Josh Winn, Jennifer Yee, Andrew Youdin, and Wei Zhu. Apologies to those I forgot to include in this list and those I forgot to cite in this review. Finally, I recognize the support from the Thomas Jefferson Chair for Space Exploration endowment from the Ohio State University, and the Jet Propulsion Laboratory. This research has made use of the NASA Exoplanet Archive, which is operated by the California Institute of Technology, under contract with the National Aeronautics and Space Administration under the Exoplanet Exploration Program.

References

Agol, E.: Mon. Not. R. Astron. Soc. **374**, 1271 (2007)
Albrow, M., Beaulieu, J.P., Birch, P., et al.: Astrophys. J. **509**, 687 (1998)
Beaulieu, J.P., Bennett, D.P., Fouqué, P., et al.: 439, 437 (2006)
Beichman, C.A., Greene, T.P.: (2018). arXiv:1803.03730
Beichman, C., Benneke, B., Knutson, H., et al.: Proc. Astron. Soc. Pac. **126**, 1134 (2014)
Bennett, D.P., Rhie, S.H.: Astrophys. J. **472**, 660 (1996)
Bennett, D.P., Rhie, S.H.: Astrophys. J. **574**, 985 (2002)
Bennett, D.P., Anderson, J., Gaudi, B.S.: Astrophys. J. **660**, 781 (2007)
Beuzit, J.L., Vigan, A., Mouillet, D., et al.: Astron. Astrophys. **631**, A155 (2019)
Bhattacharya, A., Beaulieu, J.P., Bennett, D.P., et al.: Astron. J. **156**, 289 (2018)
Bond, I.A., Udalski, A., Jaroszyński, M., et al.: Astrophys. J. **606**, L155 (2004)
Bonfils, X., Delfosse, X., Udry, S., et al.: Astron. Astrophys. **549**, A109 (2013)
Borucki, W.J., Koch, D., Basri, G., et al.: Science **327**, 977 (2010)
Boss, A.P.: Science **276**, 1836 (1997)
Bowler, B.P.: Proc. Astron. Soc. Pac. **128**, 102001 (2016)
Bowler, B.P., Blunt, S.C., Nielsen, E.L.: Astron. J. **159**, 63 (2020)
Bozza, V., Mancini, L., Sozzetti, A.: Astrophysics and Space Science Library, Volume 428 (Springer International Publishing Switzerland) (2016)
Brakensiek, J., Ragozzine, D.: Astrophys. J. **821**, 47 (2016)
Bryan, M.L., Knutson, H.A., Howard, A.W., et al.: Astrophys. J. **821**, 89 (2016)
Burke, C.J., Gaudi, B.S., DePoy, D.L., Pogge, R.W.: Astron. J. **132**, 210 (2006)
Burke, C.J., Christiansen, J.L., Mullally, F., et al.: Astrophys. J. **809**, 8 (2015)
Burt, J., Gaudi, B., Callas, J., et al., 2020, American Astronomical Society Meeting Abstracts, 236, 107.05
Campbell, B., Walker, G.A.H., Yang, S.: Astrophys. J. **331**, 902 (1988)
Carter, J.A., Yee, J.C., Eastman, J., et al.: Astrophys. J. **689**, 499 (2008)
Casertano, S., Lattanzi, M.G., Sozzetti, A., et al.: Astron. Astrophys. **482**, 699 (2008)
Cash, W.: Nature **442**, 51 (2006)
Cassan, A., Kubas, D., Beaulieu, J.P., et al.: Nature **481**, 167 (2012)
Chen, J., Kipping, D.: Astrophys. J. **834**, 17 (2017)
Chiang, E.I., Tabachnik, S., Tremaine, S.: Astron. J. **122**, 1607 (2001)
Clanton, C., Gaudi, B.S.: Astrophys. J. **791**, 90 (2014)
Clanton, C., Gaudi, B.S.: Astrophys. J. **791**, 91 (2014)
Clanton, C., Gaudi, B.S.: Astrophys. J. **819**, 125 (2016)
Clanton, C., Gaudi, B.S.: Astrophys. J. **834**, 46 (2017)
Collaboration, Gaia: Astron. Astrophys. **595**, A1 (2016)
Collaboration, Gaia: Astron. Astrophys. **616**, A1 (2018)
Crepp, J.R., Johnson, J.A., Howard, A.W., et al.: Astrophys. J. **761**, 39 (2012)
Cumming, A., Butler, R.P., Marcy, G.W., et al.: Proc. Astron. Soc. Pac. **120**, 531 (2008)
Dodson-Robinson, S.E., Veras, D., Ford, E.B., Beichman, C.A.: Astrophys. J. **707**, 79 (2009)
Dong, S., Zhu, Z.: Astrophys. J. **778**, 53 (2013)
Dong, S., Mérand, A., Delplancke-Ströbele, F., et al.: Astrophys. J. **871**, 70 (2019)
Dressing, C.D., Charbonneau, D.: Astrophys. J. **767**, 95 (2013)
Dulz, S.D., Plavchan, P., Crepp, J.R., et al.: Astrophys. J. **893**, 122 (2020)

Fernandes, R.B., Mulders, G.D., Pascucci, I., et al.: Astrophys. J. **874**, 81 (2019)
Fischer, D.A., Howard, A.W., Laughlin, et al.: In: Beuther, H., Klessen, R.S., Dullemond, C.P., Henning, T. (eds.) Protostars and Planets VI, p. 715 (2014)
Foreman-Mackey, D., Hogg, D.W., Morton, T.D.: Astrophys J **795**, 64 (2014)
Foreman-Mackey, D., Morton, T.D., Hogg, D.W., et al.: Astron. J. **152**, 206 (2016)
Gaidos, E., Mann, A.W.: Astrophys. J. **762**, 41 (2013)
Gaidos, E., Mann, A.W., Kraus, A.L., Ireland, M.: Mon. Not. R. Astron. Soc. **457**, 2877 (2016)
Gaudi, B. S., Seager, S., Mennesson, B., et al.: (2020). arXiv:2001.06683
Gaudi, B.S.: Ann. Rev. Astron. Astrophys. **50**, 411 (2012)
Gaudi, B.S., Albrow, M.D., An, J., et al.: Astrophys. J. **566**, 463 (2002)
Gaudi, B.S., Seager, S., Mallen-Ornelas, G.: Astrophys. J. **623**, 472 (2005)
Gould, A.: (2008). arXiv:0807.4323
Gould, A.: Microlensing planets. In: Bozza, V., Mancini, L., Sozzetti, A. (eds.), Methods of Detecting Exoplanets: 1st Advanced School on Exoplanetary Science, Astrophysics and Space Science Library, vol. 428, Springer International Publishing Switzerland, p. 135 (2016)
Gould, A.: Astrophys. J. **535**, 928 (2000)
Gould, A., Loeb, A.: Astrophys. J. **396**, 104 (1992)
Gould, A., Dorsher, S., Gaudi, B.S., Udalski, A.: Acta Astron. **56**, 1 (2006)
Gould, A., Udalski, A., An, D., et al.: Astrophys. J. **644**, L37 (2006)
Gould, A., Dong, S., Gaudi, B.S., et al.: Astrophys. J. **720**, 1073 (2010)
Green, J., Schechter, P., Baltay, C., et al. (2012). arXiv:1208.4012
Green, J., Schechter, P., Baltay, C., et al.: (2011). arXiv:1108.1374
Guyon, O.: Astrophys. J. **629**, 592 (2005)
Guyon, O., Pluzhnik, E.A., Kuchner, M.J., et al.: Astrophys. J. Suppl. Ser. **167**, 81 (2006)
Han, C., Gaudi, B.S., An, J.H., Gould, A.: Astrophys. J. **618**, 962 (2005)
Hartman, J.D., Gaudi, B.S., Holman, M.J., et al.: Astrophys. J. **695**, 336 (2009)
Henderson, C.B., Gaudi, B.S., Han, C., et al.: Astrophys. J. **794**, 52 (2014)
Herman, M.K., Zhu, W., Wu, Y.: Astron. J. **157**, 248 (2019)
Hinkley, S., Oppenheimer, B.R., Zimmerman, N., et al.: Proc. Astron. Soc. Pac. **123**, 74 (2011)
Ho, S., Turner, E.L.: Astrophys J **739**, 26 (2011)
Howard, A.W., Marcy, G.W., Johnson, J.A., et al.: Science **330**, 653 (2010)
Howard, A.W., Marcy, G.W., Bryson, S.T., et al.: Astrophys. J. Suppl. Ser. **201**, 15 (2012)
Hsu, D.C., Ford, E.B., Ragozzine, D., Morehead, R.C.: Astron. J. **155**, 205 (2018)
Hsu, D.C., Ford, E.B., Ragozzine, D., Ashby, K.: Astron. J. **158**, 109 (2019)
Huber, D., Silva Aguirre, V., Matthews, J.M., et al.: Astrophys. J. Suppl. Ser. **211**, 2 (2014)
Ida, S., Lin, D.N.C., Nagasawa, M.: Astrophys. J. **775**, 42 (2013)
Johnson, J.A., Howard, A.W., Marcy, G.W., et al.: Proc. Astron. Soc. Pac. **122**, 149 (2010)
Johnson, J.A., Aller, K.M., Howard, A.W., Crepp, J.R.: Proc. Astron. Soc. Pac. **122**, 905 (2010)
Johnson, S.A., Penny, M., Gaudi, B.S., et al.: Astron. J. **160**, 123 (2020)
Jung, Y.K., Gould, A., Zang, W., et al.: Astron. J. **157**, 72 (2019)
Kane, S.R.: Icarus **214**, 327 (2011)
Kennedy, G.M., Kenyon, S.J.: Astrophys. J. **673**, 502 (2008)
Kim, H.W., Hwang, K.H., Gould, A., et al.: (2020). arXiv:2007.06870
Kim, S.L., Lee, C.U., Park, B.G., et al.: JKAS **49**, 37 (2016)
Kipping, D.M.: Mon. Not. R. Astron. Soc. **392**, 181 (2009)
Kipping, D.M.: Mon. Not. R. Astron. Soc. **396**, 1797 (2009)
Kipping, D.M., Fossey, S.J., Campanella, G.: Mon. Not. R. Astron. Soc. **400**, 398 (2009)
Kopparapu, R.K., Hébrard, E., Belikov, R., et al.: Astrophys. J. **856**, 122 (2018)
Kovács, G., Zucker, S., Mazeh, T.: Astron. Astrophys. **391**, 369 (2002)
Kratter, K.M., Murray-Clay, R.A., Youdin, A.N.: Astrophys. J. **710**, 1375 (2010)
Kraus, A.L., Ireland, M.J.: Astrophys. J. **745**, 5 (2012)
Lagrange, A.M., Bonnefoy, M., Chauvin, G., et al.: Science **329**, 57 (2010)
Latham, D.W., Mazeh, T., Stefanik, R.P., et al.: Nature **339**, 38 (1989)

Laughlin, G., Bodenheimer, P., Adams, F.C.: Astrophys. J. Lett. **612**, L73 (2004)
Ma, S., Mao, S., Ida, S., et al.: Mon. Not. R. Astron. Soc. **461**, L107 (2016)
Macintosh, B., Graham, J.R., Ingraham, P., et al.: Proc. Natl. Acad. Sci. **111**, 12661 (2014)
Macintosh, B., Graham, J.R., Barman, T., et al.: Science **350**, 64 (2015)
Mann, A.W., Gaidos, E., Lépine, S., et al.: Astrophys. J. **753**, 90 (2012)
Mao, S., Paczynski, B.: Astrophys. J. Lett. **374**, L37 (2011)
Marcy, G.W., Butler, R.P.: Proc. Astron. Soc. Pac. **104**, 270 (1992)
Marois, C., Macintosh, B., Barman, T., et al.: Science **322**, 1348 (2008)
Marois, C., Zuckerman, B., Konopacky, Q.M., et al.: Nature **468**, 1080 (2010)
Mayor, M., Marmier, M., Lovis, C., et al.: (2011). arXiv:1109.2497
Mayor, M., Queloz, D.: Nature **378**, 355 (1995)
Mayor, M., Duquennoy, A., Halbwachs, J.L., Mermilliod, J.C.: Astron. Soc. Pac. Conf. Ser. **32**, 73 (1992)
Meixner, M., Cooray, A., Leisawitz, D., et al.: (2019). arXiv:1912.06213
Meyer, M.R., Amara, A., Reggiani, M., Quanz, S.P.: Astron. Astrophys. **612**, L3 (2018)
Mizuno, H.: Progress Theoret. Phys. **64**, 544 (1980)
Mochejska, B.J., Stanek, K.Z., Sasselov, D.D., et al.: Astron. J. **129**, 2856 (2005)
Montet, B.T., Crepp, J.R., Johnson, J.A., et al.: Astrophys. J. **781**, 28 (2014)
Morbidelli, A., Bitsch, B., Crida, A., et al.: Icarus **267**, 368 (2016)
Mordasini, C., Alibert, Y., Georgy, C., et al.: Astron. Astrophys. **547**, A112 (2012)
Mordasini, C., Mollière, P., Dittkrist, K.M., et al.: Int. J. Astrobiol. **14**, 201 (2015)
Mroz, P., Poleski, R., Gould, A., et al.: (2020a). arXiv:2009.12377
Mroz, P., Udalski, A., Skowron, J., et al.: Nature **548**, 183 (2017)
Mroz, P., Ryu, Y.H., Skowron, J., et al.: Astron. J. **155**, 121 (2018)
Mroz, P., Udalski, A., Skowron, J., et al.: Astrophys. J. Suppl. Ser. **244**, 29 (2019)
Mroz, P., Udalski, A., Bennett, D.P., et al.: Astron. Astrophys. **622**, A201 (2019)
Mroz, P., Poleski, R., Han, C., et al.: Astron. J. **159**, 262 (2020)
National Academies of Sciences, Engineering, and Medicine.: Exoplanet Science Strategy. Washington, DC: The National Academies Press (2018)
National Research Council: New Worlds. New Horizons in Astronomy and Astrophysics. The National Academies Press, Washington, DC (2010)
Nielsen, E.L., De Rosa, R.J., Macintosh, B., et al.: Astron. J. **158**, 13 (2019)
Pascucci, I., Mulders, G.D., Gould, A., Fernandes, R.: Astrophys. J. Lett. **856**, L28 (2018)
Penny, M.T., Henderson, C.B., Clanton, C.: Astrophys. J. **830**, 150 (2016)
Penny, M.T., Gaudi, B.S., Kerins, E., et al.: Astrophys. J. Suppl. Ser. **241**, 3 (2019)
Perryman, M.: The Exoplanet Handbook. Cambridge University Press (2018)
Perryman, M., Hartman, J., Bakos, G.Á., Lindegren, L.: Astrophys. J. **797**, 14 (2014)
Petigura, E.A., Howard, A.W., Marcy, G.W.: Proc. Natl. Acad. Sci. **110**, 19273 (2013)
Pinsonneault, M.H., An, D., Molenda-Żakowicz, J., et al.: Astrophys. J. Suppl. Ser. **199**, 30 (2012)
Pollack, J.B., Hubickyj, O., Bodenheimer, P., et al.: Icarus **124**, 62 (1996)
Quanz, S. P., Absil, O., Angerhausen, D., et al.: (2019). arXiv:1908.01316
Rafikov, R.R.: Astrophys. J. Lett. **621**, L69 (2005)
Rameau, J., Chauvin, G., Lagrange, A.M.: Astrophys. J. Lett. **772**, L15 (2013)
Rauer, H., Catala, C., Aerts, C., et al.: Exp. Astron. **38**, 249 (2014)
Ricker, G.R., Winn, J.N., Vanderspek, R., et al.: JATIS **1**, 014003 (2015)
Ryu, Y. H., Mróz, P., Gould, A., et al.: (2020). arXiv:2010.07527
Seager, S.: Exoplanets. University of Arizona Press (2010)
Seager, S., Mallén-Ornelas, G., et al.: Astrophys. J. **585**, 1038 (2003)
Shvartzvald, Y., Maoz, D., Udalski, A., et al.: Mon. Not. R. Astron. Soc. **457**, 4089 (2016)
Skemer, A.J., Hinz, P., Esposito, S., et al.: Proc. SPIE **9148**, 91480L (2014)
Sozzetti, A., Giacobbe, P., Lattanzi, M.G., et al.: Mon. Not. R. Astron. Soc. **437**, 497 (2014)
Spergel, D., Gehrels, N., Baltay, C., et al.: (2015) arXiv:1503.03757
Spergel, D., Gehrels, N., Breckinridge, J., et al.: (2013) arXiv:1305.5422

Spitzer, L., Jr.: Am. Sci. **50**, 473 (1962)
Stassun, K.G., Collins, K.A., Gaudi, B.S.: Astron. J. **153**, 136 (2017)
Stevens, D.J., Gaudi, B.S.: Proc. Astron. Soc. Pac. **125**, 933 (2013)
Stevens, D.J., Stassun, K.G., Gaudi, B.S.: Astron. J. **154**, 259 (2017)
Stevenson, D.J.: Planet. Space Sci. **30**, 755 (1982)
Sumi, T., Bennett, D.P., Bond, I.A., et al.: Astrophys. J. **710**, 1641 (2010)
Sumi, T., Kamiya, K., Bennett, D.P., et al.: Nature **473**, 349 (2011)
Sumi, T., Bennett, D.P., Bond, I.A., et al.: Astrophys. J. **778**, 150 (2013)
Suzuki, D., Bennett, D.P., Sumi, T., et al.: Astrophys. J. **833**, 145 (2016)
Suzuki, D., Bennett, D.P., Ida, S., et al.: Astrophys. J. Lett. **869**, L34 (2018)
Tabachnik, S., Tremaine, S.: Mon. Not. R. Astron. Soc. **335**, 151 (2002)
The LUVOIR Team (2019). arXiv:1912.06219
The Lynx Team (2018). arXiv:1809.09642
Thompson, S.E., Coughlin, J.L., Hoffman, K., et al.: Astrophys. J. Suppl. Ser. **235**, 38 (2018)
Torres, G.: Proc. Astron. Soc. Pac. **111**, 169 (1999)
Torres, G., Andersen, J., Giménez, A.: Astron. Astrophys. Rev. **18**, 67 (2010)
Udalski, A., Jaroszyński, M., Paczyński, B., et al.: Astrophys. J. Lett. **628**, L109 (2005)
Udalski, A., Szymański, M.K., Szymański, G.: Acta Astron. **65**, 1 (2015)
Udalski, A., Ryu, Y.H., Sajadian, S., et al.: Acta Astron. **68**, 1 (2018)
Veras, D., Raymond, S.N.: Mon. Not. R. Astron. Soc. **421**, L117 (2012)
Villanueva, S.J., Dragomir, D., Gaudi, B.S.: Astron. J. **157**, 84 (2019)
Wang, J., Meyer, M., Boss, A., et al.: Bull. Am. Astron. Soc. **51**, 200 (2019)
Wittenmyer, R.A., Butler, R.P., Tinney, C.G., et al.: Astrophys. J. **819**, 28 (2016)
Wright, J.T., Gaudi, B.S.: Exoplanet Detection Methods, p. 489 (2013)
Wright, J.T., Marcy, G.W., Howard, A.W., et al.: Astrophys. J. **753**, 160 (2012)
Yee, J.C., Gaudi, B.S.: Astrophys. J. **688**, 616 (2008)
Youdin, A.N.: Astrophys. J. **742**, 38 (2011)
Zhu, W., Dong, S.: Ann. Rev. Astron. Astrophys. submitted (2021)
Zhu, W., Wu, Y.: Astron. J. **156**, 92 (2018)
Zucker, S., Mazeh, T.: Astrophys J **562**, 1038 (2001)